Laser Guide Star Adaptive Optics for Astronomy

NATO ASI Series

Advanced Science Institute Series

A Series presenting the results of activities sponsored by the NATO Science Committee, which aims at the dissemination of advanced scientific and technological knowledge, with a view to strengthening links between scientific communities.

The Series is published by an international board of publishers in conjunction with the NATO Scientific Affairs Division

A. Life Sciences B. Physics	Plenum Publishing Corporation London and New York
C. Mathematical and Physical Sciences D. Behavioural and Social Sciences E. Applied Sciences	Kluwer Academic Publishers Dordrecht, Boston and London
F. Computer and Systems Sciences G. Ecological Sciences H. Cell Biology I. Global Environment Change	Springer-Verlag Berlin, Heidelberg, New York, London, Paris and Tokyo

PARTNERSHIP SUB-SERIES

1. Disarmament Technologies	Kluwer Academic Publishers
2. Environment	Springer-Verlag / Kluwer Academic Publishers
3. High Technology	Kluwer Academic Publishers
4. Science and Technology Policy	Kluwer Academic Publishers
5. Computer Networking	Kluwer Academic Publishers

The Partnerschip Sub-Series incorporates activities undertaken in collaboration with NATO's Cooperation Partners, the countries of the CIS and Central and Eastern Europe, in Priority Areas of concern to those countries.

NATO-PCO-DATA BASE

The electronic index to the NATO ASI Series provides full bibliographical references (with keywords and/or abstracts) to about 50,000 contributions from international scientists published in all sections of the NATO ASI Series. Access to the NATO-PCO-DATA-BASE is possible via a CD-ROM "NATO Science and Technology Disk" with user-friendly retrieval software in English, French, and German (©WTV GmbH and DATAWARE Technologies, Inc. 1989). The CD-ROM contains the AGARD Aerospace Database.

The CD-ROM can be ordered through any member of the Board of Publishers or through NATO-PCO, Overijse, Belgium.

Series C: Mathematical and Physical Sciences – Vol. 551

Laser Guide Star Adaptive Optics for Astronomy

edited by

N. Ageorges
European Southern Observatory,
Santiago, Chile

and

C. Dainty
Imperial College,
Blackett Laboratory,
London, United Kingdom

Kluwer Academic Publishers

Dordrecht / Boston / London

Published in cooperation with NATO Scientific Affairs Division

Proceedings of the NATO Advanced Study Institute on
Laser Guide Star Adaptive Optics for Astronomy
Cargèse, France
September 29-October 10, 1997

A C.I.P. Catalogue record for this book is available from the Library of Congress.

ISBN 0-7923-6381-7

Published by Kluwer Academic Publishers,
P.O. Box 17, 3300 AA Dordrecht, The Netherlands.

Sold and distributed in North, Central and South America
by Kluwer Academic Publishers,
101 Philip Drive, Norwell, MA 02061, U.S.A.

In all other countries, sold and distributed
by Kluwer Academic Publishers,
P.O. Box 322, 3300 AH Dordrecht, The Netherlands.

Printed on acid-free paper

All Rights Reserved
© 2000 Kluwer Academic Publishers
No part of the material protected by this copyright notice may be reproduced or utilized in any form or by any means, electronic or mechanical, including photocopying, recording or by any information storage and retrieval system, without written permission from the copyright owner.

Printed in the Netherlands.

TABLE OF CONTENTS

Preface xiii

List of Participants xv

Chapter 1
OPTICAL EFFECTS OF ATMOSPHERIC TURBULENCE 1
J.C. Dainty
    1   A Primer on Statistics and Random Processes   1
    2   Kolmogorov Turbulence   3
    3   Field Correlation and Phase Structure Function   4
    4   Strehl Ratio, Marechal Criterion   7
    5   Zernike Expansion of Kolmogorov Turbulence   8
    6   Angle-of-Arrival Statistics   13
    7   Long- and Short-Exposure Transfer Functions   15
    8   Temporal Behaviour of Atmospheric Turbulence   17
    9   Angular Anisoplanatism   19
    10   Conclusion   21

Chapter 2
ADAPTIVE OPTICS WITH LASER GUIDE STARS: BASIC CONCEPTS
AND LIMITATIONS 23
A. Quirrenbach
    1   Introduction   23
    2   The Formation of Artificial Guide Stars   24
    3   The Tilt Determination Problem   25
    4   Focal Anisoplanatism   30
    5   Properties of the Sodium Atom and Sodium Layer   32
    6   Sodium Guide Star Brightness   35
    7   Laser Technology for Sodium Guide Stars   37
    8   Practical Considerations   40
    9   Laser Guide Star Experiments and Operational Systems   41
    10   An Example: the ALFA System   43
    11   Astronomy with Adaptive Optics and Laser Guide Stars   46

Chapter 3
THE PHYSICS OF THE SODIUM ATOM 51
E.J. Kibblewhite
    1   Introduction   51
    2   Sodium in the Mesosphere   51

3	All You Need to Know about Atomic Physics	52
4	Saturation of the Sodium Atom in the Radiation Field	55
5	Radiation Pressure	57
6	Optical Pumping	57
7	Calculation of the Return Flux from the Layer for a Single Frequency Laser	60
8	Return Flux for Different Laser Spectral and Pulse Formats	62
9	Summary and Conclusions	65

Chapter 4
THE DESIGN AND PERFORMANCE OF LASER SYSTEMS FOR GENERATING SODIUM BEACONS 67
Edward Kibblewhite

1	Introduction	67
2	How Small Can We Make the Laser Beacon?	67
3	Saturation Properties of the Laser at the Sodium Layer	73
4	Lasers for the Generation of Sodium Beacons	74
	4.1 CW Dye Lasers - (UC/UA/Max Planck)	74
	4.2 Pulsed Dye Lasers	75
	4.3 Sum-Frequency Lasers	76
5	Laser Field Tests	81
	5.1 Laser Beacon Profile at the Layer	81
	5.2 Return Flux for the Sodium Beacon	83
	5.3 Optical Pumping Effects	84
6	Estimation of the Beacon Position for the Gemini South Telescope	85
7	System Issues	87

Chapter 5
LASER GUIDE STAR OPERATIONAL ISSUES 89
C.E. Max

1	Introduction	89
2	Operational Implications for the Laser System	90
	2.1 Rayleigh Scattering	90
	2.2 Focus Changes	92
	2.3 Variations in Sodium Column Density	93
	2.4 Requirement to Nod the Telescope for Infra-Red Observing	93
3	Calibration of the Adaptive Optics System for Sodium Laser Guide Star Operation	94
	3.1 Types of Internal Calibration Sources	96
	3.2 Static Calibration	96
	3.3 Auxiliary Wavefront Sensors	97
	3.4 Dynamic Calibration (Real-Time Point-Spread-Function Measurements)	98
4	Safety Considerations Regarding Laser Guide Star Systems	99

	4.1	Laser Eye Safety	100
	4.2	Fire Safety	100
	4.3	Aircraft Avoidance	101
	4.4	Spacecraft Damage Avoidance	102
	4.5	Laser Coordination on Multi-Telescope Summits	103
5	Conclusions		104

Chapter 6
THE CONE EFFECT 107
R. Foy

1	Introduction		107
2	What is the Cone Effect?		108
3	Parameters and Astrophysical Implications		109
4	Multiple LGS: Stitching Method		112
5	Multiple LGSs: the 3D-Mapping Method		114
	5.1	Hypotheses	115
	5.2	Telescope to Laser Spot Array Approach	115
	5.3	Laser Spot Array to Telescope Approach	118
	5.4	The Inversion	119
	5.5	Modal Approach	119
6	Spot Array Geometry and Field		119
7	Conclusion		122

Chapter 7
LASER GUIDE STAR ADVANCED CONCEPTS: TILT PROBLEM 125
Roberto Raggazoni

1	The Framework		125
2	Searching for New Ideas		126
3	Tuning the Parameters		128
4	Looking for Non-Standard Tiny Effects		129
5	LGSs in a New Perspective		134
	5.1	A Statistical Approach	135
	5.2	Gauging with a Natural Guide Star	138
6	A Few Remarks about Thinking Globally		142
7	Experimental Verifications		144
8	Future Directions		145

Chapter 8
THE TILT PROBLEM - MULTIWAVELENGTH 147
R. Foy

1	Introduction: What is the Tilt Problem		147
2	Methods to Measure the Tilt		148
	2.1	Use of a Natural Guide Star	148
	2.2	The Dual Adaptive Optics Concept	150
3	Principle of the Polychromatic Artificial Star		151

		3.1	Required Photon Flux	153
		3.2	Rayleigh and Raman Scattering	154
		3.3	The Excitation of the $4P_{3/2}$ Energy Level of Mesospheric Sodium Atoms	154
	4	Excitation of the $4P_{3/2}$ Atomic Level of Sodium		155
		4.1	The $3S_{1/2} \to 4P_{3/2}$ Absorption	156
		4.2	The $3S_{1/2} \to 4D_{5/2}$: Non-Resonant Two-Photon Absorption	156
		4.3	Saturations	157
		4.4	Integration Time	158
		4.5	Laser Power	159
	5	From the Theory to the Experiments: PASS		159
		5.1	Description of the PASS Experiment	160
		5.2	Results and Discussion	160
		5.3	The Power Balance between the Two Beams	163
	6	From the Experiment to a Full Demonstration: ELP-OA		163
		6.1	Atomic Physics	164
		6.2	Lasers	164
		6.3	Optics	164
		6.4	Infrastructure	165
		6.5	Integration	165
		6.6	Measurements and Analysis	166
	7	Conclusion		166

Chapter 9
SKY COVERAGE WITH LASER GUIDE STAR SYSTEMS ON
8m TELESCOPES 169
M. Le Louarn

	1	Introduction		169
	2	Sky Coverage Computation		170
		2.1	AO System Model	170
		2.2	Statistical Sky Coverage	175
		2.3	Sky Coverage: Object-Counts	177
	3	Conclusion		182

Chapter 10
GROUND BASED ASTRONOMY WITH ADAPTIVE OPTICS 185
Stephen T. Ridgway

	1	Introduction		185
	2	The Investment in Adaptive Optics		185
	3	Does Adaptive Optics Work?		186
	4	Where Are the Adaptive Optics Systems		188
	5	What Will Adaptive Optics Do for Astronomy?		188
	6	Sensitivity and Speed of Observations with Adaptive Optics		189
		6.1	Raw Sensitivity	189
		6.2	Multiobject Observations	192

7	The Isoplanatic Patch	192
8	The Point Spread Function	192
9	What about Photometry?	193
10	Coronography, Nulling, Super AO and Speckle	194
11	Adaptive Optics and Spectroscopy	196
12	How Adaptive Optics Will Make the Rich Richer	196
13	How Much is Adaptive Optics Used	197
14	What Does Natural Guide Star Adaptive Optics Offer Astronomers?	197
	14.1 Correction of Static Aberrations	198
	14.2 Painless Focus	198
	14.3 Tilt Correction	199
	14.4 Resolution Improvement near Sufficiently Bright Stars	200
15	Why Not Higher Order Correction?	200
16	Laser Beacons	200
17	Partial Adaptive Correction	202
18	Science Results from Adaptive Optics	203
19	A Closer Look at Adaptive Optics Science	204
	19.1 Young Stars	205
	19.2 Solar System	206
	19.3 Extragalactic and Galactic Center	207
	19.4 Binary Stars and Star Clusters	208
	19.5 Compact Nebulae and Circumstellar Shells	209
	19.6 Summary of Science Directions	210
20	Future Directions in Adaptive Optics	210
	20.1 Natural Guide Star Adaptive Optics	211
	20.2 Sodium Beacon Adaptive Optics	211
	20.3 Rayleigh Beacon Adaptive Optics	211
	20.4 Adaptive Optics for Partial Correction	212
	20.5 Low Altitude Turbulence	212
	20.6 Adaptive Secondaries	213
21	Summary	213

Chapter 11
POLARIMETRIC MEASUREMENTS AND DECONVOLUTION
TECHNIQUES 219
N. Ageorges

1	Introduction	219
2	Polarimetric Measurements	219
	2.1 Introduction	219
	2.2 Observational Mode	220
	2.3 How to Calibrate the Data	221
	2.4 Analysis of the Data	222
	2.5 Results and Conclusion	228
3	Post-Processing of Adaptive Optics Data: Recovery of Diffraction Limited Images	230

		3.1	Introduction: Why Are Deconvolution Techniques Needed?	230
		3.2	What Does Exist?	230
		3.3	Why Is It Different for Adaptive Optics?	236
		3.4	New Methods Well Adapted to Adaptive Optics Data	236

Chapter 12
MEASURING ASTEROIDS WITH ADAPTIVE OPTICS 243
Jack D. Drummond

1	Introduction		243
2	Ellipses from a Triaxial Ellipsoid		246
	2.1	Basic Formulae	246
	2.2	Useful Relationships	247
3	Fitting the Ellipse		250
	3.1	A Mean Ellipse	250
	3.2	The Fourier Domain	251
4	A Triaxial Ellipsoid from Ellipses		254
	4.1	Residuals for Least Squares Fitting	254
	4.2	Derivatives for Least Squares Fitting	256
	4.3	Final Calculations	257
5	Summary		259

Chapter 13
THE ENVIRONMENTS OF YOUNG STARS 263
T.P. Ray

1	Introduction	263
2	The Early Years	263
3	The T Tauri and Herbig Ae/Be Stars	268
4	HH Flows from Young Stars	271
5	Disks around YSOs	277
6	A Few Examples of How AO Could Help Our Understanding of YSOs	279

Chapter 14
DISTANT (RADIO) GALAXIES: PROBES OF GALAXY FORMATION 285
Huub Rottgering

1	Introduction		285
2	Lookback Time and Redshift		286
3	Main Questions		288
	3.1	Galaxy Evolution and Formation	288
	3.2	Active Galactic Nuclei: Quasars, Radio Galaxies	288
4	Normal Galaxies at $z \sim 3$		290
	4.1	Models of Galaxy Formation	290
5	Distant Radio Galaxies		293
	5.1	Cygnus A	295
	5.2	Search for High Redshift Radio Galaxies	297
	5.3	Properties of High-z Radio Galaxies	298

	5.4	HST Imaging and the Stellar Content	301
	5.5	Distant Radio Galaxies: Probing the Formation of Central Cluster Galaxies?	303
6	Issues for High Resolution Imaging	305

Chapter 15
ACTIVE GALAXIES IN THE LOCAL AND DISTANT UNIVERSE 309
Cláudia S. Rola

1	When Is a Galaxy Active?	309
	1.1	Introduction	309
	1.2	Taxonomy of Emission-Line Galaxies	310
2	Unification Models	321
3	The Physics behind an Emission-Line Galaxy Spectrum	323
	3.1	Introduction	323
	3.2	Nature of an Emission-Line Galaxy	332
4	Active Galaxies and Adaptive Optics	332

Index 339

PREFACE

Groundbased optical and infrared astronomy is on the threshold of a new era, thanks to the invention and development of *adaptive optics*. This technique enables the theoretical limit of angular resolution to be achieved from a large telescope, despite the presence of atmospheric turbulence, or seeing. Thus an eight-metre class telescope, such as one of the four that comprise the Very Large Telescope (VLT) operated by the European Southern Observatory in Chile, will in future routinely be capable of an angular resolution of almost 0.01 arc-seconds, compared to the present resolution of about 0.50 arc-seconds for conventional imaging in good conditions. Adaptive optics on groundbased telescopes will provide higher angular resolution than space telescopes, as for the foreseeable future we shall always be able to build bigger (and much less expensive) telescopes on Earth than those to be launched into space.

All the World's major telescopes either have adaptive optics (AO) installed or are in the process of building AO systems. These first generation systems will not achieve the highest angular resolution, and probably will provide a resolution in the range 0.05 to 0.10 arc-seconds, an order of magnitude improvement on conventional imaging but still significantly below the value of 0.01 arc-seconds. To understand why this is so, we need to understand how an AO system works.

An adaptive optics system consists of three essential components: a wavefront sensor, a deformable mirror, and a control system. In the early days of AO, the technology of deformable mirrors and fast control systems were the main concerns of those developing systems, but nowadays there is fairly universal agreement that the key part of the process is wavefront sensing. If one cannot measure the wavefront deformation induced by the turbulent atmosphere, there is no hope of correcting for it. To sense the wavefront within the atmospheric coherence time, a fairly bright source is required, almost certainly brighter than the astronomical object of interest. So the strategy is usually to select the nearest bright star to the science object for wavefront sensing, but the disadvantage of this is that the sensed turbulence is not exactly the same as that experienced by the science object, an effect known as angular isoplanatism. When all the calculations are complete, it turns out a reasonable fraction of the sky can be observed using adaptive optics, with moderately good imaging quality, provided that the wavelength used for imaging is in the near infra-red. Of course, the angular resolution is proportional to wavelength, so near IR images do not provide as good angular resolution as visible light.

The solution is for astronomers to create their own *laser guide star* to facilitate the wavefront sensing process. There is a layer of sodium atoms at approximately 90 km altitude, and this can be excited by a laser to produce such a source. Alternatively, or in addition, Rayleigh scattering lower in the atmosphere may be employed, although this is not the currently favoured approach.

Since the laser guide can be created anywhere in the field of view — for example, right in the middle of the science object — and in principle (and with enough money!) can be relatively bright, it might seem that this is the ideal solution and therefore that laser guide star adaptive optics would effortlessly attain the goal of 0.01 arc-second resolution at visible wavelengths.

Unfortunately, the production and use of a laser guide star is not trivial from a technical standpoint, and in addition there are fundamental issues not directly solved by the use of a laser guide star. These provided the motivation for the NATO Advanced Study Institute (ASI) on "Laser Guide Star Adaptive Optics for Astronomy", held in Cargèse, Corsica, from September 29 to October 10, 1997, upon which this book is based. At this meeting, the key issues determining the successful implementation of laser guide stars were discussed: these included the physics of the sodium atom, the cone effect, tilt determination, sky coverage and numerous potential astronomical applications. Approximately 80 scientists participated in the ASI over a two-week period. Although there has been a significant delay between this meeting and the publication of this book, all the Chapters are fully up-to-date and should be of particular value to those now building and using the first generation of astronomical laser guide star systems.

The Advanced Study Institute was co-chaired by J C Dainty (Imperial College) and N Hubin (European Southern Observatory), who were assisted by a Scientific Organising Committee consisting of R Foy (Observatoire de Lyon), C Max (Lawrence Livermore Laboratory), S T Ridgway (NOAO, Tucson) and C E Webb (Oxford). Particular thanks go to Annie Touchant and the staff of the Institut d'Etudes Scientifiques, Cargèse for their assistance with the local organisation and their warm hospitality, and to Georgette Huber of ESO for her administrative help. The Advanced Study Institute was sponsored by the Scientific Affairs Division of NATO.

The Editors are deeply grateful to the contributors of this Volume for the care and patience they exercised in the preparation and revision of their manuscripts. We are grateful also to the many participants who provided photographs taken during the meeting. We hope that this book will stimulate those interested in developing and using new astronomical observing techniques to produce even more spectacular observations, and a greater understanding, of our Universe.

Nancy Ageorges, Galway

Chris Dainty, London

February, 2000

List of participants

Nancy Ageorges
European Southern Observatory
Casilla 19001
Santiago 19 - Chile

Jeffrey Baker
Boeing/Phillips Laboratory
PO Box 5670
Kirtland Air Force Base
NM 87185 - USA

Fabio Baroncelli
via S de' Tivoli 12
57125 Livorno - Italy

Andrea Baruffolo
Osservatorio Astronomico di Padova
Vicolo dell'Osservatorio 5
35122 Padova - Italy

Jim Beletic
European Southern Observatory
Karl-Schwarzschild-Straße 2
D-85748 Garching-bei-München - Germany

Carmen Dolores Bello Figueroa
Instituto de Astrofisica de Canarias
C/ Via Lactea, s/n
C.P. 39200 La Laguna
S/C de Tenerife - Spain

Aniceto Belmonte
Polytechnic University of Catalonia
Campus Nord UPC - Modulo D3
Despacho 118 c/ Gran Capitan s/n
08034 Barcelona - Spain

Philippe Berio
Observatoire de la Côte d'Azur
Département Fresnel
CNRS UMR 6528
06460 Saint Vallier de Thiey - France

Bruce Bigelow
Observatories of the Carnegie Institute of Washington
813 Santa Barbara St
Pasadena CA 91101 - USA

Nicoletta Bindi
Arcetri
Via Carducci n.18
50121 Firenze - Italy

Domenico Bonaccini
ESO
Karl-Schwarzschild-Straße 2
D-85748 Garching - Germany

Marcel Carbillet
Osservatorio Astrofisico di Arcetri
Largo Enrico Fermi 5
50125 Firenze - Italy

Mark Chang
51 Furrow Way
Maidenhead
Berkshire SL6 3NY - UK

Jean-Christophe F Chanteloup
Lawrence Livermore National Laboratory
University of California
700 East avenue, L-447
Livermore, CA 94550 - USA

Mark Chun
Gemini 8-m Telescopes Project
670 Aohoku Place
Hilo, Hawaii 96720 - USA

Claudio Cumani
ESO
Karl-Schwarzschild Straße 2
85748 Garching bei München - Germany

Céline D'Orgeville
Gemini Observatory
670 N A'ohoku Place
Hilo HI-96720 - Hawaii

Gérard Daigne
Observatoire de Bordeaux
BP 89, Avenue P. Semirot
33270 Floirac - France

Chris Dainty
Imperial College
Blackett Laboratory
London, SW7 2BZ UK

Francoise Delplancke
ESO
Karl-Schwarzschild-Straße 2
85748 Garching b. München - Germany

Peter Doel
Optical Science Laboratory
Dept Physics and Astronomy
University College London
London WC1E 6BT - UK

Jack Drummond
Starfire Optical Range
Air Force Research Laboratory
Directed Energy Directorate
3550 Aberdeen Avenue SE
Kirtland AFB
NM 87117-5776 - USA

Yvan Dutil
Universidad Politecnica de Catalunya
Departamendo de matematica aplicada II
Pan Gargallo 5, Edificio U
E-08028 Barcelona - Spain

Jacopo Farinato
ESO
Karl-Schwarzschild-Straße 2
85748 Garching - Germany

Orla Feeney
Osservatorio Astrofisico di Arcetri
Largo Enrico Fermi 5
50125 Firenze - Italy

Marc Ferrari
Observatoire de Marseille
2 Place Le Verrier
13248 Marseille Cedex 4 - France

Renaud Foy
CRAL/ Observatoire de Lyon
9 Avenue Charles André
69561 Saint-Genis-Laval - France

Eric Gendron
Observatoire de Meudon
Bâtiment Lyot
5 Place Jules Janssen
92195 Meudon - France

Adriano Ghedina
Centro Galileo Galilei
C/Alvarez Abreu, 70
38700 Santa Cruz de la Palma
S.C. Tenerife - Spain

Chris Holstenberg
MPE-Garching
Giessenbachstr 1
D-85748 Garching - Germany

Georgette Hubert
ESO
Karl-Schwarzschild-Straße 2
D-85748 Garching-bei-München - Germany

Norbert Hubin
European Southern Observatory
Karl-Schwarzschild-Straße 2
D-85748 Garching-bei-München - Germany

Natalia Iaitskova
Moscow State University
Physical Department
International Laser Centre
119899 Vorobyevy Gory
MSU, Moscow - Russia

Neville Jones
Imperial College
Blackett Laboratory
Physics Dept.
Prince Consort Road
London SW7 2BZ

Markus Kasper
MPI für Astronomie
Königstuhl 17
69117 Heidelberg - Germany

Ed Kibblewhite
Department of Astronomy
University of Chicago
5640 Ellis Avenue
Chicago IL 60637 - USA

Etienne Le Coarer
LAOG Observatoire de Grenoble
BP 53
38041 Grenoble Cedex 9 - France

Miska Le Louarn
ESO
Karl-Schwarzschild-Straße 2
85748 Garching - Germany

Jean-Pierre Lemonnier
CRAL - Observatoire de Lyon
9, avenue Charles André
69561 Saint-Genis-Laval Cedex - France

Gary Loos
USAF Phillips Laboratory
PL/LIMI
3550 Aberdeen SE
KAFB - NM 87117-5776 - USA

Philip Lucas
Dept. of Physical Sciences
University of Hertfordshire
College Lane
Hatfield AL10 9AB - UK

Sergio Mallucci
University of Bologna
Department of Astronomy
c/o Fusi PECCI
Vio Zamboni 33
40100 Bologna - Italy

Enrico Marchetti
ESO
Karl-Schwarzschild-Straße 2
85748 Garching - Germany

Claire Max
Director of University Relations
Lawrence Livermore National Laboratory
L-413, 7000 East Avenue
Livermore CA 94550 - USA

Patrick McGuire
Center for astronomical adaptive optics
Steward Observatory
University of Arizona
933 North Cherry Avenue
Tucson - Az 85721 - USA

Vincent Michau
ONERA
Département d'optique théorique et appliquée
29 avenue de la division Leclerc
BP 72
92322 Chatillon cedex - France

Guy Monnet
European Southern Observatory
Karl-Schwarzschild-Straße 2
D-85748 Garching-bei-München - Germany

Rhys Morris
University of Cardiff
Dept. of Physics & Astronomy
PO Box 913
Cardiff CF2 3YB - UK

Créidhe O'Sullivan
Department of Physics
NUI, Maynooth
Co. Kildare - Ireland

Vadim Parfenov
S I Vavilov State Optical Institute
12 Birzhevaya Liniya
St Petersburg
199034 Russia

Patrizio Patriarchi
CNR/ Gruppo Naz Astronomia
Largo E Fermi 5
I-50125 Firenze - Italy

Guy Perrin
Département de Recherche Spatiale
Observatoire de Paris, section de Meudon
5 place Jules Janssen
92190 Meudon - France

Yuri Protasov
Professsor of BMSTU
BMSTU
2-nd Bauman Str 5
Moscow 107005 - Russia

Andreas Quirrenbach
Prof of Physics
University of California, San Diego
Department of Physics
Center for Astrophysics and Space Sciences
9500 Gilman Drive
Mail Code 0424
La Jolla, CA 92093-0424 - USA

Roberto Ragazzoni
Astronomical Observatory of Padova
vicolo dell' Osservatorio 5
I-35122 Padova - Italy

Sam Ragland
Osservatorio Astrofisico di Archetri
Largo Enrico Fermi 5
50125 Firenze - Italy

Tom Ray
Dublin Institute for Applied Studies
5 Merrion Square
Dublin 2 - Ireland

Marcos Reyes Garcia-Talavera
Instituto de Astrofisica de Canarias
C/Via Lactea, s/n
La Laguna
38200 Tenerife - Spain

Stephen Ridgway
NOAO
PO Box 26732
Tucson AZ 85726 - USA

Francois Rigaut
Gemini Observatory
670 N A'ohoku Place
Hilo HI-96720 - Hawaii

Clelia Robert
ONERA
BP 72
92322 Chatillon Cedex - France

Tom Roberts
University of Arizona
Steward Observatory
933 Cherry Avenue
Tucson - Arizona 85721 - USA

Claudia Maria Silva Rola
Institute for Astronomy
University of Cambridge
Madingley Road
Cambridge CB3 0HA - England

Huub Rottgering
Sterrewacht, Oort Gebouw P.O. Box 9513
2300 RA Leiden - The Netherlands

Juan A Rubio
Univ. Politecnica de Catalunya
Dept. TSC
Jordi Girona 1-3. Mod. D3
08034 Barcelona - Spain

Ernesto Sanchez-Blanco Mancera
Gran Telescopiio de Canarias (GTC)
Instrumentation Group
C/Via Lactea s/n
La Laguna
38200 Santa Cruz de Tenerife - Spain

Andrew Sheinis
Lick Observatory
University of California
1156 High St.
Santa Cruz - CA 95064 - USA

Beatrice Sorrente
ONERA - DOTA/OASO
29 avenue de la division Leclerc
BP 72
92322 Chatillon Cedex - France

Jérome Vaillant
Observatoire de Lyon
9 Avenue Charles Andre
69561 Saint-Genis-Laval Cedex - France

Christophe Verinaud
Observatoire de la Côte d'Azur
06460 St Vallier de Thiey - France

Elise Viard
ESO
Karl-Schwarzschild-Straße 2
85748 Garching b. München - Germany

Keith Wilson
Jet Propulsion Laboratory
4800 Oak Grove Drive
MS 161-135
Pasadena
California 91109-8099 - USA

Andrew Zadrozny
University of Durham
Physics Dept.
South Road
Durham DH1 3LE - UK

CHAPTER 1
OPTICAL EFFECTS OF ATMOSPHERIC TURBULENCE

J C DAINTY
Blackett Laboratory
Imperial College
London SW7 2BZ - UK

The aim of this Chapter is to provide a *gentle* introduction to the optical effects of atmospheric turbulence on astronomical images. A number of more detailed descriptions should be read in conjunction with this Chapter and those by Roddier[13][14], Fried[5], and Roggemann and Welsh[16] are particularly recommended.

In my view, the only way to really appreciate how atmospheric turbulence affects astronomical images is to make observations and measurements yourself: one quickly learns to appreciate the meaning of words like "random" and "average", and acquire the detailed understanding that is necessary to design and operate adaptive optics systems. Relying on computer simulations is second best and may encourage a false sense of security.

1. A Primer on Statistics and Random Processes

The value f of a random variable F is not precisely predictable: one can only talk of the *probability* of it taking a particular value (for discrete variables such as the value of the throw of a dice, or the number of photons detected), or it lying between two values (for a continuous variable such as the air temperature or an optical wavefront). The *probability density function* $p(f)$ is defined as:

$$p(f)df \equiv \text{Prob}[f \leq F < f + df] \qquad (1)$$

Rather than specify the whole function $p(f)$, it is sometimes convenient to describe the behaviour of the random variable by a few numbers, the *moments* μ_n and indeed this is sometimes a complete description: for example, a Gaussian distribution (this accurately quantifies the phase distortion introduced by atmospheric turbulence) is *completely* specified by the first two moments. The first moment μ_1 is the *mean*

$$\mu_1 = \langle f \rangle = \int_{-\infty}^{+\infty} p(f) f \, df \qquad (2)$$

and the second moment is

$$\mu_2 = \langle f^2 \rangle = \int_{-\infty}^{+\infty} p(f) f^2 \, df \qquad (3)$$

The *variance* σ^2 is defined as

$$\sigma^2 = \mu_2 - \mu_1^2 \qquad (4)$$

and the square root of the variance is called the *standard deviation*. The mean and standard deviation are the most common two parameters used to describe a random variable, and random quantities are frequently expressed as `mean ± std dev`. In adaptive optics, the variance σ^2 is often preferred, since when two independent random variables are added, their variances also add: this is useful in "error budgets", where the wavefront phase variances due to the (many) various errors add to give the total wavefront phase variance.

A *random process* $F(x)$ is a sequence of random variables in time or space, or both: the variable x can represent time, space, or both, although in this Chapter we take x to represent just space (distance). The phase of a wave that has propagated through turbulence is a space- and time-varying random process. The value of a random process at one point x is a random variable, and has a probability density function $p(f[x])$, mean $m(x)$ and variance $\sigma(x)$. If the process is *stationary*, then none of these quantities depends on x, that is, the statistics are the same everywhere. Atmospheric turbulence is a stationary process to a good approximation.

The function $p(f)$ is strictly called the *first order* probability density function (pdf) since it only describes the statistics at a single point x. If we wish to describe the spatial or temporal structure of a random process (for example, is it changing slowly or rapidly?) then we need to use a second order probability density function: like the first order pdf, this can be parameterised in terms of its moments, and by far the most important moment is the *covariance function* $C(x')$ which can be written as follows for a stationary real process:

$$C(x') \equiv \langle (f(x) - \langle f \rangle)(f(x + x') - \langle f \rangle) \rangle \qquad (5)$$

Note that often $\langle f \rangle = 0$ and also that the variance is given by

$$\sigma^2 = C(x' = 0) = \langle f^2 \rangle - \langle f \rangle^2 \qquad (6)$$

As we shall see for the case of atmospheric turbulence, there is sometimes a problem in defining the covariance function (sometimes called the autocorrelation function), and a *structure function* is defined instead:

$$D(x') = \langle (f(x) - f(x+x'))^2 \rangle \tag{7}$$
$$= 2\left[C(0) - C(x')\right] \tag{8}$$

Finally, the power spectrum $\Phi(\kappa)$ of a stationary process is the Fourier transform of the covariance function:

$$\Phi(\kappa) = \int_{-\infty}^{+\infty} C(x') \exp(-2\pi i \kappa x') dx \tag{9}$$

The power spectrum describes the structure of the process in Fourier, or frequency, space.

2. Kolmogorov Turbulence

The phase of a wave that propagates through turbulence becomes distorted because of the random refractive index fluctuations in the atmosphere. These refractive index fluctuations (and hence the imposed phase fluctuations) are Gaussian to a good approximation. Their spatial statistics were first "derived" by Kolmogorov, using a dimensional argument (see [8] and [4]) and here we simply quote the result. Within a range of separations $\Delta r = |\vec{r_1} - \vec{r_2}|$ ($\vec{r}$ is a three-dimensional vector) that are greater than the *inner scale* ℓ_0 and less than the *outer scale* L_0, the structure function obeys a power law:

$$D_n(\Delta r) = C_N^2 \Delta r^{2/3} \qquad \ell_0 < \Delta r < L_0 \tag{10}$$

This is the famous "two-thirds power law" and we shall find that all the important quantities in atmospheric turbulence have characteristic power laws different amounts of "thirds". The quantity C_N^2 is called the refractive index structure constant: its value depends very strongly with altitude z and so usually we write

$$D_n(\Delta r) = C_N^2(z) \Delta r^{2/3} \tag{11}$$

The units of $C_N^2(z)$ are m$^{-2/3}$.

The power spectrum of the refractive index fluctuations is given by

$$\Phi_n(\vec{\kappa}) = 0.033 C_N^2(z) |\vec{\kappa}|^{-11/3} \tag{12}$$

within the range $1/L_0 < |\vec{\kappa}| < 1/\ell_0$. When we need to include the effects of finite inner and outer scales, the modifed von Karman spectrum is often used:

$$\Phi(\vec{\kappa}) = \frac{0.033 C_N^2}{(|\vec{\kappa}|^2 + \kappa_0^2)^{11/6}} \exp\left(-\frac{|\vec{\kappa}|^2}{\kappa_m^2}\right) \tag{13}$$

where $\kappa_0 = 2\pi/L_0$ and $\kappa_m = 5.92/\ell_0$.

Some comments on Eqs. 11 and 12 are in order. Eq.11 states that the average refractive index difference between two points increases without bound (with a two-thirds power law) as their separation increases, a result that clearly is unphysical for large separations. The existence of an outer scale makes this result physical, the outer scale being a measure of the value at which this average difference ceases to increase: there are few reliable measurements of the outer scale but it is believed to be on the order of tens of metres[11]. Similarly, examination of Eq.11 reveals that the spectrum approaches infinity as $|\vec{\kappa}| \to 0$: the finite outer scale again prevents this from happening in practice. Eq.13 is widely used in modelling to incorporate values for the inner and outer scales.

When one is interested primarily in the *phase* of a wavefront that has propagated through turbulence (the usual situation in adaptive optics), then the outer scale L_0 is usually of much greater significance that the inner scale ℓ_0, particularly regarding issues such as wavefront tilt. On the other hand, if considering the *intensity* (scintillation) of a wave that has propagated through turbulence, it is the inner scale that is the more significant of the two parameters.

3. Field Correlation and Phase Structure Function

A key concept in adaptive optics is the *wavefront*. A wavefront is found by tracing out an equal optical path (distance × refractive index) from a source to the region of interest, for example, the entrance pupil of an optical system. For a point source and free space, wavefronts are spherical, and for starlight the distance is so large that for all practical purposes the wavefronts entering the Earth's atmosphere are plane. After propagating through the random refractive index of the atmosphere, the wavefront entering the telescope pupil is random, and it statistics determine the image quality, and also govern how an adaptive optical system might be used to compensate for the distortion.

In this Chapter, we denote the wavefront by $W(\vec{x})$, where $\vec{x}$ is a two-dimensional vector in the plane of the pupil. Since W is defined by an optical path, its units are length (metres), although often ones states the length in units of the wavelength being used. To first order, the wavefront aberration introduced by turbulence is achromatic, i.e. independent of wavelength of light, so that, for example, the number of microns of optical path retardance is the same for both visible and infrared light: however, the number of *wavelengths* of optical path is different — there are fewer waves of retardance at (the longer) infrared wavelengths — and therefore, from this elementary picture, it is clear that the distortion experienced by infrared light is less serious than that experienced by visible light. It immediately follows that optical degradation at IR wavelengths should be easier to compensate for. It should be noted that the refractive index of air is not *quite* the same for all wavelengths, but that like any of material it varies with wavelength, that is, there is dispersion. As described in a later Chapter, this dispersion can be used to advantage in a polychromatic laser guide star to determine the absolute tilt due to atmospheric turbulence.

Although the basic quantity of interest is the wavefront aberration, it is more usual to deal with the *phase fluctuation* $\phi(\vec{x})$ where

$$\phi(\vec{x}) = \frac{2\pi}{\lambda} W(\vec{x}) \tag{14}$$

where λ is the wavelength. Note that this phase distortion is specified between $-\infty$ and $+\infty$, whereas optically, at a given wavelength, we cannot distinguish between ϕ and $\phi \pm 2n\pi$, where n is any integer. We shall denote the 2π wrapped phase as $\phi|_{\text{Mod}2\pi}$. Clearly the phase distortion ϕ or $\phi|_{\text{Mod}2\pi}$ is also a strong function of wavelength.

The complex amplitude $U(\vec{x})$ of a plane wave of wavelength λ that has propagated through turbulence is given by

$$U(\vec{x}) = |U(\vec{x})| \exp i\phi(\vec{x}) = |U(\vec{x})| \exp\left(\frac{2\pi i}{\lambda} W(\vec{x})\right) \tag{15}$$

The problem of finding the properties of the complex amplitude $U(\vec{x})$ is quite involved: we basically have to solve the wave equation in a three-dimensional random medium. The first order statistics are simple: $|U(\vec{x})|$ is a log-normal process and ϕ (or W) is Gaussian. If the turbulence is not too strong — usually the case in astronomy — then a detailed analysis (the clearest derivation is given by Roddier [13]) shows that the covariance function of the complex amplitude for a wave that has propagated through Kolmogorov turbulence is given by

$$C(\vec{x}') = \langle U(\vec{x}) U^*(\vec{x} + \vec{x}')\rangle \tag{16}$$
$$= \exp\left(-\frac{1}{2} D_\phi(\vec{x}')\right) \tag{17}$$

where $\langle U \rangle = 0$ and $D_\phi(\vec{x}')$ is the *phase structure function*. This in turn is given by

$$D_\phi(\vec{x}') = 2.91\, k^2 (\cos\gamma)^{-1} |\vec{x}'|^{5/3} \int_0^\infty C_N^2(z) dz \tag{18}$$
$$= 6.88 \left(\frac{|\vec{x}'|}{r_0}\right)^{5/3} \tag{19}$$

where $k = 2\pi/\lambda$ and r_0, the Fried parameter, is given by

$$r_0 = \left(\frac{2.91}{6.88}k^2 (\cos\gamma)^{-1} \int_0^\infty C_N^2(z)dz\right)^{-3/5} \qquad (20)$$

In these equations, γ is the zenith angle (angle of observation measured from the zenith).

It is clear from Eqs.17 and 19 that the *Fried parameter* r_0 is a crucial quantity, since if this is known then both the phase structure function and field correlation are also determined. Fried defined this quantity in his 1965 paper [2], and r_0 remains the single most important parameter describing the quality of a wave that has propagated through turbulence. The Fried parameter also has two physical interpretations. First, it is the aperture over which there is approximately one radian of root-mean-square phase aberration. Second, it is the aperture which has the "same resolution" (as defined by Fried) as a diffraction-limited aperture in the absence of the turbulence.

A typical value for r_0 in the visible region of the spectrum lies in the range of 10–20cm: this means therefore that there is approximately one radian of *rms* phase error over a (say) 10cm aperture and that the image resolution, in the presence of this amount of turbulence, is equivalent to that of a *diffraction-limited* optical system (no turbulence) of 10cm aperture. In other words, if the "seeing" is 10cm at the Keck 10m telescope, the image resolution is no better than that given by a 10cm amateur telescope!

If one measures r_0 (using, for example, a differential image motion monitor [17]), it is often found to be a highly fluctuating quantity from minute to minute and can easily vary by a factor of two over a few minutes. It is a feature of human nature that astronomers always remember the best moments of seeing!

There are two important functional dependences of r_0 that should be noted: first, r_0 is a simple integral over the $C_N^2(z)$ profile:

$$r_0 \propto \left(\int_0^\infty C_N^2(z)dz\right)^{-3/5} \qquad (21)$$

We shall see that all the important atmospheric parameters for adaptive optics are integrals (some are weighted integrals) over $C_N^2(z)$, as thus the profile of the refractive index structure constant is a very important quantity. Second, r_0 is proportional to the six-fifths power of the wavelength:

$$r_0 \propto \lambda^{6/5} \qquad (22)$$

Table 1 gives some typical values of the Fried parameter at several wavelengths. The wavelength dependence is of crucial importance, since the ratio of $\mathcal{D}/r_0$ is important.

$\lambda = 0.55\mu m$	$\lambda = 1.2\mu m$	$\lambda = 1.6\mu m$	$\lambda = 2.2\mu m$
10	25	36	53
15	38	54	79
20	50	72	106

TABLE 1. Values of the Fried parameter r_0, in cm, for the given wavelengths, assuming a six-fifths power law dependence.

The power spectrum of the phase fluctuation corresponding to the structure function of Eq.19 is

$$\Phi_\phi(\vec{\kappa}) = \frac{0.023}{r_0^{5/3}} |\vec{\kappa}|^{-11/3} \qquad (23)$$

where $\vec{\kappa}$ is a two-dimensional frequency vector.

4. Strehl Ratio, Marechal Criterion

In this section, we address the relationship between the phase aberration and image quality. This is traditionally quantified, in a high quality imaging system, by the *Strehl ratio* S, defined as the ratio of the central intensities of the aberrated point spread function and the diffraction-limited point spread function:

$$S = \frac{I(0,0)_{\text{aberrated}}}{I(0,0)_{\text{d-l}}} \qquad (24)$$

where $I(\xi, \eta)$ is the intensity point spread function, and (ξ, η) are image plane coordinates (both equal to zero in this case).

For *small* arbitrary aberrations, the Strehl ratio is related to the *variance* of the phase aberration by

$$S \approx 1 - \sigma_\phi^2 \qquad \sigma_\phi^2 \ll 1 \qquad (25)$$

A system is said to be "well-corrected", i.e. close to diffraction-limited in practical terms, when $S \geq 0.8$, the equality being called the Strehl or Marechal criterion. At the Marechal limit,

$$\sigma_\phi^2 = \left(\frac{2\pi}{\lambda}\right)^2 \sigma_W^2 = 0.2 \qquad \text{rad}^2 \qquad (26)$$

corresponding to an rms wavefront aberration of $\sigma_W \approx \lambda/14$ or a wavefront variance $\sigma_W^2 \approx \lambda^2/200$. Note that the Marechal residual phase value of 0.2

rad² is smaller than the residual phase of approximately one rad² in the definition of the Fried parameter r_0. As we shall see in the next section, the diameter of telescope required to achieve a Strehl ratio of 0.8 (in the absence of any adaptive optics), is approximately $0.4r_0$ (i.e. significantly smaller than r_0).

If the phase aberration is *not* small, then the Strehl ratio depends on the specific form of the aberration. However, for a random Gaussian aberration, such as that given by atmospheric turbulence, provided that the telescope diameter $\mathcal{D}$ is very much greater than r_0, it can be shown that

$$\mathcal{S} \approx \exp(-\sigma_\phi^2). \tag{27}$$

This result is a property of Gaussian random processes. Since it subsumes the previous expression, we will take Eq.27 as the definitive form of the Strehl ratio in terms of the variance of the phase aberration. Naïvely, one might expect the goal of an adaptive optics system would to reach the Marechal limit of $\mathcal{S} = 0.8$ by reducing the residual phase variance, after correction, to 0.2 rad²: in practical terms, this is a rather ambitious goal, at least for imaging through the atmosphere.

5. Zernike Expansion of Kolmogorov Turbulence

The Zernike polynomials $Z_j(\rho, \theta)$ are an orthogonal expansion over the unit circle, and have a long tradition of use in classical optical aberration analysis. They have been adopted by the AO community as the *de facto* standard rightly or wrongly (probably the latter). They are defined as [10],[9]:

$$\begin{align} Z_j &= \sqrt{(n+1)} \ R_n^0(\rho), & m &= 0 \\ Z_{\text{even } j} &= \sqrt{(n+1)} \ R_n^m(\rho)\sqrt{2}\cos m\theta & m &\neq 0 \\ Z_{\text{odd } j} &= \sqrt{(n+1)} \ R_n^m(\rho)\sqrt{2}\sin m\theta & m &\neq 0 \end{align} \tag{28}$$

where

$$R_n^m(\rho) = \sum_{s=0}^{(n-m)/2} \frac{(-1)^s (n-s)!}{s![(n+m/2-s)!][n-m/2-s]!} \rho^{(n-2s)} \tag{29}$$

The orthogonality of the Zernike polynomials is expressed by

$$\int_0^R \int_0^{2\pi} \mathcal{W}(\rho) Z_j(\rho,\theta) Z_k(\rho,\theta) \rho d\rho d\theta = \delta_{jk} \tag{30}$$

where the aperture weighting function $\mathcal{W}(\rho)$ is defined by

$$\begin{align} \mathcal{W}(\rho) &= 1/\pi & r &\leq 1 \\ &= 0 & r &> 1. \end{align}$$

j	n	m	Zernike Polynomial	Name
1	0	0	1	Constant
2	1	1	$2\rho\cos\theta$	Tilt
3	1	1	$2\rho\sin\theta$	Tilt
4	2	0	$\sqrt{3}(2\rho^2 - 1)$	Defocus
5	2	2	$\sqrt{6}\rho^2 \sin 2\theta$	Primary Astigmatism
6	2	2	$\sqrt{6}\rho^2 \cos 2\theta$	Primary Astigmatism
7	3	1	$\sqrt{8}(3\rho^3 - 2\rho)\sin\theta$	Primary Coma
8	3	1	$\sqrt{8}(3\rho^3 - 2\rho)\cos\theta$	Primary Coma
9	3	3	$\sqrt{8}\rho^3 \sin 3\theta$	
10	3	3	$\sqrt{8}\rho^3 \cos 3\theta$	
11	4	0	$\sqrt{5}(6\rho^4 - 6\rho^2 + 1)$	Primary Spherical
12	4	2	$\sqrt{10}(4\rho^4 - 3\rho^2)\cos 2\theta$	
13	4	2	$\sqrt{10}(4\rho^4 - 3\rho^2)\cos 2\theta$	
14	4	4	$\sqrt{10}\rho^4 \cos 4\theta$	
15	4	4	$\sqrt{10}\rho^4 \sin 4\theta$	
16	5	1	$\sqrt{12}(10\rho^5 - 12\rho^3)\cos\theta$	
17	5	1	$\sqrt{12}(10\rho^5 - 12\rho^3)\sin\theta$	
18	5	3	$\sqrt{12}(5\rho^5 - 4\rho^3)\cos 3\theta$	
19	5	3	$\sqrt{12}(5\rho^5 - 4\rho^3)\sin 3\theta$	
20	5	5	$\sqrt{12}\rho^5 \cos 5\theta$	
21	5	5	$\sqrt{12}\rho^5 \sin 5\theta$	
22	6	0	$\sqrt{7}(20\rho^6 - 30\rho^4 + 12\rho^2 - 1)$	Secondary Spherical

TABLE 2. First 22 Zernike polynomials.

Table 2 lists the first 22 Zernike polynomials, with the numbering convention that the first one is $j = 1$.

Figure 1 shows the form of the first few Zernike polynomials.

Any wavefront or wavefront phase can be expanded in terms of Zernike polynomials. For a circular pupil of radius R, the expansion is

$$\phi(R\rho,\theta) = \sum_j a_j Z_j(\rho,\theta) \tag{31}$$

where the coefficients a_j are given by

$$a_j = \int_0^R \int_0^{2\pi} \mathcal{W}(\rho)\phi(R\rho,\theta)Z_j(\rho,\theta) \tag{32}$$

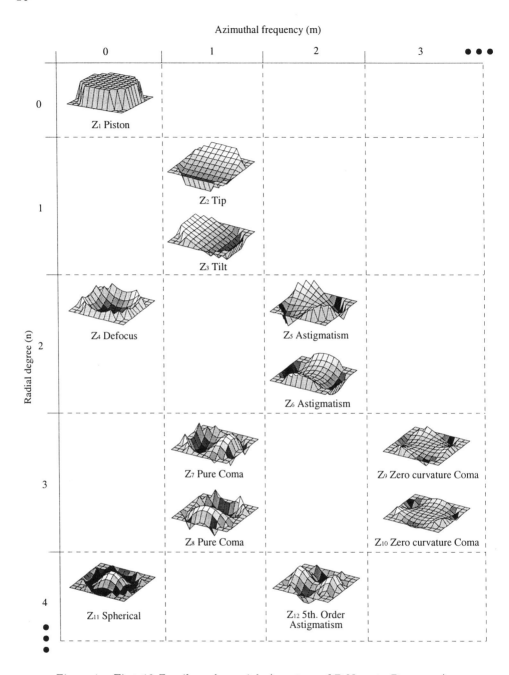

Figure 1. First 12 Zernike polynomials (courtesy of P Negrete-Regagnon).

This last equation states that the a_j are linear combinations of the pupil phase $\phi(\rho,\theta)$ and therefore the a_j are Gaussian random variables if the phase is Gaussian.

Noll [12] was the first person to apply Zernike polynomials to the description of atmospheric turbulence. He showed that the Zernike coefficients are weakly correlated (Eq.25 of [12]), the correlation existing only when both m and the parity are different. He also calculated the residual mean square phase error that resulted when the first J Zernike terms are corrected and the results of this calculation are given below.

The wavefront phase resulting from the summation of the first J Zernike terms is

$$\phi_C(R\rho, \theta) = \sum_{j=1}^{J} a_j Z_j(\rho, \theta) \qquad (33)$$

so that the mean squared phase error σ_ϕ^2, or to use Noll's terminolgy Δ_J, when these J terms are exactly corrected is

$$\begin{aligned} \sigma_\phi^2 &= \Delta_J = \int_{-\infty}^{+\infty} W(\rho, \theta) \langle [\phi(R\rho, \theta) - \phi_C(R\rho, \theta)]^2 \rangle \rho d\rho d\theta \\ &= \langle \phi^2 \rangle - \sum_{j=1}^{J} \langle |a_j|^2 \rangle \end{aligned}$$

As noted earlier, the infinity of $\langle \phi^2 \rangle$ is contained in the $j=1$ term (piston), so that Δ_1 (correction for piston) is finite. Table 3 lists the residual phase variances for $J=1$ to $J=21$, for a telescope of diameter $\mathcal{D}$ and a Fried parameter r_0.

It can now be seen from Table 3 that if $\mathcal{D} = r_0$, then then piston-removed wavefront variance is 1.03 rad^2: to achieve a Strehl ratio of 0.8, the residual phase variance would have to equal 0.2 rad^2, implying $\mathcal{D} \approx 0.4 \times r_0$. Note the large effect on the residual variance of the correction of piston + tip + tilt (i.e. Δ_3 in Table 3): approximately 87% of the wavefront variance is contained in these terms, and it is tempting to think that correction of only these terms might give significant increase in the Strehl value. Obviously the degree of correction required to achieve a certain Strehl value depends very strongly upon $\mathcal{D}/r_0$: using the approximate expression given by Noll,

$$\Delta_J \approx 0.29 J^{-\frac{\sqrt{3}}{2}} \left(\frac{\mathcal{D}}{r_0}\right)^{\frac{5}{3}} \qquad \text{rad}^2 \qquad J \gg 1 \qquad (34)$$

we obtain the relationship

$$\begin{aligned} S &= \exp(-\Delta_J) & & (35) \\ &\approx 1 - \Delta_J & \Delta_J \ll 1 & (36) \end{aligned}$$

so that we can plot graphs of the required number of Zernike polynomials that have to be corrected as function of $\mathcal{D}/r_0$ to achieve given Strehl values.

j	Squared error	j	Squared Error
1	1.030	12	0.0352
2	0.582	13	0.0328
3	0.134	14	0.0304
4	0.111	15	0.0279
5	0.0880	16	0.0267
6	0.0648	17	0.0255
7	0.0587	18	0.0243
8	0.0525	19	0.0232
9	0.0463	20	0.0220
10	0.0401	21	0.0208
11	0.0377		

TABLE 3. The residual phase variance (in rad^2) when the first J Zernike coefficients of Kolmogorov turbulence are corrected exactly (after Noll). The values given are for $\mathcal{D} = r_0$ and scale as $(\mathcal{D}/r_0)^{5/3}$.

This is done in Fig. 2, for Strehl ratios of 0.1, 0.4 and 0.8. For example, if $\mathcal{D}/r_0 = 40$ (a typical value for a 4m diameter telescope in the visible part of the spectrum in average seeing), then >1000 Zernike terms must be corrected, if the Marechal criterion is to be reached! However, if the wavelength is increased to 2.2μm, the number of Zernikes to be corrected reduces dramatically to about 70.

Of course, an adaptive optical system does not work by compensating for the lowest J Zernike terms exactly: in practice, it is not terms in a Zernike expansion that are corrected, and anyway *nothing* is corrected exactly. However, the above calculations are useful as they give an idea of the "degrees-of-freedom" required to obtain good correction and give an idea of the overall magnitude of the correction problem.

As remarked above, the Zernike coeeficients are weakly correlated, and therefore the Zernike expansion is *not* the best expansion to use for Kolmogorov turbulence: the optimum expansion is the Karhunen-Loevé expansion. Roddier [15] shows that these can be expressed in terms of Zernike polynomials and a comparison of the two can be found in the book by Roggemann and Welsh [16].

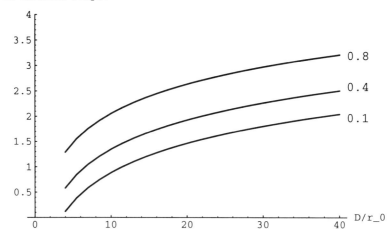

Figure 2. The logarithm (base 10) of the number of Zernike polynomials that must be precisely compensated for, in order to reach Strehl ratios of 0.8 (upper), 0.4 (middle) and 0.1 (lower), as a function of $\mathcal{D}/r_0$.

6. Angle-of-arrival Statistics

Determining the (average) slope of a wavefront is important for two reasons in adaptive optics. First, the correction of average slope ("tip" and "tilt") is, in itself, an important correction, and one can view it as the first step in the implementation of an AO system, although as we saw in the last section, it is rarely sufficient. Second, wavefront sensing is often done by sensing the tilt over sub-apertures, for example, the lenslets in a Shack-Hartmann sensor. Noting the relationship between the wavefront and the phase:

$$W(x,y) = \frac{\lambda}{2\pi}\phi(x,y)$$

we define the slopes along the x and y directions as:

$$\alpha(x,y) \equiv \frac{\partial}{\partial x}W(x,y) = \frac{\lambda}{2\pi}\frac{\partial}{\partial x}\phi(x,y) \tag{37}$$

$$\beta(x,y) \equiv \frac{\partial}{\partial y}W(x,y) = \frac{\lambda}{2\pi}\frac{\partial}{\partial y}\phi(x,y) \tag{38}$$

If $\phi(x,y)$ is Gaussian, then $\alpha(x,y)$ and $\beta(x,y)$ are also Gaussian. This means that the image moves around the image plane with a Gaussian probability distribution.

Using the differentiation property of Fourier transforms, the power spectra of the slopes can be written as:

$$\Phi_\alpha(\vec{\kappa}) = \lambda^2 \kappa_x^2 \Phi_\phi(\vec{\kappa}) \tag{39}$$
$$\Phi_\beta(\vec{\kappa}) = \lambda^2 \kappa_y^2 \Phi_\phi(\vec{\kappa}) \tag{40}$$

where κ_x and κ_y are the components of κ, $\Phi_\phi(\vec{\kappa})$ is the power spectrum of the phase (Eq.23) and $\Phi_{\alpha,\beta}(\vec{\kappa})$ are power spectra of the wavefront slopes. The variance of the slope along say x is therefore

$$\sigma_\alpha = \lambda^2 \int_{\text{bandpass}} \kappa_x^2 \Phi_\phi(\vec{\kappa}) d\vec{\kappa}$$

and the total variance is

$$\sigma_{\alpha\beta}^2 = \lambda^2 \int_{\text{bandpass}} |\kappa|^2 \Phi_\phi(\vec{\kappa}) d\vec{\kappa} \tag{41}$$

A rigorous analysis of this problem was carried out by Fried [2] and the result is:

$$\sigma_{\alpha\beta}^2 = 0.36 \lambda^2 \mathcal{D}^{-1/3} r_0^{-5/3} \tag{42}$$
$$= 0.36 \left(\frac{\lambda}{\mathcal{D}}\right)^2 \left(\frac{\mathcal{D}}{r_0}\right)^{5/3} \tag{43}$$

Three comments on Eq.43 are worthwhile. The apparent wavelength dependence does not exist, as it should be remembered that r_0 is proportional to $\lambda^{6/5}$. Secondly, note the weak dependence on the telescope diameter, which is also unphysical as $\mathcal{D} \to 0$: the finite *inner scale* takes care of this problem. Finally, note that $(\lambda/\mathcal{D})$ is approximately the angular width of the diffraction-limited image, so that the standard deviation of the image motion is approximately $0.6(\mathcal{D}/r_0)^{5/6}$ in units of the width of the *diffraction-limited* point spread function.

Consider an example of a $\mathcal{D} =$1m telescope, wavelength 0.5μm and $r_0 =$10cm (typical seeing in the mid-visible): in that case,

$$\sigma_{\alpha\beta}^2 \approx 4.2 \times 10^{-12} \quad \text{rad}^2$$
$$\approx 0.18 \quad (\text{arcsec})^2$$

or a standard deviation of approximately 0.42 arcsec.

Repeating this calculation for $\mathcal{D} =$10cm, a value appropriate for the lenslets of a Shack-Hartmann array in the visible, gives a standard deviation of approximately 0.62 arcsec.

7. Long- and Short-exposure Transfer Functions

The performance of an optical imaging system is often quantified by either a *point spread function* or a *transfer function*. In the present case of astronomical imaging through turbulence, we are interested in the imaging of spatially incoherent objects, so it is the *intensity* point spread function that is the basic quantity of interest, or alternatively, its Fourier transform, the Optical Transfer Function (OTF).

Over a time that is short compared to the coherence time of the atmosphere (for example, less than a millisecond in the visible region), the object intensity $O(\xi,\eta)$ and image intensity $I(\xi,\eta)$ are related by

$$\begin{aligned} I(\xi,\eta) &= \int\int_{-\infty}^{+\infty} O(\xi',\eta')\mathcal{P}(\xi'-\xi,\eta'-\eta)\,d\xi'd\eta' \\ &= O(\xi,\eta) \odot \mathcal{P}(\xi,\eta) \end{aligned} \quad (44)$$

where $\mathcal{P}(\xi,\eta)$ is is intensity point spread function and $\odot$ represents convolution. In Fourier space this can be represented as

$$\tilde{I}(u,v) = \tilde{O}(u,v) \cdot \mathcal{H}(u,v) \quad (45)$$

where the tilde denotes a Fourier transform and $\mathcal{H}(u,v)$ is the Fourier transform of $\mathcal{P}(\xi,\eta)$; $\mathcal{H}(u,v)$ is called the instantaneous optical transfer function (OTF).

The instantaneous OTF $\mathcal{H}(u,v)$ is given by the normalised autocorrelation of the pupil function:

$$\mathcal{H}(u,v) = P(x,y) \star P(x,y) \quad (46)$$

where $\star$ denotes spatial autocorrelation, distances in the pupil x,y are related to spatial frequencies u,v of the image by

$$x = \lambda f_T u \qquad y = \lambda f_T v$$

f_T is the focal length of the telescope and the pupil function is defined by

$$P(x,y) = \exp\left(ikW(x,y)\right)$$

Both the instantaneous point spread function $\mathcal{P}(\xi,\eta)$ and the instantaneous OTF $\mathcal{H}(u,v)$ are random functions because of the random pupil function. The simple *average* point spread function is obtained by summing all the point spread functions in the normal way, as would happen in a long-exposure image. We therefore denote it $\langle\mathcal{P}(\xi,\eta)\rangle_{LE}$ and the corresponding long-exposure OTF by $\langle\mathcal{H}(u,v)\rangle_{LE}$. However, if each instantaneous point spread function were *centroided* prior to averaging, (e.g. by an adaptive tip-tilt mirror), then one would obtain an average point spread function called the short-exposure point spread function, $\langle \mathcal{P}(\xi,\eta) \rangle_{SE}$ and the corresponding short-exposure OTF is denoted $\langle \mathcal{H}(u,v) \rangle_{SE}$.

The long-exposure transfer function is given by [7]:

$$\langle \mathcal{H}(u,v) \rangle_{LE} = C(\lambda f_T u, \lambda f_T v) T_o(u,v) \qquad (47)$$

where $T_o(u,v)$ is the optical transfer function of the telecope optical system, and the quantity $C(\lambda f_T u, \lambda f_T v)$ was defined in Eq.(16): it is the correlation of the complex amplitude across the telescope aperture. Substituting from Eq.(19),

$$\langle \mathcal{H}(u,v) \rangle_{LE} = T_o(\vec{u}) \cdot \exp\left(-3.44 \left(\frac{\lambda f_T |\vec{u}|}{r_0}\right)^{5/3}\right) \qquad (48)$$

The two parts of the long-exposure OTF are plotted in Fig. 3 for the case where the optical system (telescope) is diffraction-limited — this is a rather unlikely scenario — and $\mathcal{D}/r_0 \approx 10$. Notice the enormous loss in resolution, a factor of 10 (i.e. $\mathcal{D}/r_0$) approximately.

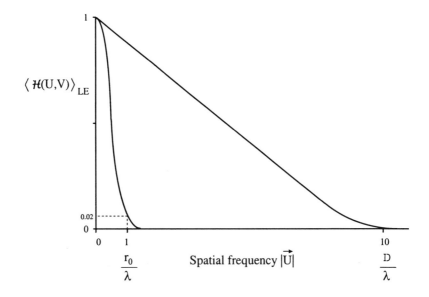

Figure 3. The two terms that make up the long-exposure transfer function $\langle \mathcal{H}(u,v) \rangle_{LE}$ for a value of $\mathcal{D}/r_0 \approx 10$.

The short-exposure transfer function was first derived by Fried [3] (again!), using the results of §2.6, specifically Eq.(43):

$$\langle \mathcal{H}(u,v) \rangle_{SE} = T_o(\vec{u}) \cdot \exp\left(-3.44 \left(\frac{\lambda f_T |\vec{u}|}{r_0}\right)^{5/3}\right) \left[1 - \left(\frac{\lambda f_T |\vec{u}|}{\mathcal{D}}\right)^{1/3}\right] \qquad (49)$$

The extra term in Eq.(49) means that, for given seeing conditions, there is an optimum telescope diameter $\mathcal{D}$ for maximum resolution. This can be seen as follows. First we define the resolution $\mathcal{R}$ (a *bandwidth*) by

$$\mathcal{R} \equiv \int_{-\infty}^{+\infty} \langle \mathcal{H}(\vec{u}) \rangle \, d\vec{u} \qquad (50)$$

Now consider a fixed value of r_0: then a plot of $\mathcal{R}$ versus telescope diameter $\mathcal{D}$ looks as shown on the lower curve in Fig. 4, i.e. it increases to an asymptotic value denoted by $\mathcal{R}_\infty$. If we now plot the ratio of $\mathcal{R}/\mathcal{R}_\infty$ for the short-exposure case, as shown in the upper curve of Fig. 4, then we see that a maximum of $\mathcal{R}/\mathcal{R}_\infty \approx 3.5$ occurs at $\mathcal{D}/r_0 \approx 3.8$.

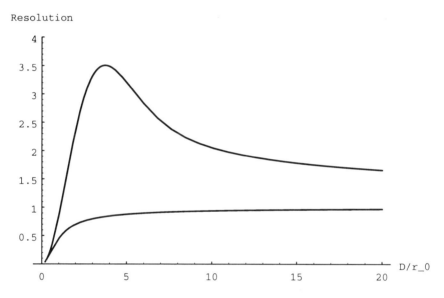

Figure 4. The resolution (bandwidth) $\mathcal{R}$ normalised to the value $\mathcal{R}_\infty$ at $\mathcal{D}/r_0 \to \infty$, for long-exposure (lower curve) and short-exposure (upper curve) imaging. Note the peak value of $\mathcal{R}/\mathcal{R}_\infty \approx 3.5$ at $\mathcal{D}/r_0 \approx 3.8$.

Thus, for a simple tip-tilt system, the optimum value of $\mathcal{D}/r_0$ is approximately 3.8, and an improvement in resolution of between three and four is obtained. However, beware: the improvement in resolution depends strongly on it how resolution is defined, and the value of ≈ 3.5 above is only for the particular definition of $\mathcal{R}$.

8. Temporal Behaviour of Atmospheric Turbulence

A description of the temporal behaviour of the phase in the telescope pupil is rather complicated. The basic model is that there are several strong layers of turbulence in the atmosphere, and each layer (height z) is moving at some velocity perpendicular to the Earth's surface $\vec{v}_\perp(z)$. If the layer simply moves rigidly, without changing its refractive index distribution, then that layer contributes a phase in the pupil which also just moves rigidly across the pupil: this is called the Taylor hypothesis. In reality, of course, the layer evolves in a statistical fashion as it moves. The result of several layers moving at different velocities gives rise to a complicated phase evolution in the telescope pupil.

A good discussion of temporal effects is given by Conan et al [1] and here we simply cite some useful results. Assuming the Taylor hypothesis to be true, and assuming there to be just *one* layer moving at velocity $\vec{v}_\perp$, then the temporal power spectrum $\Phi_\phi(f)$ of the phase in the telescope pupil is given by:

$$\Phi_\phi(f) \propto (C_N^2 \; dz) \frac{1}{|\vec{v}_\perp|} \left(\frac{f}{|\vec{v}_\perp|}\right)^{-8/3} \tag{51}$$

That is, there is -8/3 power law dependence of the phase on the frequency f at any point in the pupil.

In wavefront sensors, we are often interested in the time-dependence of the spatial derivative of the phase, such as $\partial\phi/\partial x$. For a single point (no spatial integration effects) the power spectrum of the phase derivative $\Phi_{\partial\phi/\partial x}(f)$ is:

$$\Phi_{\partial\phi/\partial x}(f) \propto (C_N^2 \; dz) \frac{1}{|\vec{v}_\perp|} \left(\frac{f}{|\vec{v}_\perp|}\right)^{-2/3} \tag{52}$$

That is, there is -2/3 power law dependence of the phase derivative on the frequency f at any *point* in the pupil. In reality, we find the average slope over a dimension d, and the results become very much more involved and vary depending upon the sub-aperture shape (circular or square) and the wind direction with respect to the direction of the phase derivative. At very low temporal frequencies, there is the -2/3 power law dependence of Eq.(52), since these frequencies arise from low spatial frequencies which are barely affected by spatial integration. At the high frequency limit, Conan et al [1] found a power law between -11/3 and -14/3 for a square aperture, depending whether the derivative is parallel or perpendicular to the wind velocity.

For adaptive optics, a useful parameter is the Greenwood frequency [6], f_G, defined for an arbitrary $C_N^2(z)$ and $\vec{v}_\perp(z)$ profile as

$$f_G = \left[0.102 \left(\frac{2\pi}{\lambda}\right)^2 (\cos\gamma)^{-1} \int_0^\infty C_n^2(z) \, (\vec{v}_\perp(z))^{5/3} \, dz \right]^{\frac{3}{5}} \tag{53}$$

For a single layer of turbulence moving at velocity $\vec{v}_\perp$, using Eq.(20), this reduces to simply

$$f_G = 0.43 \frac{|\vec{v}_\perp|}{r_0} \tag{54}$$

As a simple example, for r_0=10cm (typical value in visible) and $|\vec{v}_\perp|$=10ms^{-1}, then $f_G \approx$40Hz, this value falling to only $\approx$8Hz at a wavelength of 2.2μm in the infrared. Typical values from real multilayer turbulence are likely to be higher than these values.

The practical utility of the Greenwood frequency is that, in a first order loop compensator, the residual phase error is related to the Greenwood frequency by

$$\sigma^2_{\text{closed-loop}} = \left(\frac{f_G}{f_{3dB}}\right)^{5/3} \quad \text{rad}^2 \tag{55}$$

where f_{3dB} is the 3dB closed-loop bandwidth of the compensator.

9. Angular Anisoplanatism

Our final topic in atmospheric turbulence is the subject of angular isoplanatism. Figure 5 shows the problem:

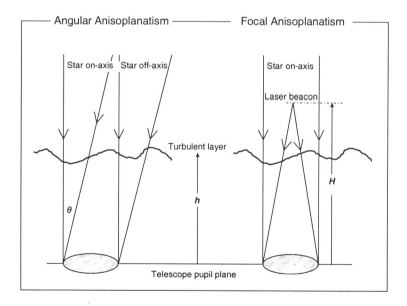

Figure 5. Angular isoplanatism (left) and focus anisoplanatism, or the cone effect (right). Lack of isoplanatism is one of the greatest restrictions to the effectiveness of adaptive optical systems.

As we shall see when we discuss the fundamental photon noise error in a Shack-Hartmann wavefront sensor, in the case of astronomical adaptive optics the "guide star" used to determine the wavefront is unlikely to be the science object under study, whose image is to be improved. The guide star will, in general, be a bright star angularly displaced from the science object, and clearly, from the left hand side of Fig. 5, the wavefront sampled by the guide star is not the same as that from the science object because of this angular displacement. When the wavefronts are, for practical purposes, the same, then we say that the system is "isoplanatic", and when they are not the same we say it is "anisoplanatic". There is an angle Θ_0, called the isoplanatic angle, which quantifies the region over which there is approximate isoplanatism, and it can be shown that this angle is given by:

$$\Theta_0 = \left[2.91\left(\frac{2\pi}{\lambda}\right)^2 (\cos\gamma)^{-1} \int_0^\infty C_n^2(z) z^{5/3} dz\right]^{-\frac{3}{5}} \quad \text{rad} \quad (56)$$

and the mean squared wavefront error σ_Θ^2 for a science object observed Θ away from the guide star is

$$\sigma_\Theta^2 = \left(\frac{\Theta}{\Theta_0}\right)^{5/3} \quad \text{rad}^2 \quad (57)$$

Note that like the isoplanatic angle, like the Fried parameter r_0 is proportional to $\lambda^{6/5}$, that is, it is approximately 5.3 times larger (28 times in solid angle) at 2.2μm than at 0.55μm. If we have a single layer of turbulence at height z, then,

$$\Theta_0 \approx 0.31 \frac{r_0}{z} \quad \text{rad} \quad (58)$$

Remembering there are $\approx$206265 arcseconds in a radian, then for a typical value of r_0 in the visible of r_0 =10cm and for a layer at altitude z =5km, then the isoplantic angle is only $\approx$1.3 arcseconds — this is a very small angle indeed. This is a somewhat pessimistic estimate (it is too small!) but clearly the isoplanatic angle is very small and this has a profound influence on the effectiveness of adaptive optics.

The isoplanatic angle is also a very important issue for some potential non-astronomical applications of adaptive optics. We have to immediately dispel the notion that large fields of view (many degress) in imaging systems can be corrected using adaptive optics, at least if the distorting medium is extended in three dimensions. If the distortion is all introduced in one plane then there is the possibility of conjugating the correction to that plane.

10. Conclusion

In this brief survey, I have highlighted some of the key issues in understanding the optical effects of atmospheric turbulence. For a more thorough understanding, I urge the interested reader to delve into the literature — there is *lots* of it — and to embark on simple measurements and observations. The analysis of practical observations is deeply rewarding, not only for a better understanding of atmospheric effects but also for a real appreciation of random processes in general.

References

1. J-M Conan, Gérard Rousset, and P-Y Madec. Wave-front temporal spectra in high resolution imaging through turbulence. *J Opt Soc Am A*, 12:1559–1570, 1995.
2. D L Fried. Statistics of a geometric representation of wavefront distortion. *J Opt Soc Am*, 55:1427–1435, 1965.
3. D L Fried. Optical resolution through a randomly inhomogeneous medium for very long and very short exposures. *J Opt Soc Am*, 56:1372–1379, 1966.
4. D L Fried. Atmospheric turbulence optical effects: Understanding the adaptive-optics implications. In D M Alloin and J-M Mariotti, editors, *Adaptive Optics in Astronomy*, pages 25–57. Kluwer Academic Publishers, 1994.
5. D L Fried and J F Belsher. Analysis of fundamental limits of artificial guide star adaptive optics system performance for astronomical imaging. *J Opt Soc Am A*, 11:277–287, 1994.
6. D P Greenwood. Bandwidth specifications for adaptive optics. *J Opt Soc Am*, 67:390–392, 1977.
7. R E Hugnagel and N R Stanley. Modulation transfer function with image transmission through turbulent media. *J Opt Soc Am*, 54:52–61, 1964.
8. A N Kolmogorov. The local structure of turbulence in incompressible fluids for very large Reynolds numbers. *Proc Roy Soc Lond A*, 434:9–13, 1991.
9. V N Mahajan. Zernike annular polynomials and optical aberrations of systems with annular pupils. *Opt Phot News*, 5:Engng Lab Notes, 1994.
10. V N Mahajan. Zernike circle polynomials and optical aberrations of systems with circular pupils. *Opt Phot News*, 5:Engng Lab Notes, 1994.
11. F Martin, A Tokovinin, A Ziad, R Conan, J Borgnino, R Avila, A Agabi, and M Sarazan. First statistical data on wavefront outer scale at La Silla Observatory from the GSM instrument. *Astron Astrophys*, 336:L49–L52, 1998.
12. R J Noll. Zernike polynomials and atmospheric turbulence. *J Opt Soc Am*, 66:Suppl 14-1, 307–310, 1975.
13. F Roddier. Effects of atmospheric turbulence in optical astronomy. In E Wolf, editor, *Progress in Optics*, volume 19, chapter 5, pages 281–376. Elsevier, 1981.
14. F Roddier, editor. *Adaptive Optics in Astronomy*. Cambridge University Press, 1999.
15. N Roddier. Atmospheric wavefront simulation using Zernike polynomials. *Opt Engng*, 29:1174–1180, 1990.
16. M Roggemann and B M Welsh. *Imaging Through Turbulence*. CRC Press, 1996.
17. M Sarazin and F Roddier. The ESO differential image motion monitor. *Astron Astrophys*, 227:294–300, 1990.

Coffee break for C. Dainty (left) & E. Lecoarer (right)

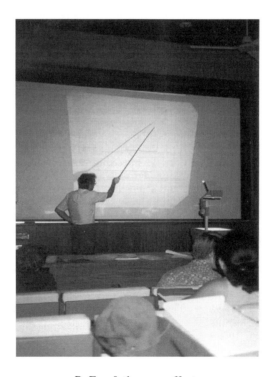

R. Foy & the cone effect

CHAPTER 2
ADAPTIVE OPTICS WITH LASER GUIDE STARS: BASIC CONCEPTS AND LIMITATIONS

A. QUIRRENBACH
University of California, San Diego
Physics Department
Center for Astrophysics and Space Sciences
Mail Code 0424
La Jolla, CA 92093-0424 - USA

Artificial guide stars generated by backscattering of laser light in the Earth's atmosphere are intended to solve one of the main problems of astronomical adaptive optics: the limited sky coverage due to the requirement of having a bright object within the isoplanatic angle, which is needed for wavefront sensing. This article gives a brief introduction into the formation of laser guide stars, either through Rayleigh backscattering in the lower atmosphere, or through resonant backscattering from the sodium layer at an altitude of 90 km. The main limitations of these methods are explained: the tilt determination problem and focal anisoplanatism. The most important properties of the sodium layer are summarized, and some technical issues relevant to the generation of sodium guide stars are discussed.

1. Introduction

Adaptive optics systems need a bright guide star within the isoplanatic field for the real-time analysis of the wavefront; this requirement limits the use of adaptive optics for astronomy. It is possible to obtain good correction in the vicinity of bright stars, which is useful e.g. for searches of faint companions, but the sky coverage for observations of faint objects is poor: for any "interesting" source located in a random position on the sky, the chances of finding a suitable guide star are only of order 1 % for observations in the infrared, and even worse in the visible.

It was therefore suggested in classified reports to the US military (see Happer et al. 1994) and independently in the astronomical literature (Foy & Labeyrie 1985) to use a laser to generate artificial guide stars at arbitrary positions: the laser is focused into the atmosphere at a small angle from the target object, and backscattered light — the "laser guide star" or "laser beacon" — is used by the wavefront sensor to analyze the atmospheric turbulence. A few modest experiments carried out in the US (Thompson & Gardner 1987) and in France (Foy et al. 1989) demonstrated that laser guide stars could in principle be very useful for astronomical applications, but the field really took off when results from the US military program were declassified in May 1991 and subsequently published (Fugate et al. 1991, Primmerman et al. 1991). Adaptive optics systems with laser guide stars are now included in the planning for nearly all major telescopes, and several prototype systems have been commissioned.

Several good reviews of adaptive optics have been published in the past few years (e.g. Beckers 1993, Fugate & Wild 1994, Thompson 1994, Tyson 1997, Hardy 1998); they contain a wealth of introductory material about laser guide stars and information about the history of the subject. A number of detailed technical accounts have been collected in Volume 11 of the Journal of the Optical Society of America A (pp. 257 – 451 and 783 – 945). Workshops and conferences have been organized by Starfire Optical Range (Albuquerque 1992), the European Southern Observatory (Garching 1993, 1995, 1997), the Optical Society of America (Maui 1996) and the Society of Photo-Optical Instrumentation Engineers SPIE (Kona 1994, San Diego 1995, 1997, Kona 1998); the proceedings of these conferences contain information about individual instruments, newly developed components, and novel concepts.

The present article is intended to give an overview of the basic principles of laser guide stars, and to set the stage for the later contributions in this volume, which will treat many aspects in more detail.

2. The Formation of Artificial Guide Stars

The fundamental principle of the formation of artificial guide stars is fairly simple: *the beam from a laser located near the telescope is transmitted into the atmosphere and pointed close to the object of interest; the backscattered light returned from a column of a specific diameter, length, and altitude is observed as the artificial star.* Two techniques have been tried for the formation of such laser guide stars: Rayleigh scattering off air molecules at moderate altitudes (up to $\sim 20\,\mathrm{km}$), and resonance scattering from sodium atoms at an altitude of about 90 km.

Because of the strength of Rayleigh scattering, it is fairly easy to generate bright guide stars with the first of these methods. Furthermore, the exact wavelength of the laser does not matter; it is therefore possible to choose a laser technology which gives high power output at moderate cost.

Copper vapor (Fugate et al. 1994) and excimer lasers (Thompson & Castle 1992) have been used for this purpose. These are pulsed systems, so that the altitude h and length l of the column from which backscattered light is accepted can be selected through "range gating": for each laser pulse, a shutter in front of the wavefront sensor is opened when the light returning from the bottom end of the desired column reaches the detector (i.e., after a time $\tau = 2(h - l/2)/c$ has elapsed), and closed again when it reaches its upper end (at $\tau = 2(h + l/2)/c$). The main drawback of Rayleigh guide stars is their low altitude h, which causes rather large residual errors in the wavefront reconstruction, in particular for large telescopes (see the section on focal anisoplanatism below). It is for this reason that this technique has not been favored by most groups interested in astronomical applications of adaptive optics.

The second method takes advantage of a layer of sodium atoms, which are deposited by meteors at a height of ~ 90 km. A laser tuned to the D_2 transition at 589.2 nm is used to excite the sodium atoms; the light emitted when they return to the ground state forms the artificial guide star. The sodium layer has a thickness of about 10 km, so that range gating is not required to select the backscattering range; both pulsed and continuous-wave lasers can be used to create sodium guide stars. The height h and thickness l of the sodium layer also give a limit on the maximum separation d between the laser projector and the telescope: the sodium "spot" is actually a long thin column, which appears elongated when viewed from a distance (see Fig. 1). The elongation at zenith ε is given by

$$\varepsilon = \frac{d \cdot l}{h^2} \sim 0.''24 \cdot d\,[\mathrm{m}] \ . \tag{1}$$

Therefore d must not be larger than a few meters; in practice this means that the beam projector has to be attached directly to the telescope structure.

Since resonant backscattering from the mesospheric sodium layer is the method chosen for most laser guide star projects at astronomical observatories, the remainder of this article will concentrate mostly on this technique.

3. The Tilt Determination Problem

It is not straightforward to use laser guide stars to determine the tip-tilt (image wander) component of the atmospheric turbulence. The reason for this important limitation is that the laser itself is projected up through the turbulent atmosphere, which causes the laser spot to bounce back and forth. The effects of the atmosphere on the laser beam (the "uplink") and on the backscattered light (the "downlink") can easily be understood in terms of

Figure 1. Sodium "spot" at Calar Alto, as seen from the 1.23 m telescope at a distance of 276 m. The length of the strip is $140''$, from which a total thickness of the sodium layer of about 20 km can be derived. (The FWHM is about half this value.)

their influence on the spots on a Shack-Hartmann wavefront sensor. The perturbations in the uplink cause the Hartmann spots to dance around and — for sufficiently large subapertures — to break up into speckles; this pattern is *the same* for all Hartmann spots, because they all observe the same guide star generated by the same laser beam. Turbulence in the downlink affects the individual Hartmann spots differently, because the downlink paths are different. Measuring tip and tilt means determining the average *absolute position* of all Hartmann spots; this information is corrupted because of the uplink turbulence. The higher-order information, which is contained in the *relative* motion of the Hartmann spots, remains intact, however, and can be determined from the laser guide star.

The simplest solution of the tilt determination problem is splitting the task of wavefront sensing in two: a separate tip-tilt sensor looks at a natural guide star, while the laser spot is used to measure only the higher orders of the wavefront deformation. A conceptual drawing of this setup is shown in Fig. 2. This may sound as if nothing had been gained over an adaptive optics system with a natural guide star, but now the requirements on the brightness of the natural guide star are much less stringent because light from the whole aperture can be used for the tip-tilt determination. The sky coverage of such systems is not 100%, but sufficient for many astronomical programs (at least at near-infrared wavelengths).

It would clearly be desirable to devise a totally autonomous system, which could be pointed at any location on the sky without any need of a natural guide star. The key to the implementation of such systems is the development of a method to obtain the tilt information from the laser guide star. One such concept, the "polychromatic sodium star" has been suggested by Foy et al. (1995). It is based on dispersion, i.e., the variation of the index of refraction in air with λ, which causes slightly larger image wander at shorter wavelengths. By measuring the *difference* of the tilts θ_1, θ_2 at two widely separated wavelengths λ_1, λ_2, it is possible to retrieve the total tilt θ from

$$\theta = \frac{n-1}{n_1 - n_2} \cdot (\theta_1 - \theta_2) \; , \qquad (2)$$

where n_1 and n_2 are the refractive indices at λ_1 and λ_2, and θ and n correspond to any wavelength of interest. The main difficulty with this method is the fact that the measured tilt difference has to be multiplied by the constant $\mathcal{D} = (n-1)/(n_1 - n_2)$, which is fairly large because the dispersion of air is quite small. To minimize $\mathcal{D}$, λ_1 should be as close to the ultraviolet cutoff of the atmospheric transmission as possible, and λ_2 should be in the red or near-IR; with this combination one can reach $\mathcal{D} \sim 30$. Still, this means that θ_1 and θ_2 have to be measured with a precision that is a factor of $\sqrt{2} \cdot 30$ better than the desired tilt accuracy. Furthermore, it is obviously important that the photons at both wavelengths come from the same spot. Sending up two laser beams with different colors to create two independent beacons would not work: due to dispersion, turbulence in the uplink path would cause a random offset between them. It is possible, however, to observe two transitions originating from the same upper level in an atom. In this case the two guide stars at λ_1 and λ_2 are at the same position in the sky, because they originate from the same population of atoms, irrespective of the excitation mechanism and the uplink wavelength. The article by R. Foy (Chapter 8 in this volume) describes in detail how such a two-color guide star can be generated in the sodium layer. The main challenge for a practical implementation of this technique remains the high

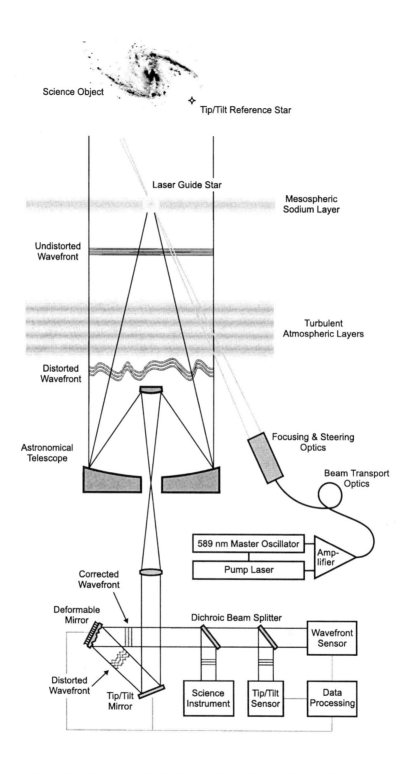

Figure 2. Principle of adaptive optics with a sodium guide star.

laser power, which is required because of the complicated excitation scheme and because of the factor $\sqrt{2}\cdot\mathcal{D}$ in the error propagation according to Eq. 2.

It should also be noted in this context that all large modern telescopes use guide cameras in a closed control loop to correct the telescope tracking. In an adaptive optics system, the tip-tilt loop is the primary tracking mechanism, which keeps the observed field in a stable position on the detector. It corrects both for telescope vibrations up to the cutoff frequency of the tip-tilt loop and for slower errors due to imperfections in the telescope tracking. In contrast, the polychromatic guide star concept — or any other method that seeks to derive the tip-tilt information from the laser beacon without any reference to natural guide stars — implicitly assumes that the laser provides a stable absolute reference. This means that vibrations in the laser projector have to be controlled to a precision that corresponds to the diffraction limit of the main telescope, and that the projector has to be pointed with that same precision. This requires a radically new approach to the telescope guiding control, for example a gyro-based pointing system.

The next best thing to a laser guide star system which does not use any natural reference is a system that can use a very large area of the sky to find a suitable guide star. One such scheme proposed by Ragazzoni et al. (1995) takes advantage of the elongation of the laser beacon when it is viewed with an auxiliary telescope at a substantial distance d from the main telescope (Eq. 1, see also Fig. 1). If there is a star near the sodium beacon, it can be used to monitor the one-dimensional motion of the laser beam by measuring its projected distance from the strip. A similar observation with a second auxiliary telescope at an approximately orthogonal direction gives the required information in the second dimension. If the separation d of the auxiliary telescopes is sufficiently large, the elongation ε is much larger than the isoplanatic angle θ_0, and the area of sky which can be searched for suitable reference star is approximately $2\varepsilon\theta_0$, which is much larger than the $\pi\theta_0^2$ area around the laser spot that can be used from the main telescope. Unlike the polychromatic guide star approach, this technique works with the "standard" excitation of the sodium D_2 line. The complexity is considerable, however, because of the two auxiliary telescopes and the changes in the geometry during the observation due to the Earth's rotation. The laser power required for the tilt determination depends on the diameter of the auxiliary telescopes and on their separation d, because the surface brightness of the strip decreases with increasing d. So there are tradeoffs between the search area, which is proportional to ε and therefore to d, the cost of the auxiliary telescopes, and the laser power requirement. Alternatively, the use of two movable auxiliary telescopes would give a further increase of the search area, but add to the complexity and cost of the system. Further information about this method and related techniques can be found in R. Ragazzoni's contribution to these proceedings.

A different approach to improving the sky coverage has been proposed by Rigaut & Gendron (1992): rather than trying to increase the search area, they seek to improve the sensitivity of the tip-tilt sensor. The accuracy of the tilt determination depends critically on the instantaneous size of the image of the guide star; the sharper an image is, the easier is it to determine its centroid. The high-order correction provided by a laser guide star system produces diffraction-limited short-exposure images of the tip-tilt guide star, provided that the adaptive optics system is good enough to give substantial correction at the operating wavelength of the tip-tilt sensor. (This is not the case for IR-optimized systems with a visible-wavelength sensor). This means that much fainter tip-tilt guide stars can be used. However, the search radius is further limited now, because the isoplanatic angle for the high-order wavefront aberrations is much smaller than the tip-tilt coherence angle. This problem can be circumvented by using two independent high-order systems, each one complete with its own laser guide star. One of the systems is used for correction of the target object, the other system for the tip-tilt guide star; only the tip-tilt information is carried over from the guide star to the science target. The cost and complexity of such a dual adaptive optics system would be quite high, since it requires duplication of all adaptive optics components, including the laser system, wavefront sensor, and deformable mirror.

4. Focal Anisoplanatism

Now we turn to a very important limitation of the laser guide star technique, known as "focal anisoplanatism" or the "cone effect". It is due to the finite height of the artificial guide star above the telescope. Whereas the light rays from the astronomical target, which is at infinite distance, form a parallel bundle and trace a cylinder through the atmosphere, the rays from the guide star trace out a cone (see Fig. 3). The turbulence above the beacon is not sampled at all, and the difference in the propagation paths leads to an incorrect sampling of the turbulence below the beacon. For sodium guide stars, which are formed well above the troposphere, the first of these effects is negligible, and the mismatch for the lower turbulence is less severe than for Rayleigh guide stars (Fig. 3).

A quantitative analysis of focal anisoplanatism shows that the phase variance σ_ϕ^2 for a telescope with diameter D can be written in the form

$$\sigma_\phi^2 = \left(\frac{D}{d_0}\right)^{5/3}, \qquad (3)$$

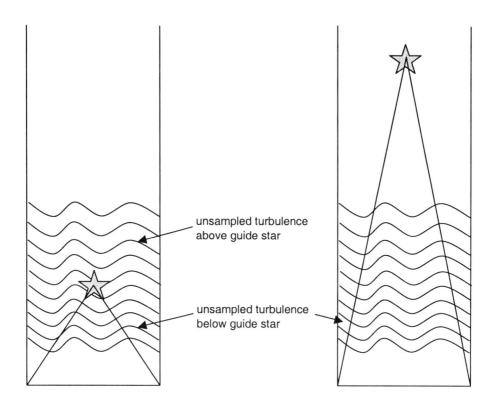

Figure 3. Focal anisoplanatism (the "cone effect").

with $d_0 \propto \lambda^{6/5}$. This important relation means that there is a maximum telescope size that can be corrected with a laser guide star, and that this size scales with $\lambda^{6/5}$. The value of d_0 for a given guide star altitude h can be calculated from profiles of the atmospheric turbulence. For a sodium guide star it is typically 5 times larger than for a Rayleigh guide star at 10 km height, as expected qualitatively from Fig. 3. At a good astronomical site, a sodium guide star system can provide good correction (i.e., $\sigma_\phi^2 \lesssim 0.5$) at a 10 m class telescope in the near-IR, and at a 2 m class telescope in the visible. Rayleigh systems cannot provide high Strehl ratio even at moderately large telescopes, and have therefore not been widely considered for astronomical applications (but see Thompson & Castle 1992). In any case, the phase variance due to focal anisoplanatism is one of the error terms which have to be taken into account in performance analyses of laser guide star adaptive optics systems.

Focal anisoplanatism sets a fundamental limitation to the quality of the correction that can be obtained with a laser guide star system. In particular, it precludes the use of laser beacons for work in the visible wavelength range at large telescopes. The only promising concept to overcome this restriction appears to be the generation of multiple guide stars in a geometric pattern on the sky. The light from each guide star maps a cone, but the guide stars can be arranged such that the union of these cones includes the cylinder traced by the rays from the star. It is therefore possible in principle to measure the stellar wavefront exactly, but much conceptual work remains to be done on optimizing the guide star geometry and devising the best wavefront sensing technique (Tallon & Foy 1990, Tyler 1992, Ragazzoni et al. 1999). As an added bonus, sensing of multiple guide stars might give the information required to extend the corrected field-of-view beyond the isoplanatic angle, e.g. in a multi-conjugate adaptive optics system. As we learn more about the details of atmospheric turbulence through SCIDAR and related methods, we will get better inputs for a realistic assessment of the potential of multiple guide star techniques.

5. Properties of the Sodium Atom and Sodium Layer

Before one can embark on the construction of a sodium guide star system, one has to understand the properties of the sodium layer and the basic physics of the sodium atom. For reference, the most important numerical values have been collected in Table 1.

The sodium is deposited in the upper atmosphere by meteors along with other metals. Iron, aluminum, and calcium are actually more abundant than sodium, but the D_2 transition of sodium is so strong that it provides the largest *product* of column density and transition cross section. There are considerable geographic, seasonal, and short-term variations in all parameters of the sodium layer (Megie et al. 1978, Papen et al. 1996, Ageorges et al. 1999, O'Sullivan et al. 1999), which have to be taken into account by the system design. Most annoying is the variability of the column density, which can reach a full order of magnitude in extreme cases. (This alone gives a variation in the guide star brightness by 2.5 magnitudes.) At mid-latitudes there are seasonal variations with a maximum in winter, which is about three times higher than the minimum in the summer; the amplitude of this seasonal effect seems to be smaller closer to the equator. On top of these systematic variations there are erratic night-to-night changes and even appreciable effects on time scales as short as ten minutes. We should actually think about the sodium column density c_{Na} as a parameter similar to the

TABLE 1. Basic data about the mesospheric sodium layer and the sodium atom.

layer altitude	92 km
layer thickness	10 km
temperature in layer	215 ± 15 K
column density	$2\ldots 9 \times 10^{13}$ m^{-2}
peak sodium density	4×10^9 m^{-3}
molecular (N$_2$, O$_2$) density at 90 km	7.1×10^{19} m^{-3}
sodium – molecule collision time	150 μs
natural width of D line components	10 MHz
natural lifetime	16 ns
oscillator strength of D$_1$ line	0.33
oscillator strength of D$_2$ line	0.66
D$_2$ line wavelength	589.2 nm
hyperfine splitting of D$_2$ line	1.77 GHz
relative strength of D$_2$ hyperfine components	5:3
Doppler width of D$_2$ line at 215 K	1.19 GHz
peak D$_2$ cross section for natural width	1.1×10^{-13} m^2
peak D$_2$ cross section for Doppler width	8.8×10^{-16} m^2
energy of 589.2 nm photon	3.38×10^{-19} J
saturation intensity for single velocity group	64 Wm^{-2}
optical depth of Na layer	$\lesssim 0.1$

seeing: its wild fluctuations have a serious effect on the performance of a laser guide star system, and we should make all efforts to monitor it, and respond quickly to changes, in order to optimize the observing program.

The effective height h of the sodium layer shows similar variations. They do not affect the performance of the system per se, but they have to be taken into account in the focusing procedures. For a large telescope, the sodium layer is very far out of focus (see Fig. 6); this means that the wavefront sensor has to be mounted on a translation stage to follow the focus of the laser guide star, whose distance $h/\cos z$ changes with h and with the zenith angle z. If the sodium layer is stationary, the focus change is fully deterministic, and the motion of the translation stage can be preprogrammed easily. Unanticipated changes in h, however, will move the focus away from its predicted position, and the wavefront sensor will see a defocus term. The adaptive optics system will interpret this as a wavefront error and command the deformable mirror to compensate for it, thus defocusing the astronomical target on the scientific instrument. To prevent this from happening, the height h has to be recalibrated regularly. If the fluctuations of h are not too fast, this can be done by periodic observations of a bright star. It may be better, however, to modify the tip-tilt sensor such that it can also measure the focus using the natural guide star. With this setup it is possible to split the focus correction in a fast and a slow control loop: the bright laser guide star is used to measure the focus term due to the fast fluctuations of the atmosphere, which are corrected with the deformable mirror, while slow changes in the focus of the much fainter natural guide star are attributed to changes of h and corrected by moving the translation stage with the wavefront sensor.

The D_2 line corresponds to the transition from the $3\,^2S_{1/2}$ ground state of the sodium atom to the $3\,^2P_{3/2}$ excited state. To first approximation, the sodium atom can be treated as a two-level system with a transition wavelength of 589.2 nm. The radiative lifetime τ_n of the upper level is 16 ns, corresponding to a natural line width of $1/(2\pi\tau_n) = 10\,\text{MHz}$. The Doppler effect due to the thermal motion of the atoms in the mesospheric sodium layer causes inhomogeneous broadening of the line, which is much larger, $\sim 1.2\,\text{GHz}$. The thermal broadening is called "inhomogeneous" because individual atoms have different radial velocities and therefore interact with light of slightly different frequency. In other words, a photon of a given frequency within the Doppler width of the line can interact with only one "velocity group", i.e., with about 1% of the sodium atoms. This effect has been taken into account in the computation of the peak cross section σ_d of the D_2 transition quoted in Table 1; the peak cross section σ_n for the natural line width is about 120 times higher. We note that the probability for a D_2 photon to be absorbed in the sodium layer, which is given by the product $c_{\text{Na}} \cdot \sigma_d$, is of order $0.02 - 0.1$: the sodium layer is optically thin.

The sodium guide star is created by "resonant backscattering" of light from a laser tuned to the D_2 line. This means that the sodium atoms are excited by absorption of a laser photon and return to the ground state after 16 ns (on average), emitting another D_2 photon. One should expect that the flux F observed from the sodium beacon is proportional to the laser power P. This is true as long as the laser intensity I (i.e., the laser power per unit area) is so small that most of the sodium atoms are still found in the ground state. If I is sufficiently high, however, it changes the level populations, and the number of sodium atoms in the excited state becomes appreciable. These atoms interact with the laser photons through stimulated emission, which goes in the forward direction and does not contribute to the guide star brightness. In this regime, the guide star flux increases slower than linearly with I; it approaches a constant value for very large intensities:

$$F \propto \frac{I/I_{\text{sat}}}{1 + I/I_{\text{sat}}} \ . \tag{4}$$

The saturation intensity I_{sat} for a single velocity group can be approximately calculated from the simple condition that the scattering rate $I \cdot \sigma_n$ should not exceed one photon per natural lifetime:

$$I_{\text{sat}} \cdot \sigma_n = 1/\tau_n \ . \tag{5}$$

This estimate gives $I_{\text{sat}} = 5.7 \cdot 10^{20}$ photons m^{-2} s^{-1} = 192 W m^{-2}; from a more detailed calculation one obtains $I_{\text{sat}} = 64$ W m^{-2}. For a typical spot size of $1''$, which corresponds to 45 cm at an altitude of 92 km, saturation is not a major concern for continuous-wave lasers with output powers of a few Watts. For pulsed lasers, however, the situation is very different. Depending on the pulse format, their peak power can be many orders of magnitude higher than the average power. For example, a laser system with a pulse duration of 20 ns, a pulse repetition frequency of 10 kHz, and an average output power of 10 W would have a peak power of 50 kW during the pulses, which would badly saturate the D$_2$ line for any useful spot diameter. This problem can be alleviated by broadening the frequency spectrum of the laser, in order to match the Doppler profile of the line and to spread the laser power over all velocity groups. Nevertheless, saturation is a major concern for the design of pulsed sodium guide star systems.

At the next level of detail in the treatment of the sodium atom, we must take into account the hyperfine structure of the $3\,^2S_{1/2}$ and $3\,^2P_{3/2}$ levels caused by the magnetic interaction of the angular momentum of the electron with the nuclear spin. The ground state is split into two levels 1.77 GHz apart; the upper state consists of four states with a much smaller separation (~ 100 MHz total). The two hyperfine components originating from the two levels of the ground state have a 5:3 intensity ratio; together with the Doppler broadening this gives rise to an asymmetric double-peaked profile of the D$_2$ line with a total width of about 2 GHz. Each hyperfine level is further split into magnetic sublevels; there are a total of 54 allowed transitions between the 8 sublevels of the $3\,^2S_{1/2}$ ground state and the 16 sublevels of the $3\,^2P_{3/2}$ state. Detailed calculations of the interaction of the sodium atom with the laser light are therefore very complicated (e.g. Morris 1994). A fairly general result of practical importance is that circular polarization gives a better backscattering efficiency than linear polarization. With pulsed lasers, which provide high peak power, it is also possible to redistribute the atoms between the magnetic sublevels; this optical pumping has been demonstrated experimentally (Jeys et al. 1992) and can further enhance the brightness of the laser beacon. More data with a variety of laser pulse formats and line shapes are required, however, to test the theoretical predictions and to assess the practical importance of polarization effects and of optical pumping.

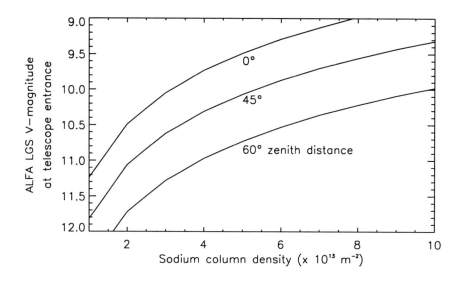

Figure 4. Guide star magnitude predicted for a 3 W continuous-wave dye laser.

6. Sodium Guide Star Brightness

From the properties of the sodium atom and the mesospheric sodium layer tabulated in the previous section, we can now estimate the laser power needed in a guide star system for astronomical purposes. The photon flux F of the laser beacon observed with the wavefront sensor is given by

$$F = \eta \cdot c_{Na} \cdot \frac{\sigma_d}{4\pi h^2} \cdot \frac{P}{E_{phot}} \cdot \cos z \, , \qquad (6)$$

where c_{Na} is the sodium column density, σ_d the peak D_2 cross section, h the height of the sodium layer, P the average laser power, E_{phot} the energy of a 589.2 nm photon, and z the zenith angle. η is an overall efficiency factor, which includes the instrumental and atmospheric transmission as well as the quantum efficiency of the detector in the wavefront sensor. If we assume $\eta = 10\%$, use an intermediate value $c_{Na} = 5 \cdot 10^{13}$ m^{-2}, and require $F = 300\,000$ m^{-2} s^{-1}, we find that an average laser power of 2.4 W is needed. A few simplifications are implicit in Eq. 6, such as the assumption of isotropic emission, which is somewhat pessimistic; on the other hand, and more importantly, saturation is not taken into account. However, this rough estimate demonstrates that the generation of laser guide stars is feasible in principle with currently available laser technology.

For any practical system, the fluctuations of the laser guide star brightness due to temporal variations of c_{Na} are a major concern, furthermore the guide star brightness depends strongly on the zenith angle z. (The $\cos z$ scaling results from the combination of a factor $\cos^2 z$ for the distance to the guide star, and a factor $1/\cos z$ for the length of the sodium column. The zenith-angle dependence of the atmospheric extinction has been neglected in Eq. 6.) Figure 4 shows the expected brightness for a 3 W continuous-wave dye laser as a function of sodium column density and zenith angle. Here the laser guide star brightness has been converted to "equivalent V magnitude", which is defined as the magnitude of a natural star which gives the same photon flux as the laser guide star when the natural star is observed through a V filter. It should be noted here that the bandpass has to be carefully defined when the brightness of a sodium star is given in magnitudes. When we compare the brightness of the laser guide star to a natural star, the result depends on the width of the bandpass, because the laser star is a monochromatic source, whereas the natural star is a continuum emitter. The magnitude scale is a logarithmic measure of flux density (flux per spectral interval). This means that a continuum source has a fixed magnitude, no matter how wide the filter is; in contrast, the magnitude of a line source is smaller for narrower bandpasses. It is therefore advisable to use the equivalent V magnitude only for qualitative arguments and rough estimates; the photon flux should be used in careful system analyses.

It is apparent from Fig. 4 that the sodium guide star brightness can vary by several magnitudes over fairly short times. Clearly, these variations have a strong influence on the performance of the adaptive optics system; they have to be taken into account for an optimization of the operating parameters of the system such as the number of subapertures and the control loop bandwidth (Parenti & Sasiela 1994, Welsh & Gardner 1989). This optimization is a bit more complicated for laser guide stars than for natural guide stars, because focal anisoplanatism and the separation of the tip-tilt measurement and the higher-order wavefront sensing have to be taken into account. The performance degradation with increasing zenith angle is stronger because of the $\cos z$ term in the laser guide star brightness, and the performance depends critically on the sodium column density and on the seeing.

7. Laser Technology for Sodium Guide Stars

Finding a suitable laser for the generation of a sodium guide star is a highly non-trivial task, because of the highly demanding and partially conflicting requirements. The need of operating at the sodium D_2 wavelength limits the choices to tunable or specially designed lasers, which are difficult to obtain with high power output. Good beam quality is needed for the generation of small guide stars, whose diameter is limited only by the seeing.

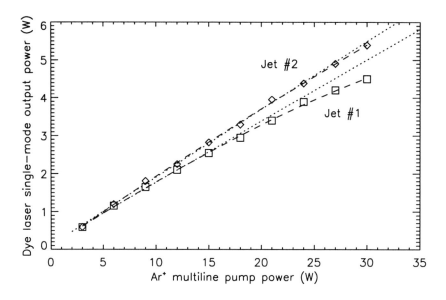

Figure 5. Output of the ALFA dye laser as a function of the pump power, with two different dye solutions. The dotted lines are extrapolations of the output at low pump power.

Pulsed systems have to have a pulse format and spectral characteristics that minimize saturation and are suitable for efficient excitation of the sodium layer. Their pulse repetition frequency has to be at least as high as the desired read rate of the wavefront sensor. Furthermore, the ideal laser system should be affordable and easy to maintain in an observatory environment. No laser has been built so far that would fulfill all these requirements simultaneously; compromises in one or the other area have to be made. Consequently, there is no agreement yet about the most promising laser technology, and a number of different approaches have been taken in the existing and planned sodium guide star systems (e.g. Fugate 1996).

Among the commercially available laser types, the one that comes closest to the requirements of sodium guide star systems is a continuous-wave dye laser, pumped by an argon ion laser (see e.g. Quirrenbach et al. 1998). These lasers are in widespread use for laboratory spectroscopy; they can be tuned over a wide wavelength range with a number of different dyes. Fortunately, at 589 nm one can use Rhodamine 6G, which is the most efficient of all dye molecules. The dye is dissolved in water or ethylene glycol, and the dye solution is pumped through a nozzle under high pressure, so that a free dye jet is generated. The pump laser is focused into the dye jet, which converts the pump photons into somewhat redder output light. This process can generate an output over a fairly wide wavelength range;

the exact dye laser frequency and the line width are then selected by tuning the laser cavity. This can fairly easily be done under computer control. Since saturation of the sodium layer is not a problem for continuous-wave lasers, it is possible to use a narrow line width and to tune the laser to the peak of the D_2 line, where the transition cross section is largest. The power output of continuous-wave lasers is a nearly linear function of the pump power, but there is an upper limit of a few Watts, due to the tendency of the dye jet to become unstable when the pump laser deposits too much heat (see Fig. 5). Argon ion lasers are well suited as pump lasers, because they are fairly robust and can deliver up to about 30 W of pump power in blue and green lines. The main drawback of this laser type is its very low efficiency, it takes about 55 kW of electrical power to generate 30 W of output light. Most of the input power gets converted into heat in the laser head, and the main problem for laser guide star applications is to get rid of this heat without generating additional turbulence near the telescope. In the near future, frequency-doubled continuous-wave Nd:YAG lasers may become an attractive alternative to argon ion lasers; they cannot yet deliver quite the same output power at 532 nm, but they are much more efficient and generate much less heat.

Instead of continuous-wave dye lasers, one can also use pulsed dye lasers. They can be pumped either by copper vapor lasers (Hogan & Webb 1995) or by pulsed Nd:YAG lasers (Friedman et al. 1994). Pulsed dye lasers can deliver higher output power than their continuous-wave counterparts, because the pulses are much shorter than the time needed for the development of thermal instabilities, and the disturbed dye solution gets flushed out of the active volume before the next pulse arrives. It is difficult, however, to produce a pulse format that avoids saturation of the sodium layer; it is always necessary to use a broad spectral format matched to the Doppler width of the mesospheric D_2 line. Therefore pulsed lasers create dimmer guide stars per Watt of laser power than continuous-wave lasers, so that their power advantage is somewhat reduced.

A laser that was specifically developed for sodium guide star systems is the sum-frequency Nd:YAG laser (Jeys 1992). Nd:YAG lasers normally operate at 1.064 μm, but it is also possible to get them to lase at 1.319 μm. By a quirk of nature, the nonlinear mixing of these two wavelengths produces light very close to the sodium D_2 wavelength ($1/1.064+1/1.319 \sim 1/0.589$). This laser type has the advantage of being all solid-state; it avoids the "dirty chemistry" of dye solutions. On the other hand, the technologies for obtaining stable laser action at 1.319 μm and for the non-linear mixing of the two Nd:YAG wavelengths are far from trivial, and the robustness of this laser for routine observations has yet to be demonstrated.

The search for better laser technologies continues. A number of solid-state techniques to obtain 589 nm light have been proposed in the past few years, and corresponding demonstration systems are under development. Some of the most promising approaches involve Raman-shifting of a Nd:YAG laser to 1.178 µm and subsequent frequency-doubling to 589 nm. It will be interesting to see which new ideas will come to fruition over the next few years. In the mean time, more practical experience can be gained with the existing dye and sum-frequency lasers. It is quite clear, however, that the widespread success of sodium guide stars will be intimately linked to the invention of improved lasers.

8. Practical Considerations

There are a number of practical aspects that have to be taken into account for the construction of a laser guide star system. First, it is imperative to make the sodium spot as small as possible, because the centroiding accuracy — and therefore the quality of the wavefront sensing — depends critically on the spot size. The spot size should only be limited by the atmospheric turbulence, which is characterized by the Fried parameter r_0. The optimum size of the beam projector is about $3r_0$; the existing systems use projection telescopes with diameters in the range 30 cm – 50 cm. Because of the spot elongation (see Eq. 1), the beam projector has to be attached directly to the telescope, either at the side, or behind the secondary mirror. Attaching the beam projector to the side of the telescope produces a geometry which is not circularly symmetric; the spots are elongated in one direction, and the amount of elongation varies strongly over the telescope aperture. This can potentially lead to systematic aberrations in the adaptive optics system: some Zernike modes can be sensed — and corrected — better than others. Projection from behind the secondary avoids this problem, but the laser beam has to be transmitted across the telescope aperture. If the beam is propagated freely in front of the aperture, the stray light level is enhanced; if it is enclosed in a pipe, part of the aperture is blocked and the infrared background increased. Furthermore, it is often difficult to put a fairly heavy beam projector behind the secondary of an existing telescope, because that changes the vibration frequencies and leads to additional bending of the telescope structure. So both methods, projection from the side or from behind the secondary, have their pros and cons; in many cases space limitations or other considerations regarding telescope engineering drive the decision towards one or the other solution.

The wavefront sensor must be able to separate the sodium spot from the much brighter Rayleigh backscattering in the lower atmosphere. In pulsed laser systems with suitable pulse repetition rates this can be done by range gating: a shutter in front of the wavefront sensor is closed while the Rayleigh light arrives, and opened only for the light from the sodium layer. This method is not applicable to continuous-wave lasers or pulsed lasers with high repetition rates. However, the projection geometry helps in this case: when viewed slightly from the side, the Rayleigh light forms a cone, which points towards the sodium spot (see Fig. 6). The Rayleigh cone corresponds to heights from 0 to about 20 km, the sodium spot to 90 km, and the gap between the cone and the spot to the region between 20 km and 90 km, which does not generate any return signal. Since there is a gap of a few arcseconds, it is possible to reject the Rayleigh light by putting a field stop in front of the wavefront sensor, which allows only the light from the sodium spot to enter the camera. Note that this technique works also when the laser is projected from behind the secondary, because each individual subaperture still sees the laser beam from the side.

In principle it is also possible to feed the laser beam into the optical train of the main telescope itself and use it as the beam projector. This "monostatic" configuration has severe disadvantages, however: the enormous amount of scattered light requires gating not only of the wavefront sensor, but also of the scientific camera. Furthermore, phosphorescence of optical coatings or other materials in the telescope may produce a glow that persists long after the laser pulse, which is unacceptable for observations of faint objects. This approach is therefore not recommended for astronomical applications.

The use of high-power lasers raises a number of safety issues which are normally not encountered at astronomical observatories, due to the flammability of laser dye solutions and the toxicity of some additives that improve the dye laser efficiency. Most importantly, however, the eye hazards of laser radiation require careful shielding of the beam, and interlocks that restrict access to the laser room and to the dome. The standard safety procedures for working with lasers obviously have to be followed by the persons working directly with the laser and beam projection systems. The laser could also dazzle pilots of aircraft in the vicinity of the observatory if they happen to look directly down the beam. It is therefore necessary to close a shutter in the laser beam when a plane comes too close. This can be done either manually by observers on the catwalk, or automatically by a system that can detect approaching aircraft, such as radar, thermal IR cameras, or CCD cameras that can identify the position lights.

9. Laser Guide Star Experiments and Operational Systems

During the past few years, a number of experiments have been carried out to demonstrate the feasibility of laser guide stars, and the first systems intended for astronomical observations have been put into operation. The

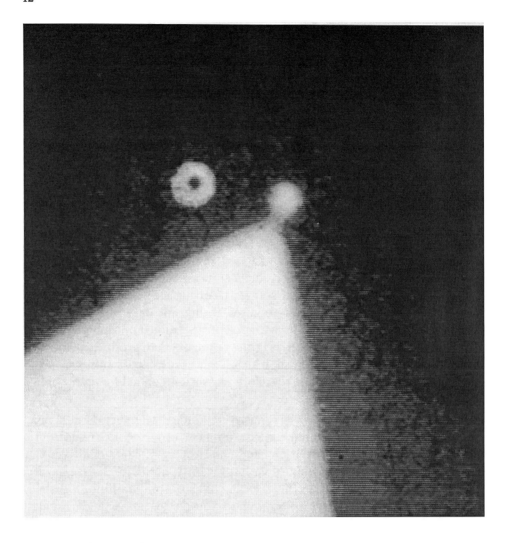

Figure 6. Sodium spot on Calar Alto, observed with the guide camera of the 3.5 m telescope. The sodium spots is located at the tip of the bright Rayleigh cone. This image was taken with the telescope focused to a height of 90 km; the doughnut-shaped object is a defocused image of a star.

pioneering work that was done by the Starfire Optical Range at the US Air Force Phillips laboratory has been summarized by Fugate et al. (1994). The ongoing laser guide star program at Starfire Optical Range supports a number of astronomical projects (e.g., McCullough et al. 1995), but the emphasis is on the needs of the defense community.

The experiments at Lawrence Livermore National Laboratory started with a 1100 W dye laser, which had been developed for atomic vapor laser isotope separation (AVLIS); they produced a sodium spot which was bright enough to be seen with the naked eye (Max et al. 1994). That same group has now built a laser guide star system for the 3 m Shane Telescope at the University of California's Lick Observatory (Max et al. 1997), which is based on a pulsed dye laser with a Nd:YAG pump. A similar system has recently been delivered to the Keck Observatory; it will form part of the new Keck adaptive optics system (Wizinowich et al. 1996).

A group at the University of Arizona has used continuous-wave dye lasers at the Multiple Mirror Telescope (MMT) on Mt. Hopkins. Because of the particular construction of this telescope, which (until recently) had six separate mirrors in a single mount, the configuration of the adaptive optics system was somewhat unusual (Lloyd-Hart et al. 1995). However, this did not preclude the use of laser guide stars. In the near future, the MMT will be refurbished with a single 6.5 m mirror, and the telescope secondary will be the deformable mirror for the adaptive optics system (Sandler et al. 1995). This configuration, which minimizes the number of optical elements between the telescope primary and the scientific instrument, is very attractive for observations of faint objects, especially in the infrared.

After initial experiments with a continuous-wave dye laser, the University of Chicago group decided to use the sum-frequency laser developed by Tom Jeys at Lincoln Laboratories for the adaptive optics system for the 3.5 m ARC Telescope at Apache Point Observatory (Kibblewhite et al. 1994). This system known under the acronym ChAOS (Chicago Adaptive Optics System) has been tested successfully at the telescope.

The first — and so far only — astronomical laser guide star system outside the US is ALFA (Adaptive optics with Laser guide star For Astronomy), a joint project of the Max-Planck-Institutes for Astronomy and for Extraterrestrial Physics (Quirrenbach et al. 1997, Glindemann & Quirrenbach 1997). ALFA uses a commercial continuous-wave dye laser with argon ion pump; the dye laser was modified and optimized for high power output. The system has been installed at the 3.5 m telescope at the German-Spanish Astronomy Center on Calar Alto, Spain. ALFA will be described in somewhat more detail in the next section, in order to give a specific example of a laser guide star installation at a telescope.

Three of these systems have so far obtained sharpened images with sodium guide stars, namely those at Lick (Max et al. 1997), at the Multiple Mirror Telescope (Lloyd-Hart et al. 1998), and on Calar Alto (Davies et al. 1999).

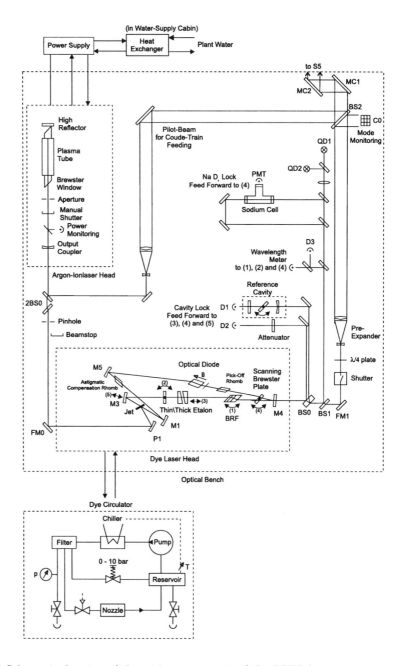

Figure 7. Schematic drawing of the main components of the ALFA laser system.

Figure 8. The ALFA laser cabin in the coudé room of the 3.5 m telescope on Calar Alto. The optical table on the right holds the pump and dye lasers, and auxiliary optics. The box on top of the laser table produces a laminar flow of air, which provides a stable environment for the lasers. The dye pump is in the box under the table. The rack in the center contains laser electronics and diagnostics.

10. An Example: the ALFA System

ALFA's laser system is based on a Coherent Model 899-21 dye ring laser using Rhodamine 6G, noncollinearly pumped by an argon ion laser with 25 W multiline output power (Coherent Innova 400). The oscillation mode of the dye laser is selected by a three-plate birefringent filter followed by electronically controlled thin and thick etalons. To achieve tunability within the mode separation, the cavity length can be continuously changed by rotating a single galvo-driven Brewster plate mounted at the vertex of the ring. This position minimizes output power modulation and beam movement during a frequency scan. For active frequency stabilization a part of the dye laser output is sent into an external reference cavity. The error signal from the reference cavity controls the scanning Brewster plate and a piezo-mounted folding mirror for fast adjustments of the cavity length. For absolute wavelength tuning and long-term frequency stabilization the reference cavity is locked to the Lamb dip in the saturated fluorescence signal of the sodium D_2 line from a laboratory sodium cell. This setup is drawn schematically in Fig. 7; Figure 8 shows a photo of the laser bench.

To achieve the power output shown in Fig. 5, some modifications of the dye laser were required; the main improvement was achieved by optimizing the optical quality of the active laser medium at high pump powers. This involved replacing the jet nozzle with a much more precisely manufactured one and fine-tuning the properties of the dye solution. To achieve the required high flow velocities, a high pressure dye circulation system was constructed out of materials compatible with Rhodamine 6G. Great care was taken to provide effective damping of the vibrations of the dye pump. The dye chiller had also to be upgraded to cope with the heat deposited in the dye by the pump laser.

The lasers are located in the coudé room of the 3.5 m telescope, and the coudé path of the telescope is used for the beam-delivery system, shown schematically in Fig. 9. The mirrors S4 and S5 are part of the coudé train of the telescope; all other mirrors were installed for the laser system. Two mirrors located on the laser bench are used to feed the collimated 20 mm diameter beam into the coudé train. S5 is located on the right ascension axis of the telescope, it is fully steerable to follow the telescope motion. S5 is driven by the telescope control system, which operates independently of the laser control system. The mirrors S4 and MT1 are located on the declination axis. MT1, at a position above and just off the edge of the primary mirror, picks the beam off the coudé path; MT1 through MT4 direct it to the laser launch telescope. The launch telescope is a reflective system with 0.5 m diameter and an actively controlled secondary mirror. It is mounted to the mirror cell of the 3.5 m telescope, at approximately 1 m distance from the primary mirror. A breadboard at the entry of the launch telescope provides space for beam diagnostics, such as power meter and beam profiler, which can be inserted into the laser beam during alignment and maintenance.

The beam position is measured in several places by position-sensitive devices (lateral-effect photodiodes). The mirrors MT1, MT3, and MT4 are driven by fast, high precision dc motors, and one of the mirrors on the laser bench is mounted on a tip-tilt piezo stage. A closed-loop control system driving these mirrors keeps the laser beam aligned during telescope slews and tracking. The secondary mirror of the beam projection telescope is used to focus the laser into the sodium layer, whose distance varies with the zenith angle, and to move the guide star around on the sky. The whole system is controlled by a networked VMEbus computer running EPICS, which provides real-time communication with the control computers of the telescope, the scientific instruments, and of the adaptive optics system.

DSAZ 3.5 m Telescope and the ALFA LGS-System

Figure 9. View of the Calar Alto 3.5 m telescope, with a schematic drawing of the laser path from the coudé room overlaid.

11. Astronomy with Adaptive Optics and Laser Guide Stars

During the past few years, a number of interesting results have been obtained with adaptive optics systems using natural guide stars (for overviews see e.g. Quirrenbach 1998, 1999, and several contributions in this volume). Among the subjects represented are the Solar system, binary stars, circumstellar and interstellar matter, dense stellar clusters, the nuclei of normal and active galaxies, and a few initial observations of high-redshift objects. There has been a recent notable increase in the number of astronomical publications from adaptive optics observations, due to the impact of the two operational facility systems – ADONIS at ESO's 3.6 m telescope on La Silla, Chile, and PUEO at the CFHT on Mauna Kea, Hawaii – and of the instruments at the University of Hawaii and at Starfire Optical Range, which are also in regular use. These programs have been most successful in "niche applications", i.e. for special observations that could not have been done in any other way, such as high-resolution infrared imaging in the time before NICMOS was installed on the Hubble Space Telescope.

To bring adaptive optics into the mainstream of astronomy, user-friendly facilities that provide good sky coverage will be needed at the major observatories with large telescopes. The Lick, Keck, MMT Upgrade and ALFA systems show that the basic technology is available today, but a substantial amount of systems engineering will be required before observations with laser guide stars will become a routine affair. The next step will then be pushing towards shorter wavelengths, with advanced techniques such as multiple guide stars. Last but not least, it should not be forgotten that adequate focal plane instrumentation is needed to get the most out of adaptively corrected large telescopes. The development of specialized spectrographs — long-slit, integral field, and ultra-high (spectral) resolution — that can take full advantage of the nearly diffraction-limited point spread function will be key to the success of adaptive optics over the next few years.

Ground-based observations that offer a spatial resolution of $0.''1$ or better in the near-infrared, combined with spectroscopic capabilities, will open new opportunities for many fields of astronomy. The biggest impact, however, may be on studies of faint objects in the early Universe. For objects that are smaller than a seeing disk, improved spatial resolution results in lower sky background, which is essential for observations of high-redshift objects and for deep source counts. With the combination of adaptive optics at a 10 m telescope and an integral field spectrograph, which can be used to suppress the airglow emission, limiting magnitudes of 26.5 and 27.5 should be possible for point-source detection in the H and J atmospheric bands. Important projects that can be carried out with these systems include studies of young galaxies and distant radio galaxies, investigation of quasar host galaxies, damped Lyman α absorption systems, and Lyman limit galaxies, studies of the members of galaxy clusters and of gravitational lenses, and observations of supernovae at high redshift. These observing programs can address fundamental questions about the geometry of the Universe, about the evolution of galaxies and the global history of star formation. With adaptive optics and laser guide stars, it will become possible to carry out observations that would otherwise be feasible only with space telescopes.

References

1. Ageorges, N., Hubin, N., & Redfern, R.M. (1999), in *Astronomy with Adaptive Optics — Present Results and Future Programs*, European Southern Observatory and Optical Society of America, p. 3
2. Beckers, J. (1993), ARAA 31, 13
3. Davies, R.I., Hackenberg, W., Ott, T., Eckart, A., Rabien, S., Anders, S., Hippler, S., Kasper, M., Kalas, P., Quirrenbach, A., & Glindemann, A. (1999). A&AS, in press
4. Foy, R., & Labeyrie, A. (1985), A&A 152, L29
5. Foy, R., Migus, A., Biraben, F., Grynberg, G., McCullough, P.R., & Tallon, M. (1995), A&AS 111, 569
6. Foy, R., Tallon, M., Séchaud, M., & Hubin, N. (1989), Proc. SPIE 1114, 174
7. Friedman, H.W., Erbert, G.V., Kuklo, T.C., Salmon, J.T., Smanley, D.A., Thompson, G.R., & Wong, J.N. (1994), Proc. SPIE 2201, 352
8. Fugate, R.Q. (1996), in *Adaptive Optics*, OSA Technical Digest Series 13, 90
9. Fugate, R.Q., Ellerbroek, B.L., Higgins, C.H., Jelonek, M.P., Lange, W.J., Slavin, A.C., Wild, W.J., Winker, D.M., Wynia, J.M., Spinhire, J.M., Boeke, B.R., Ruane, R.E., Moroney, J.F., Oliker, M.D., Swindle, D.W., & Cleis, R.A. (1994), JOSA A 11, 310
10. Fugate, R.Q., Fried, D.L., Ameer, G.A., Boeke, B.R., Browne, S.L., Roberts, P.H., Ruane, R.E., & Wopat, L.M. (1991), Nature 353, 144
11. Fugate, R.Q., & Wild, W.J. (1994), S&T 87, 5/25 & 6/20
12. Glindemann, A., & Quirrenbach, A. (1997), SuW 36, 950 & 1038
13. Happer, W., MacDonald, G.J., Max, C.E., & Dyson, F.J. (1994), JOSA A 11, 263
14. Hardy, J.W. (1998), *Adaptive Optics for Astronomical Telescopes*, Oxford University Press
15. Hogan, G.P., & Webb, C.E. (1995), in *Adaptive Optics*, ESO Conf. Proc. 54, 257
16. Jeys, T.H. (1992), in *Laser Guide Star Adaptive Optics Workshop Proceedings*, Albuquerque, 196
17. Jeys, T.H., Heinrichs, R.M., Wall, K.F., Korn, J., Hotaling, T.C., & Kibblewhite, E. (1992), OptL 17, 1143
18. Kibblewhite, E.J., Wild, W., Carter, B., Chun, M.R., Shi, F., Smutko, M.F., & Scor, V. (1994), Proc. SPIE 2201, 458
19. Lloyd-Hart, M., Angel, J.R.P., Groesbeck, T.D., Martinez, T., Jacobsen, B., McLeod, B.A., McCarthy, D., Hooper, E.J., Hege, K., & Sandler, D.G. (1998), ApJ 493, 950
20. Lloyd-Hart, M., Angel, J.R.P., Jacobsen, B., Wittman, D., Dekany, R., McCarthy, D., Kibblewhite, E., Wild, W., Carter, B., & Beletic, J. (1995), ApJ 439, 455
21. Max, C.E., Avicola, K., Brase, J.M., Friedman, H.W., Bissinger, H.D., Duff, J., Gavel, D.T., Horton, J.A., Kiefer, R., Morris, J.R., Olivier, S.S., Presta, R.W., Rapp, D.A., Salmon, J.T., & Waltjen, K.E. (1994), JOSA A 11, 813

22. Max, C.E., Olivier, S.S., Friedman, H.W., An, J., Avicola, K., Beeman, B.V., Bissinger, H.D., Brase, J.M., Erbert, G.V., Gavel, D.T., Kanz, K., Liu, M.C., Macintosh, B., Neeb, K.P., Patience, J., & Waltjen, K.E. (1997), Science 277, 1649
23. McCullough, P.R., Fugate, R.Q., Christou, J.C., Ellerbroek, B.L., Higgins, C.H., Spinhire, J.M., Cleis, R.A., & Moroney, J.F. (1995), ApJ 438, 394
24. Megie, G., Bos, F., Blamont, J.E., & Chanin, M.L. (1978), Planet. Space Sci. 26, 27
25. Morris, J.R. (1994), JOSA A 11, 832
26. O'Sullivan, C.M.M., Redfern, R.M., Ageorges, N., Holstenberg, H.-C., Hackenberg, W., Rabien, S., Ott, T., Davies, R., & Eckart, A. (1999), in *Astronomy with Adaptive Optics — Present Results and Future Programs*, European Southern Observatory and Optical Society of America, p. 333
27. Papen, G.C., Gardner, C.S., & Yu, J. (1996), in *Adaptive Optics*, OSA Technical Digest Series 13, 96
28. Parenti, R.R., & Sasiela, R.J. (1994), JOSA A 11, 288
29. Primmerman, C.A., Murphy, D.V., Page, D.A., Zollars, B.G., & Barclay, H.T. (1991), Nature 353, 141
30. Quirrenbach, A. (1998), in *Laser Technology for Laser Guide Star Adaptive Optics Astronomy*, European Southern Observatory, p. 6
31. Quirrenbach, A. (1999), in *Astronomy with Adaptive Optics — Present Results and Future Programs*, European Southern Observatory and Optical Society of America, p. 361
32. Quirrenbach, A., Hackenberg, W., Holstenberg, H.-C., & Wilnhammer, N. (1997), in *Adaptive Optics and Applications*, Proc. SPIE 3126, 35
33. Quirrenbach, A., Hackenberg, W., Holstenberg, H.-C., & Wilnhammer, N. (1998), in *Laser Technology for Laser Guide Star Adaptive Optics Astronomy*, European Southern Observatory, p. 126
34. Ragazzoni, R., Esposito, S., & Marchetti, E. (1995), MNRAS 276, L76
35. Ragazzoni, R., Marchetti, E., & Rigaut, F. (1999), A&A 342, L53
36. Rigaut, F., & Gendron, E. (1992), A&A 261, 677
37. Sandler, D.G., Lloyd-Hart, M., Martinez, T., Gray, P., Groesbeck, T., Angel, R., Barrett, T., Bruns, D., Stahl, S. (1995), in *Adaptive Optics*, ESO Conf. Proc. 54, 49
38. Tallon, M., & Foy, R. (1990), A&A 235, 549
39. Thompson, L.A. (1994), PhT 47, 12/24
40. Thompson, L.A., & Castle, R.M. (1992), OptL 17, 1485
41. Thompson, L.A., & Gardner, C.S. (1987), Nature 328, 229
42. Tyler, G.A. (1992), in *Laser Guide Star Adaptive Optics Workshop Proceedings*, Albuquerque, 405
43. Tyson, R.K. (1997), *Principles of Adaptive Optics, 2nd Edition*, Academic Press
44. Welsh, B.M., & Gardner, C.S. (1989), JOSA A 6, 1913
45. Wizinowich, P., Acton, D.S., Gleckler, A., Gregory, T., Stomski, P., Avicola, K., Brase, H., Friedman, H., Gavel, D., & Max, C. (1996), in *Adaptive Optics*, OSA Technical Digest Series 13, 8

CHAPTER 3
THE PHYSICS OF THE SODIUM ATOM

E J KIBBLEWHITE
Department of Astronomy
University of Chicago
5640 Ellis Avenue
Chicago IL 60637 - USA

1. Introduction

The central topic of this workshop is how to use and generate artificial beacons as the reference source for adaptive optics systems. Beacons generated by resonant scattering of sodium atoms in the mesosphere are attractive for large telescopes because the size of telescope that can be corrected at a given wavelength with a single beacon scales as its height. However, efficient generation of a compact sodium beacon is harder to achieve than for a Rayleigh beacon generated at lower altitudes due to the physics of the sodium atom in the layer and the difficulty of building suitable lasers. I will discuss the physics of the sodium atom in this talk, and the physics of lasers in the next.

2. Sodium in the mesosphere

Although the total amount of sodium in the mesosphere is very small, about 600 kg in the entire mesosphere, its resonant cross-section is sufficiently large that it can scatter a few percent of the energy of a laser beam tuned to its D_2 line. We do not completely know how the layer got there in the first place. The usual theory is that it comes from meteorites which boil off sodium as they pass through the atmosphere. Meteorites are undoubtedly a source of sodium and lidar observations sometimes see very intense short-lived local emission of the layer during meteor showers. Nevertheless, the column density of potassium, which is also present in the mesosphere, does not change with season whereas that of the sodium density does [8]. Although we may posit that this reflects fundamental differences in the atomic chemistry of the two elements, in my view these observations are difficult to square with an entirely cosmic origin of the sodium metal and it is possible that at least some of the sodium comes from the oceans in an, as yet unidentified, transport mechanism.

The layer itself has been extensively studied over the last 50 years. The average height of the layer is at about 90 to 95 kilometers and it usually about 10 km thick. Lidar studies show that its vertical height and column density distribution changes with time. The height distribution of sodium atoms is normally Gaussian in shape, but it can become stratified and, as we will see in the next talk, this stratification can cause problems in the AO system. The abundance profile changes on time scales of order hours with a trend towards increasing atomic sodium abundance throughout the night. All in all, there is a factor of at least 6 between the maximum and minimum abundance. This has important consequences for the laser power required - you can probably demonstrate a sodium beacon AO system with much less power than you need for a fully operational system. I will come back to this topic in my second lecture. In this lecture I will concentrate on the physics of the sodium atom an its interaction with laser light.

3. All you need to know about atomic physics

It is necessary to understand many of the details of the physics of the sodium atom before you can design efficient lasers to produce compact, bright beacons. The simplest quasi-classical model of the simplest atom hydrogen is the Bohr atom, which assumes that the electron orbits the proton without dissipating energy except during a change of state. Bohr proposed that the angular momentum of the electron is quantized in units of $h/2\pi$. The state of the electron in this model was specified by the principal quantum number, n, and the azimuthal number, k (which defined the eccentricity of the orbit). Selection rules allowed only transitions with $\Delta k = +1$. The orbital motion of the charged electron was allowed to create a magnetic dipole that could interact with an external magnetic field to cause precession of the orbit. This interaction was also quantized into (2k+1) sublevels and used to explain the Zeeman line splitting when the atom is placed in a magnetic field.

Later experiments suggested that the electron itself also possessed its own intrinsic angular and magnetic moment, called spin, which had a value of 1/2 the angular momentum unit of the atom. The nucleus was also found to have intrinsic angular and magnetic moment. The nuclear and electron moments interact to hyperfine split the energy levels. Since angular momentum is itself a vector, these numbers add vectorially to provide a total angular momentum for the atom which is itself quantized. We will see that these details are important.

The Bohr theory was ad hoc and the full theory was derived in the Individual electrons are now characterized by four numbers:

n=1, 2., principal quantum number
ℓ=0,1,......n-1 azimuthal quantum number
j=ℓ-1/2, ℓ+1/2 total angular momentum quantum number
m=-j,-j+1, +j magnetic quantum number

If we assume the Pauli exclusion principle, we can predict the properties of the periodic table by assuming each electron in an atom has its own 4 numbers. The sodium atom consists of 10 electrons in 2 spherically symmetric closed shells and a single valence electron which, in the ground state, has n=3 and ℓ=0, (written as (3,0)). To first order, the closed shells of electrons do not effect the valence electron and we may consider the sodium atom is being similar to the hydrogen atom with a single electron of spin 1/2 orbiting a nucleus of charge +1, atomic mass 11 and a nuclear spin, I, of 3/2.

The ground state of the atom must include the effect of the nuclear spin I. When nuclear spin is included, the total angular momentum of the atom is given by F= I + J. Since I = 3/2 and j can be ± 1/2 (depending on whether the spins of the nucleus and valence electron are aligned parallel or anti-parallel), the ground state is hyperfine split into two levels, with values of F=1 or 2. The separation of the two levels is 1.772 GHz, similar to the hyperfine splitting of the ground state of the hydrogen atom (1.42 GHz). The first excited state of the valence electron has ℓ=1 and is a fine structure doublet since the mechanical spin of the electron can increase or decrease its total angular momentum. This effect is responsible for the splitting of the D lines. The separation between the D lines is 0.6 nm (513 GHz). Each D line is further split by hyperfine magnetic interactions with the nuclear spin. For j>i the number of levels is (2i+1) and for j<i the number of levels is 2j+1.. For j=1/2 there are only two ways for I and j to combine vectorially, so the D_1 line is split into two levels with F=1 and 2. The j=3/2 transition can be split 4 ways with F=0, 1, 2 and 3 and the strength of the D_2 line is twice that of the D_1 line. The hyperfine splitting of these upper energy levels are much less than the ground state hyperfine splitting.

The energy levels of the sodium atom and line strengths are given in Fig. 1. Due to selection rules only transitions with ΔF=-1, 0 or +1 are permitted. Even this structure is too simple because it is known that the population of atomic states depends on the polarization of the light. Each hyperfine level has (2j+1) magnetic states which introduce further selection rules between allowed transitions. For linearly polarized light ΔM= 0 and

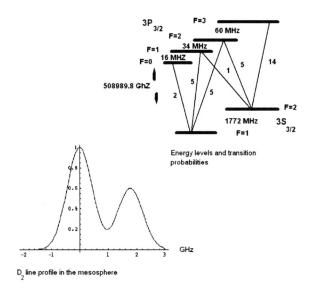

Figure 1. Energy levels and line profile of the D₂ line

the transition between two M=0 levels is forbidden. If circularly polarized light is incident on the atoms only transition with $\Delta M = \pm 1$ are allowed, depending of whether the light is left or right-handed.

What does this all mean? First of all it means that the detailed structure can only be worked out with large computer programs, so that most papers about the sodium atom are difficult to understand and require a certain trust that the program is indeed correct. However, with some thought and very simple models we can get a very good idea of what is happening in the interactions between the sodium atom and the radiation field. Apart from the energy diagrams we need three pieces of information:

(1) We need a high resolution spectra of the D_2 absorption line in the mesosphere. We find it is well fitted by two gaussians of FHWM equal to about 1.07 GHz separated by the hyperfine splitting levels of the ground state of 1.772 GHz. This Doppler width corresponds to a temperature of about 200°K. This is shown in Fig. 1.

(2) We can then calculate the density of air at the mesosphere. This density varies from about 2×10^{14} molecules/cm³ at the bottom of the layer to about 3×10^{13} molecules/cm³ at the top. Knowing the average velocity of air molecules from the temperature we can calculation the mean time between collisions, which is about 100 microseconds in the middle of the layer.

(3) We can measure in the laboratory the natural lifetime, τ, of the atom in the upper state. If the atoms are excited with a very short pulse,

the lifetime is defined as the time it takes for the electric field amplitude to decay by 1/e. For the D lines this lifetime is 16.1 nanoseconds.

The natural linewidth of the atom (which is its spectrum at 0°K) is the Fourier transform of the electric field intensity decay and is given by

$$I(\nu) = (\delta\nu_0/\pi)(1/[(\nu - \nu_0)^2 + \delta\nu_0^2]) \tag{1}$$

where $\delta\nu_0 = 1/(2\pi\tau\lambda)$ = FWHM linewidth, ν_0 = center line frequency.

Since we know τ = 16.1 nsec, the FWHM linewidth is 10MHz and is therefore much narrower than either the Doppler linewidth or the separation between the hyperfine levels. Because the time between collisions is also much longer than the natural lifetime, we can make an immediate and important physical insight; that if the sodium layer is illuminated with a single frequency laser tuned to the peak of the D_2 line only a few per cent(of order 10 MHz/1GHz) of the atoms travel in the right direction to interact at all with the radiation field. These atoms interact strongly with the radiation field until they collide with another atom or molecule or change direction. The flux scattered by the atoms are limited by three effects, saturation, optical pumping and radiation pressure, which will be discussed in turn below.

4. Saturation of the sodium atom in the radiation field

Consider a two level atom which initially has a ground state n containing N atoms and an empty upper state m. The atom is then excited by a radiation field tuned to the transition

$$\nu = W_m - W_n/h, \qquad h\nu \gg kT \tag{2}$$

In equilibrium

$$B_{nm}U(\nu)N_n = A_{mn}N_m + B_{mn}U(\nu)N_m \tag{3}$$

where A_{mn} is Einstein's coefficient A equal to 1/lifetime in the upper state, $B_{nm} = B_{mn}$ = Einstein's coefficient B and, $U(\nu)$ is the radiation density in units of Joules/cm^3 Hz.

Hence

$$N_m = B_{nm}U(\nu)N_n/(B_{nm}U(\nu) + A_{mn}) \tag{4}$$

If we define the fraction of atoms in level m as f and the fraction in level n as (1-f) we can rewrite this equation as

$$f = B_{mn}U(\nu)(1-f)/(B_{mn}U(\nu) + A_{mn}) \tag{5}$$

$$f = 1/(2 + A_{mn}/(B_{mn}U(\nu))) \tag{6}$$

This formula shows us that, at low levels of U(ν) the fraction of atoms in the upper level is $B_{mn}U(\nu)/A_{mn}$ and the number of photons radiated in spontaneous emission/sec given by N B_{mn} U(ν) and is proportional to U(ν). As the pump radiation increases, the fraction of atoms in the upper levels saturates to a maximum level of 1/2 for an infinite value of U(ν). If we define a saturation level as the radiation field generating half this maximum value we get

$$U_{sat}(\nu) = A_{mn}/2B_{mn} \tag{7}$$

The ratio A_{mn}/B_{mn} is known from Planck's black body formula and is equal to $8\pi h\nu^3/c^3$ joules cm^{-3} Hz.

The intensity of the radiation field I(ν) is related to U(ν) by

$$I(\nu) = U(\nu)c \quad watts/cm^2.Hz \tag{8}$$

so that we can estimate the saturation intensity for a natural linewidth of $\Delta\nu$ as $I_{sat} = 4\pi(h\nu)\Delta\nu/\lambda^2$

Writing this in terms of the transitional probability A ($= 2\pi\Delta\nu$) we obtain

$$I_{sat} = 2(h\nu)A/\lambda^2 \quad watts/cm^2 \quad (12.1 mW/cm^2) \tag{9}$$

In terms of photons N_{sat} = 3.6 x 10^{16} photons/sec [1]. More refined calculations, taking into account the hyperfine splitting and the polarization of the exciting radiation field, give

$$I_{sat} = (\pi/2)(h\nu)A/\lambda^2 \quad watts/cm^2 \tag{10}$$

for linearly polarized light (9.48 mW/cm^2)

$$I_{sat} = (\pi/3)(h\nu)A/\lambda^2 \quad watts/cm^2 \tag{11}$$

for circularly polarized light (6.32 mW/cm^2).

Experiments at Sac Peak and the MMT have already demonstrated that sub-arcsecond beacons can be generated at astronomical sites implying that a typical beacon diameter at the layer will be less than 250 cm^2 For a single frequency CW laser the central intensity reaches I_{sat} for a laser power of only 2 watts.

---

[1] We can use this number to estimate the natural linewidth cross section. Since both A_{mn} and B_{mn} U(ν) have dimensions of 1/lifetime, $1/A_{mn}$ is the spontaneous emission lifetime τ and $1/(B_{mn}U(\nu))$ is the time for the atom to absorb a photon, τ_a. We can now estimate the natural linewidth cross-section since for I_{sat}, $\tau_a = 2\tau$ and $\tau_a = 1/B_{mn}U(\nu) = 1/\sigma N_{sat} = 2\tau = 32$ nanoseconds

Which gives σ = 8.7 x 10^{-10} cm^{-2}. The correct value of the cross-section to the upper level is 6.9 x10^{-10} cm^{-2}.

5. Radiation pressure

The cycle time of a sodium atom is the average time that it takes the absorb and reemit a photon. Since the lifetime in the upper state is 16 nanoseconds the cycle time is 16.1/f nanoseconds, where f is the fraction of atoms in the upper state. Solving for I_{sat} in Eqs. 7 and 8, will can write the fraction of atoms in the upper level as

$$f = 1/(2(1 + I_{sat}/I)) \qquad (12)$$

where I is the intensity of the laser/m^2 for a single frequency laser. For infinitely high pumping (I = ∞) no more than 1/2 of the atoms can be in the upper state and the minimum cycle time of the sodium atom is therefore 32 nanoseconds; if I=I_{sat} the cycle time is 64 nanoseconds. If the laser is tuned to the line center of a sodium atom at I_{sat} moved in a given direction we would expect it to make about 1500 cycles in the collision lifetime of 100 microseconds. However each photon absorbed by the atom carries momentum in the direction of the beam and, on average the atom is given a Doppler shift of 50kHz for every photon it absorbs. After only 100 cycles it is already 5 MHz from the linecenter and the absorption cross-section has been reduced by a factor of 2. By integrating along the line-profile we can compute how many cycles occur between collisions and compare this with the expected number of photons returned without radiation pressure effects. This is shown in Fig. 2 for two cases representing the maximum and minimum photon returns under different conditions [2]. These calculations suggest that the return signal is significantly effected by radiation pressure when I > 0.2 I_{sat} so that radiation pressure may be more important than saturation for a single frequency laser although the effect can be reduced by chirping the laser frequency [1].

6. Optical Pumping

Study of the energy level diagram of Fig. 1 shows that, if only a single frequency laser, tuned to the peak of the D$_2$ absorption line, is used to excite the sodium atoms, it may be possible to optically pump the atom from the F=2 to the F=1 ground state. While atoms travelling approximately normal to the laser beam can only cycle between the F=2 ground state and the F=3 upper state until they change direction by collision, atoms moving in other directions can be excited to the F=2 or F=1 upper state. Once excited to

---

[2] The minimum return occurs when the laser frequency coincides with the center of the natural linewidth at the start of its path. The maximum return occurs when the initial laser frequency is blueward of the atomic line, the atom is then pushed towards and through the center frequency.

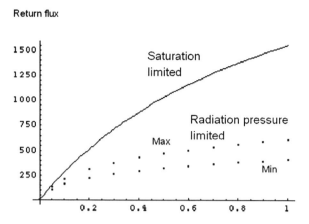

Figure 2. Return signals as a function of I/I_{sat} for single frequency excitation of a sodium atom in the mesosphere for the F=2 ground to F=3 upper state

these levels, they can fall back either to the F=2 or F=1 ground state, and after only a few cycles will end up trapped in the ground state. About 1% of all atoms get pumped to the F=1 ground state every collision lifetime (100 microseconds). If this were the only process the layer would become optically transparent in a time scale of 0.01 seconds. The significance of this effect depends on how quickly the ground level is rethermalized by collision or the atoms replaced by the mesospheric wind.

Once the atom is in the lower ground state it is only rethermalized with great difficulty in collisions with most molecules, because the ground state is spherical symmetric and typical rethermalization cross sections are $<10^{-23}$ cm^2. Happer [3] points out that spin-exchange with paramagnetic molecules such as O or O_2 is possible. In spin exchange the spin of the interacting atoms is conserved and the spin exchanged during collision. Since O_2 makes up 20% of the atmosphere and half the time the spin of the molecule will be the same as the sodium atom, the best we can hope for is relaxation times on order 10 collision lifetimes (1 millisecond). However, since the spin exchange cross section of Na-Na interactions is 10^{-14} cm^2 and the magnetic susceptibility of oxygen molecules are a factor of ten lower than sodium, so it seems unlikely that the spin exchange cross section is greater than 10^{-15} cm^2, which would give a relaxation time of order 10^{-2} seconds. The exact cross-section for oxygen molecules is therefore critical for this analysis and is not known. The other mechanism is by physical movement of new atoms into the beam through wind. Mesospheric wind speeds are very high, typically 30 m/s, so for a beam 30cm wide all the atoms are replaced in 0.01 seconds.

However we can use optical pumping effects to overcome this problem,

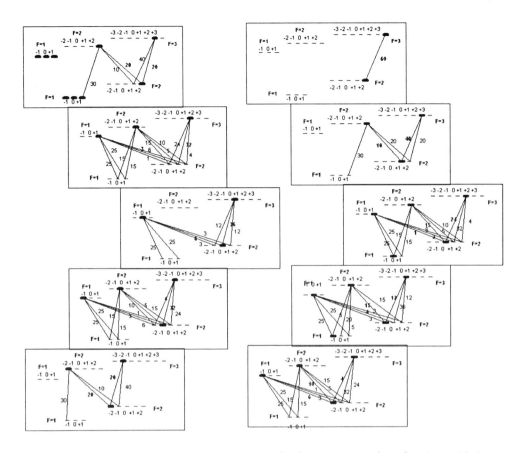

Figure 3. Transition probabilities for linear (left) and circular (right) polarized light

and potentially increase the signal return by over a factor of 2 if we use circularly polarized light and a second frequency displaced by 1.772 GHz. The transition probabilities of allowed transitions are shown in Fig. 3 for both linear and circularly polarized light [10].

Each of the F levels splits up into 2l+1 magnetic quantum levels in the presence of a magnetic field or linearly or circularly polarized light. For linearly polarized light only transitions with $\Delta m=0$ are permitted (see Fig. 3), whereas for circularly polarized light, Δm must be either +1 or -1, depending on the handedness of the light. Atoms in a given m level in the upper state can fall back according to the rule $\Delta m = -1, 0, +1$. If we look at the F=2 ground state → F=3 upper state transition we see that the bottom ground state consists of 5 M levels and the upper consists of 7 M levels. Study of the left hand side of Fig. 3 shows that, if an atom is cycled many times, the selection rules tend to move the atoms towards the m=0, m=0 state for linearly polarized light (Fig. 3). If however we use circularly polarized light, the selection rule for absorption is Δm= +1 (or -1 for the other polarization). This forces the levels to the right and after about 6-10 cycles the atoms is pumped into the m=2, m=3 state. The only radiation that can be emitted by the F=3, M=3 → F=2, M=2 transition is a dipole transition (circular polarization) which backscatters 50% more light than the average light scattered over 4π radians. A single frequency CW laser which is circularly polarized can therefore increase the backscattered return by a factor of 50%. The backscattered signal is also circularly polarization in the same sense as the laser beam.

We have only been able to use atoms which start of in the F=2 ground state, which, in thermal equilibrium comprise only 5/8 of he total number of atoms. As a further refinement we can introduce a second frequency 1.772 GHz above the primary frequency and thereby excite atoms from the F=1 ground state into upper levels F=0, F=1 and F=2. Atoms in the F=1 and F=2 upper state have a reasonable probability of falling into the lower F=2 state and then being pumped to the m=+3 state at the far right. We can therefore in principle pump all of the atoms into this state. This enables us to use potentially 100% of the atoms rather than the 62.5% that populate the F=2 ground state in thermal equilibrium. Models which use the rate equations between all the levels predict that you should be able to obtain a factor of 2.4 more light for the same power levels with circularly polarized light and a second frequency containing about 10% of the total energy. More detailed calculations show that the light has to be accurately polarized and the Earth's magnetic field precess the atoms (Larmor precession) which destroys the effect in time scales of a few microsecond. However this does not seem to be as serious as might be expected [9]. It takes about 20 cycles to obtain the full effect of optical pumping so the cycle time of the sodium atom should be short, preferably less than 500 nanoseconds and the intensity of the laser beam and the pulse format are important.

Some experimental evidence for significant optical pumping is available [6]. The beacon profiles for a sum frequency laser, are shown in Fig. 4 for both linear and circular polarized light and shows the advantage of optical pumping.

7. Calculation of the return flux from the layer for a single frequency laser

If we neglect saturation and changes in the populations in the ground state, and we know the column density, n, of the sodium atoms, the return flux is given by:

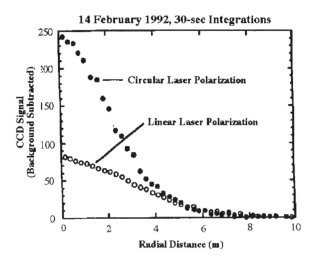

Figure 4. Beacon profiles obtained with linear and circularly polarized light using a 10 Hz 2.5 watt sum frequency laser at Lincoln Labs, Bedford MA

$$N = 3 X 10^{18} n\sigma T_a^2 \kappa sec(z)/(4\pi h^2) \ photons/cm^2/sec/watt \quad (13)$$

where σ is the atomic cross section (10×10^{-12} cm^2 for single frequency at a temperature of 200 °K), T_a is the atmospheric transmission, κ depends on optical pumping effects (≈ 1), z is the Zenith distance and, h is the vertical height of the layer [3]

Most measurements of the column density place it between 1.5 to 15×10^9 atoms/cm^2 (depending on the season and time of night). A column density of 2×10^9 atoms/cm^2 should be considered as a reasonable lower level in designing an AO system and we may expect a factor of 3 higher abundance for at least 3 months of the year. If we assume $\kappa = 1.5$ then 1 watt of single frequency power at the layer will return of 88 T_a photons/cm^2/sec provided saturation effects are not important. This is equivalent to a 10th magnitude star in the V band, or an 11.5 magnitude G star in the bandpass usually used for natural guide star AO.

Direct measurements of the return using a single mode laser and direct measurements of the sodium column density [2], by the Arizona's group at the University of Arizona give a return of 65 T_a photons/cm^2/sec when scaled to a column density of 2×10^9 atoms/cm^2. This is reasonable agree-

---

[3] The return flux scales as sec (z) rather than sec^2 (z) because, although the range scales as sec^2(z), which reduces the flux by this factor, the effective thickness of the layer increases as sec (z).

ment with the simple calculation which does not include the effect of optical pumping from the F=2 to F=1 ground state, which must exist for a single frequency laser. Peter Milloni at Los Alamos has calculated the return putting in the detailed effects of optical pumping, collisional mixing and the effect of the earth's magnetic field and obtains a very similar estimate of the return flux to that measured by Roger Angels group [9], [2].

To investigate the effect of pumping atoms from the f=1 ground state, Rodolphe Vuillemier [6], undertook to evaluate the return assuming that a second frequency is available which will optically pump atoms back from the F=1 to F=2 ground state. He using the full rate equations of each state of the sodium atom and found that a power level of only 10% of the main frequency is sufficient to achieve pumping of almost all atoms to the two level state. The calculated return flux under these conditions is 113 photons/cm^2/sec/watt of laser power at the mesosphere. These calculations do not take into the effects of the earth's magnetic field and the sodium beacon return flux under this condition has not been measured experimentally. However simple theory, complex modeling and direct observation all give similar estimates of the return flux of a single frequency, unsaturated CW laser of 65 T_a sec(z) photons/cm^2/sec/watt at the layer when scaled to a column density of 2×10^9 atoms/cm^2 This value seems reliable for calculations of AO performance. The return scales linearly with column density.

8. Return flux for different laser spectral and pulse formats

If a CW laser has a finite linewidth is its effective cross-section can be calculated and used in Eq. 13 to determine the appropriate return flux. The effective cross section is shown in Fig. 5 as a function of the laser linewidth, assuming that the laser profile is Gaussian and that the laser is tuned to the peak absorption frequency of the sodium layer.

Pulsed lasers can significantly effect the returns, depending on their format. The formats of two types of lasers currently being used in astronomy are shown in Figs. 6 and 7.

(i) LLNL pulse format

This laser was developed by LLNL for astronomy as a spin-off from its laser isotope separation program [4]. A CW master oscillator generates a double peaked spectral line width, one peak centered on the F=2 ground state transitions and the other the F=1 ground state and this signal is amplified by a series pulsed dye cells. If the line widths were infinitely narrow the effective cross-section of the sodium atom would be the mean of the two peaks or (8+5/16) of a single line tuned to the F=2 ground state transitions. The lines in the LLNL oscillator are electronically broad-

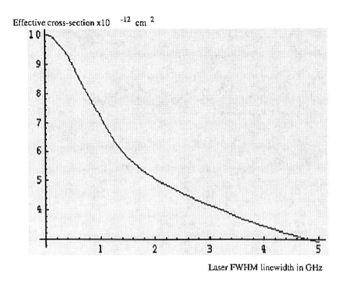

Figure 5. Effective cross-section of sodium in the mesosphere as a function of laser FWHM linewidth in GHz

Figure 6. Pulse format of the LLNL laser for the Keck AO system

ened to about 1.2 GHz FWHM so the effective cross section of the laser is $(13/16) \times 6.5 \times 10^{-12}$ cm^2 or 5.3×10^{-12} cm^2.

If we assume that the peak intensity of the pulses is at I_{sat}, the cycle time of an atom will be 64 nsec and on average an atom will absorb and emit 2-3 photons/pulse. The next pulse occurs after 33 microseconds so that an atom only cycles 6-10 times during a collision life-time and each pulse occurs after some 10 Larmor precession times. The net result of both effects is that the atom will not be pumped to the two level M=3 upper, M=2 lower, state and the factor κ will be approximately 1, rather than 1.5. The expected return flux/watt from this laser is therefore be 0.35 of the single frequency CW laser or 23 T_a $(n/2 \times 10^9)$ sec(z). However, because the linewidth of the laser is continuous, we should not expect significant additional losses due to radiation pressure.

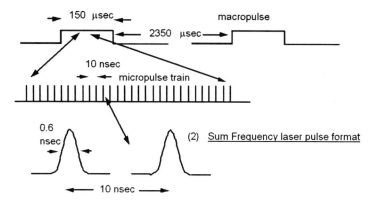

Figure 7. Pulse format of a sum-frequency laser

(ii) Sum Frequency pulse format

The laser used by our group is a sum frequency laser [5] [7], built by Tom Jeys at Lincoln Labs. This laser is described in more detail in the next talk but its pulse format is shown in Fig. 7. The format consists of a regular train of pulses lasting 150 microseconds with a repetition frequency of 400 Hz. These pulses are called macropulse and each macropulse consist of a train of very short pulses (micropulses) which are about 0.6 nsec wide and have a repetition frequency of 100 MHz.

This laser format is substantially more difficult to evaluate. The classical way to investigate optical pumping effects is to determine rate equations which describe the rate at which atoms are excited to each level and the rate each level decays via spontaneous and stimulated emission using the transition probabilities and selection rules given in Fig. 3. These equations are solved to form a steady state solution for all the levels and sublevels in the atom.

In the case of the sum frequency pulse format micropulse train, the period of the pulses is shorter the decay time of the atom in its upper state and the transition probability of an atom depends on the past history of the radiation field. For this case it is necessary to obtain a full solution of the Optical Block equations which do to relate the population levels to a periodic driving force. The Block equations are three interrelated first order differential equations derived from the Schrodinger field equations and in the steady state they simplify to the classical rate equation. For the driving field consisting of a series of short pulses with duration $t_p \ll 16$ nanoseconds the Block solution predict a population inversion of 1 for a strong enough fields (unlike the rate equations, have a maximum value of 1/2) and at even higher power levels the return flux can be reduced to low levels. However it has been shown [1] that, for a pulse train period of about

$0.65\tau_s$ (i.e. about 10 nsec) and power levels appropriate for AO systems, the sodium atom does not see the individual pulses but rather the mean power level, averaged over all the pulses, so that the atom acts like a low pass filter.

Under these conditions, the linewidth of the laser is set by the transform limited width of the micropulses with a typical FWHM linewidth of 0.9 GHz. Because the pulse length of the macropulse is of the same order as the collision lifetime, we would expect strong optical pumping with κ =1.5, so a simple model of the interaction of this laser with the sodium atom would give a cross section of 8×10^{-12} cm^2 and a return flux of 52 T_a $(n/2\times10^9)$ sec(z) photons/cm^2/watt at the layer. More detailed calculations by Milonni solving the Optical Block Equations directly give a somewhat higher flux than a CW laser [4].

These numbers are consistent with an experimental return of 37 T_a $(n/2\times10^9)$ sec(z) photons/cm^2/watt at the layer with a 3 GHz laser bandwidth ($\sigma_{eff} = 4.2\times10^{-12}$cm^2).

9. Summary and Conclusions

We have been able the derive many of the characteristics of the sodium atom interaction with a laser beam using elementary physics. The numbers obtained from these simple calculations agree surprising well with more detailed computer calculations and such experimental data that is available. The abundance of sodium in the mesosphere varies by a factor of 6 over the year from a level of about 2×10^9 to over 10×10^9 atoms/cm^2 so that low power lasers can sometimes be expected to generate a bright laser beacon. Laser beacon systems for facility AO systems should be designed to meet the overall system specification in an abundance of 2×10^9 atoms/cm^2.

Saturation effects, especially due to radiation pressure, are significant and impose challenging constraints in the laser designer which are discussed in more detail in the next lecture. Optical pumping can increase the return signal by a factor of at least 2.4 and this has been demonstrated using a sum-frequency laser with I = 0.2 mW/cm^2. This indicates that the effect of Lamor precession is not as large as might be expected. Pulsed lasers with a format of 30-100 nanosecond pulses every collision lifetime do not optically pump the atoms and have a serious saturation problem. Pulsed lasers consisting of pulse trains with a period of order the upper state lifetime and a

---

[4]The simple calculation gives a return flux of 27 T_a $(n/2\times10^9)$ sec(z) photons/cm^2/watt. There is therefore some experimental and theoretical evidence to suggest that the return of a sum-frequency laser may be higher than a single frequency CW laser since the pulse and spectral format is able to pump a significant fraction of the F=1 ground state into the two level F=2 ground < – > F=3 upper state.

pulse duration of $\ll$ this lifetime are approximately equivalent to a CW laser with a broad spectral envelop.

At a sodium abundance of 2×10^9 atoms/cm^2, a single frequency CW or transform limited sum frequency laser should give returns of order 65 T_a (n/2$\times10^9$) sec(z) photon/cm^2/watt at the mesosphere, which is equivalent to about a 10th magnitude star in the V band. The return of an ideal laser with a LLNL pulse and spectral format is about 23 T_a (n/2$\times10^9$) sec(z) photon/cm^2/watt.

References

1. Lee.C. Bradley. Pulse-train excitation of sodium for use as a synthetic beacon. *J Opt Soc Am A*, 9:1931–1944, 1992.
2. Jian Ge, B.P.Jacobsen, J.R.P. Angel, P.C.McGuire, T.Roberts, B.McLeod, and M.Lloyd-Hart. Simultaneous measurement of sodium column density and laser guide star brightness. *SPIE*, 3353:242–253, 1998.
3. W. Happer, G.J. MacDonald, C.E. Max, and F.J. Dyson. Atmospheric-turbulence compensation by resonant optical backscattering from the sodium layer in the upper atmosphere. *J Opt Soc Am A*, 11:263–276, 1994.
4. H.W.Friedman, G.V.Erbert, T.C.Kuklo, J.T.Salmon, D.A.Smauley, G.R.Thompson, J.G.Malik, N.J.Wong, K.Kanz, and K.Neeb. Sodium beacon laser system for the lick observatory. *SPIE*, 2534:150–160, 1995.
5. T.H. Jeys, A.A. Brailove, and A.Mooradian. Sum frequency generation of sodium resonance radiation. *Appl Opt*, 28:2588, 1989.
6. T.H. Jeys, R.M. Heinrichs, K.F. Wall, J. Korn, T.C. Hotaling, and E.J. Kibblewhite. Observation of optical pumping of mesospheric sodium. *Opt Lett*, 17:1143–1145, 1992.
7. E.J. Kibblewhite and Fang Shi. Design and field test of an 8 watt sum-frequency laser for adpative optics. *SPIE*, 3353:300–309, 1998.
8. G. Megie, F. Bos, J.E. Blamont, and M.L. Chanin. Simultaneous nighttime lidar measurements of atmospheric sodium and potassium. *Planet Space Sci*, 26:27–35, 1978.
9. P.W.Milonni, R.Q.Fugate, and J.M.Telle. Analysis of measured photon returns from sodium beacons. *J Opt Soc Am A*, 15:217–233, 1998.
10. P.J. Ungar, D.S. Weiss, E. Riis, and Steven Chu. Optical molasses and multilevel atoms: theory. *J Opt Soc Am B*, 6:2058–2071, 1989.

CHAPTER 4
THE DESIGN AND PERFORMANCE OF LASER SYSTEMS FOR GENERATING SODIUM BEACONS

EDWARD KIBBLEWHITE
Department of Astronomy
University of Chicago
5640 Ellis Avenue
Chicago IL 60637 - USA

1. Introduction

In a previous lecture we discussed the basic physics of the sodium atom, the various types of saturation effects which might limit the return flux and some of the optical pumping techniques which could be used to maximize the return. In this talk we discuss techniques to optimize the design of lasers used to generate sodium beacons and measure their performance.

2. How small can we make the laser beacon?

The size of the laser beacon at the mesosphere is a critical issue, since it defines the saturation effects of the laser and the power needed to reach a given system performance. The laser beam is usually focused onto the sodium layer using a dedicated telescope. There is an optimum diameter of the laser beam leaving the telescope, because, if the diameter is too small, the beam will be spread out by diffraction and if it is too large it will be distorted due to atmospheric turbulence. The simplest case involves the use of a laser with a uniform beam power density (a 'top-hat') distribution since, for this distribution and for a telescope of moderate aperture, the Strehl ratio of the laser beam will be given by:

$$\Gamma = exp[-0.134(d/r_o)^{5/3}]\Gamma_0 \qquad (1)$$

where

- d is the diameter of the launch telescope

- r_o is Fried Parameter at 0.589μm
- Γ_o is the Strehl ratio of the laser + optics

The peak intensity at the layer is then given by

$$I_{pk} = 0.88\Gamma/(\lambda h/d)^2 \qquad (2)$$

where

- λ is the wavelength (0.589 μm)
- h is the height of the layer above the telescope

In 1 arcsecond seeing conditions (r_0=0.12 m), I_{pk} is a maximum when d=0.42 meters. Assuming good optics and laser beam quality (Strehl ratio of $\approx$ 0.6), we get a peak intensity of 1 mw/cm^2/watt at the sodium layer and an effective spot size of 28 cm FWHM, corresponding to an angular size of 0.6 arcseconds.

Most laser beams have a Gaussian radial profile and the mtf of the unsaturated beacon at the layer is given by [9]:

$$mtf(r) = \frac{exp[-3.43(r/r_0)^{5/3}(1 - 0.965(r/u_0)^{1/3} + 0.067(r/u_0)^{7/3})) - (r^2/2u_0^2)])}{(1 + 0.52(r^2/u_0^2)(u_0/r_0)^{5/3} + 0.742(r^4/u_0^4)(u_0/r_0)^{10/3})^{1/2}} \qquad (3)$$

where

- $mtf(r)$ is the mtf of the beacon at the layer
- r_0 is the Fried parameter at 0.589 μm
- u_0 is the $1/e^2$ intensity radius of the beam at the telescope pupil

The intensity profile can then be calculated using a Fourier-Bessel transform. The Strehl ratio (normalized peak intensity on axis) is plotted as a function of u_o/r_0 in Fig. 1. For a launch telescope of infinite aperture the optimum $1/e^2$ beam radius is about 2 r_o with a peak Strehl ratio of 0.22.

The theoretical laser beacon profile at the sodium layer for u_0=0.2 meter, r_0=0.105 meter is shown in Fig. 2 for a unaberrated laser and telescope of infinite diameter..

This profile is well represented by the sum of two Gaussians, a core of $e^{-1/2}$ radius of 0.16 arcseconds containing 30% of the total energy and a halo of $e^{-1/2}$ radius 0.48 arcseconds [1]. In practice, the telescope aperture is finite and there will be aberrations in both the laser beam and telescope optics which reduce the Strehl ratio. This profile therefore represents the smallest beacon which can be focused onto the layer in 1 arcsecond seeing without adaptive optic correction of the wavefront.

[1]For comparison, a diffraction-limited image would have an $e^{-1/2}$ radius of 0.096 arcsecond.

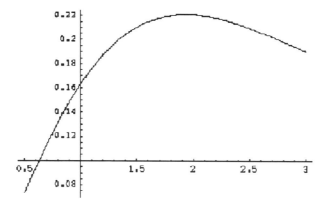

Figure 1. Strehl ratio as a function of u_0/r_0 for an unaberrated telescope and laser. Optimum $u_o/r_0 = 1.92$

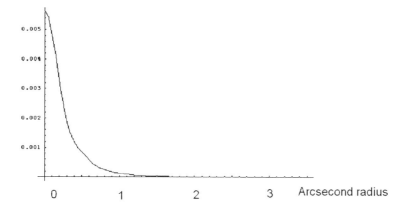

Figure 2. Theoretical laser beacon profile for a $1/e^2$ intensity diameter of 0.4 meters and $r_0 = 0.105$ m at 0.59 μm.

Two additional factors will increase the size of the beacon :
(1) The finite depth of focus of the beam at the layer.
The sodium layer has an approximate Gaussian density profile with a 1/e density half width of about 4 km. The beam diameter is given by:

$$w^2 = w_0^2 + (z\lambda/2\pi w_0)^2 \qquad (4)$$

For a diffraction limited $1/e^2$ beam radius of 0.20 meters at the launch telescope, the waist radius, w_0, at the mesosphere is 0.084 meters. For comparison, the depth of focus term $(z\lambda/2\pi w_0)$ is only 0.0045 meters at the 4 km half width of the mesospheric sodium layer, so this effect is small and is usually neglected.

(2) Obliquity of the beam

A more serious problem is caused by the obliquity of the beam when seen by subapertures some distance from the optical axis. The thin laser pencil beam passing through the layer is seen as a line in the radial direction by the outer subapertures on the pupil with an rms angular size of ($\Delta h\, r/h^2$). For a typical 4 km rms layer thickness and a beacon altitude of 90 km, this effect introduces a FWHM broadening in a radial direction equal to (0.24d) arcseconds (where d is the displacement of the laser beam in meters from a given subaperture).

This effect is minimized if the beam is launched along the optical axis of the telescope, but, even so, since we have seen that the FWHM of the beacon can be as small as 0.6 arcseconds at the layer, this more than doubles the size of the beacon in the radial direction at the edge of an 8 meter telescope. If the laser telescope is mounted at the side of an 8 meter telescope the spot size changes by a factor of 3 across the pupil. This not only requires more laser power (to make up for the effect of the larger laser beacon), but optimum reconstruction of the wavefront from the wavefront sensor data is more difficult (and as yet unsolved), since there is now significant position dependent noise across the pupil.

This is not the only problem associated with obliquity. The sodium layer can sometimes become stratified and this stratification can introduce systematic errors into the wavefront determination. If the laser telescope is located at the middle of the telescope pupil, this effect causes a change of focus, which is not very important because the laser beacon already has to be automatically focused by its own control loop, but if the laser is launched from the side, stratification introduces astigmatism into the wavefront measurement which must be corrected using data obtained from a natural guide stars located within the isoplanatic patch. This effect is independent on the spot size in the beacon layer. Both these effects are minimized if the beacon is launched along the optical axis, a factor that must be traded against the added complexity and cost of placing the launch telescope in this position. There are, however, strong theoretical arguments for putting the laser telescope behind the secondary mirror of the telescope.

(3) Accuracy of estimating the wavefront

The laser power required to measure the wavefront to a given accuracy depends ultimately on the accuracy with which we can measure the position of the beacon in the presence of photon and detector noise.

If we had a truly photon limited detector which could determine the position of every photon hitting the focal plane, each photon position would be an independent estimate of the position of the image so that, if the image shape is Gaussian, the variance on the position would be given by:

$$Var\{position\} \equiv <x_s^2> = \sigma^2/N_s\tau \tag{5}$$

where
- N_s is the number of photons per second detected from the beacon
- τ is the integration time
- σ is the standard deviation of the Gaussian image on the detector

This is the minimum variance attainable with any position estimation algorithm. For the more usual quad cell estimator, using a CCD with read noise n_R, the variance is greater and is given by:

$$<x_s^2> = (\pi/2)\sigma^2(N_s\tau + 4n_R^2)/(N_s\tau)^2 \tag{6}$$

We can transform wavefront sensor image displacements across the detector into angle of arrivals on the sky using the equation

$$\alpha_s = (\sigma/f)(lm/D) = \sigma.g \tag{7}$$

where
- f is the focal length of the lenslet array
- l is the size of a subaperture on the lenslet array
- m is the number of subapertures across the pupil
- D is the telescope diameter

For a stellar beacon, the variance of the angle of arrival α is given by:

$$<\alpha_s^2> = (\pi/2)(\theta_s^2)(N_s\tau + 4n_R^2)/(N_s\tau)^2 \tag{8}$$

where
- θ_s is the short exposure
- $e^{-1/2}$ angular radius of a star viewed through an individual subaperture

If the laser beacon has a Gaussian profile of $e^{-1/2}$ radius, σ_b, at the mesosphere, the effective angular size of the laser beacon viewed through the atmosphere is given approximately by $(\theta_s^2 + \theta_b^2)^{1/2}$ where $\theta_b = \sigma_b/h$ and the position variance of the laser beacon is given by:

$$<\alpha_b^2> = (\pi/2)((\theta_s^2 + \theta_b^2)(N_b\tau + 4n_r^2)/(N_b\tau)^2 \tag{9}$$

$$<\alpha_b^2> = (1 + (\theta_b/\theta_s)^2) <\alpha_s^2> \tag{10}$$

where $<\sigma_s^2>$ is the angle of arrival variance in a subaperture of a star with the same flux as the beacon.

Equation 10 suggests that we should try to make the laser beacon angular diameter about equal or slightly smaller than to the short exposure resolution of the subaperture, since there is a high penalty in accuracy if θ_b is much larger than the stellar image. However, if the beacon is very small (e.g. by using adaptive optics in the launch telescope), its size will change very significantly with the radial position of its subaperture. This change is size will affect the variance of the gradient measurements across the pupil and this effect must be factored into the computation of the optimum reconstruction algorithms.

If we impose two additional conditions:

(1) The wavefront sensor is not detector noise limited, i.e. the total flux/subaperture²/integration time $\geq 4n_R^2$.

(2) The sodium beacon is not saturation limited, i.e. $I(0) \leq I_{sat}$.

then the angle of arrival variance in individual subapertures is:

$$<\alpha_b^2> \leq (2\pi\theta_b^2)/N_b\tau \quad rad^2 \qquad (11)$$
$$\leq (4\pi\sigma_b^2)/(h^2 N_b\tau) \quad rad^2 \qquad (12)$$

Now the central intensity of a Gaussian profile beacon at the mesosphere with an integrated flux N_0 is:

$$n(0) = N_0/2\pi\sigma_b^2 \quad photons/cm^2/sec \qquad (13)$$

so that, since the flux collected by a wavefront sensor subaperture of side L is $N_0 T_a L^2 \tau / 4\pi h^2$, we can write the variance of estimating the angle of arrival of the wavefront is a subaperture as:

$$<\alpha_b^2> \leq 8\pi/(n(0)T_a L^2 \tau) \qquad (14)$$

and the variance of the wavefront at a wavelength λ_s, derived from these data using a least squares reconstruction matrix is given approximately by:

$$<\phi_b^2> \approx 0.7 <\alpha_b^2> L^2 (2\pi/\lambda)^2 = <\alpha_b^2> L^2 \quad rad^2 \qquad (15)$$
$$\leq 22\pi^2/(\lambda_s^2 n(0) T_a \tau) \qquad (16)$$

Equation 16 shows that the accuracy of determining the wavefront depends only on the central intensity of the laser beacon, which is ultimately set by the characteristics of the laser, and the integration time, which is set by the overall system design requirement on the servo bandwidth.

3. Saturation properties of the laser at the sodium layer

The peak power/natural linewidth incident at the sodium layer is given by

$$I_{pk} = 0.88\Lambda(PT_s/\eta)(\Gamma/(\theta_{laser}h)^2) \qquad (17)$$

where

- I_{pk} is the peak power density in watts/m^2/natural linewidth
- P is the laser power
- T_s is the transmission of the optics and the atmosphere
- η is the duty cycle of the laser
- θ_{laser} is the FWHM angular diameter of the beacon
- h is the height of the beacon
- Γ is the Strehl ratio through the optics and atmosphere
- Λ is the spectral filling factor

This equation allows us to set a maximum usable power for the laser for which saturation will not be a problem. This power is given by the inequality:

$$P_{usable} < 1.14 I_{sat}(\theta_{laser}h)2\eta/(T_s\Lambda\Gamma) \qquad (18)$$

Assuming $I_{sat} \approx 20$ watts/m^2 for a single frequency laser (radiation pressure limited), a FWHM beacon diameter of 0.20 meters and an overall transmission to the layer of 0.5, we get significantly loss of efficency if the laser power > 2 watts. Operation at higher powers requires broadening the spectral coverage, until, at the limit of a completely filled linewidth 3 GHz wide laser spectral profile, $\Lambda = 10/3000$ and $P_{usable} = 1560\eta/\Gamma$ watts [2].

Equation 18 provides a straightforward means of assessing some of the trade-offs required in designing a laser for astronomical AO. If the laser is pulsed, then the duty cycle should be as high as possible and the spectral filling factor adjusted to provide a cycle time of a few hundred nanoseconds (see previous Chapter).

A true figure of merit estimates the wavefront error due to photon noise for different lasers and can be calculated directly once the measured profile and flux return has been measured. An example of this calculation for a sum frequency laser using measured flux and profile data is given in section 6. Significant factors making up this figure of merit are the spectral and temporal properties of the laser and its ability to optical pump the sodium atoms in the mesosphere.

---

[2] The cross section of the sodium atom depends on the linewidth of the laser so this equation cannot be used directly to compare lasers of different types. Other factors are also important, for instance, radiation pressure is less significant for a fully filled laser linewidth than will a laser with discrete modes.

Type of Laser	Power (watts)	Linewidth (MHz)	Duty Cycle	Optical pumping
CW dye	1-5	0 to 500	100%	Yes
LLNL	10-1000	3000	0.1-1%	No
Sum frequency	10-20	600 to 3000	6-20%	Yes

TABLE 1. Characteristics of three different lasers.

Figure 3. This image shows a 12th magnitude beacon generated by a 1/2 watt CW laser at Yerkes in October 1991 [3]. The sodium beacon is about 2 arcseconds in diameter. The laser beam was displaced by 5 meters from the optical axis of the telescope and Rayleigh backscattering from the atmosphere is seen on the right hand side. The bright spot towards its end being caused by scattering from dust from the Mount Pinatubo volcanic eruption in 1990. It is usual to displace the laser beam to reduce the effect of Rayleigh backscattering. Elongation of the beacon along the direction of the laser beam can be clearly seen.

4. Lasers for the generation of sodium beacons

There are currently three types of laser which have been successfully used to generate compact laser beacons. Details of the lasers are summarized in Table 1.

4.1. CW DYE LASERS - (UC/UA/MAX PLANCK)

The use of a CW dye laser for infra-red AO systems was first proposed in 1989 [6], based on the calculation the surface brightness of the laser beacon would be comparable with that due to low atmospheric Rayleigh backscattering and that displacement of the CW beacon from the optical axis of the telescope would enable a CW beacon to be used for adaptive optics. A photograph of an early beacon is shown in Fig. 3.

CW dye lasers are well developed and their design is discussed in the literature [e.g. [1]]. They rely on pumping of a small volume of dye (typically ≈ 10 microns diameter) with a diffraction limited laser beam, typically generated by an argon ion laser, and its power and performance is limited by thermal effects. The dye is formed into a planar jet in air by a nozzle and the thermal effects minimized by using a mix of AMOX (a type of detergent), glycol and water at a temperature of about 10°C. At this temperature, the change of refractive index with temperature, dN/dT, undergoes an inversion, reducing the thermal effects. Introduction of a poisonous, smelly substance, COT, reduces the build up of a triplet state of the dye and enables operation at higher power though poses a health and safety issue. CW dye lasers give outputs of about 3 watts in either single or narrow band operation with reasonable beam quality. A travelling wave laser cavity configuration is important for single frequency operation, for wider band operations standing wave lasers are also suitable and are simpler to run [1].

The great advantage of this type of laser is that it is commercially available. Its major disadvantage is the low power; it is just very difficult to get the power above 3-4 watts and still maintain good beam quality and this is not enough for high order correction even at infra-red wavelengths. Less serious problems are its generally messy nature and the high power needed for the argon ion laser (40 kW) [4].

4.2. PULSED DYE LASERS

To increase the power of a dye laser significantly, it is necessary to pump the dye laser with a pulse duration so that $\tau_p \ll$ interaction length/sound velocity. This ensures that the laser pulse is over before the fluid knows that it has been heated by the pump beam. The interaction length is larger for pulsed than for CW lasers, typically being 100-1000 microns. If we assume a sound speed of 1000m/sec, this means the pulse lengths must by shorter than about 100 nanoseconds. The velocity of the dye should be such that the dye is completely changed in the active region before the next pulse. Velocities of order 10-30m/sec are possible, so that pulse reppetition rate can be of order 10-30KHz.

These lasers can be made to be both very reliable and to give enormous amounts of power (the LLNL isotope separation plant gives over a 1000 watts!). However, their effectiveness is limited by the maximum duty cycle (< 1 %) that the laser can operate under without using either multiple lasers, to increase the effective pulse rate or optical delay lines, to stretch out the pulse length. Both options increase the complexity and cost of the laser. A field version of the LLNL system operated at Lick was able to

---

[4] The argon ion laser can in principal be replaced with a frequency doubled YAG lasers. Such a laser will have much lower power.

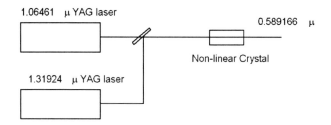

Figure 4. Schematic diagram of a sum frequency laser.

power the first AO system to correct images using a sodium beacon and operated with a duty cycle of 0.001 [2]. The duty cycle of the Keck system is about 0.5% [3].

4.3. SUM-FREQUENCY LASERS

The sum frequency laser was designed specifically for the generation of sodium beacons and uses a fortunate accident of nature, which is that the sum of two infra-red Nd: YAG laser frequencies, combined in a non-linear optical crystal, can be tuned to the D_2 line transition [5]. Sum frequency generation can be very efficient, provides an inherently high quality wavefront and its spectral and temporal properties can be matched to the sodium layer.

The laser is shown schematically in Fig. 4. It consists of two diode-pumped mode locked lasers, one operated at 1.06 microns and the other at 1.32 microns which are combined in a non-linear crystal to generate yellow light at 0.589 microns. Strictly speaking it is not a sum frequency laser, since the sum frequency conversion, at least in this implementation, is achieved outside the laser resonators.

YAG lasers can be made to give high duty cycles with good beam quality. This beam quality is further improved by the non-linear conversion process of the crystal which "cleans up" the beam. Diode laser pumped YAG lasers are compact, reliable and highly efficient and can be made to produce infra-red light of over 10 watts with an ideal pulse and spectral format.

The key challenge for the sum frequency laser is to obtain efficient conversion of the infra-red light using the non-linear crystal. Since the conversion is a non-linear process (the electric field of the yellow light being proportional to the produce of the electric field of the two laser beams) the power in yellow light generated in the crystal is given by

$$P_{0.59} = \eta P_{1.06} P_{1.32} L^2 [sin^2(\Delta kl/2)/(\Delta kl/2)^2]/d^2 \tag{19}$$

where

- $P_{0.59}$, $P_{1.06}$, $P_{1.32}$ are the powers at $0.59\mu m$, $1.06\mu m$ and $1.32\mu m$
- η is the conversion efficiency (which depends on the material)
- l is the length of the non-linear crystal
- $\Delta k = k_{1.06} + k_{1.32} - k_{0.59}$ (k is the wavenumber $= 2\pi n/\lambda$)
- d is the diameter of the laser beams focused in the crystal.

It is necessary that the phase of the infra red and yellow light remain in step. Because the polarization of the yellow light is orthogonal to that of the infra-red light, the orientation and temperature of the crystal can be adjusted so that the refractive index of both the infra-red light and the yellow light make Δk =0. The length of the crystal is then limited by the depth of focus of the beam in the crystal and by the finite bandwidth of the YAG lasers. Although multiple pass configurations through the crystal can increase the conversion efficiency, the only way to dramatically increase the power at 0.59 microns is to use a high power pulsed laser, which immediately runs up against the saturation problem discussed in section 3. The approach taken by its inventors Tom Jeys and Aram Mooradian was to use mode locked YAG lasers to overcome this problem.

- Mode locking of a laser

A laser normally operates in a number of longitudinal modes separated by a frequency of ν_{cavity} (c/2L, where L is the cavity length), shown schematically in Fig. 5. These modes normally operate independently with a random, time dependent, phase shift between each mode. For a laser with a FWHM linewidth of ν_{YAG} there are N ($\approx \nu_{YAG}/\nu_{cavity}$) modes.

In a mode locked laser, an acousto-optic modulator is placed in the cavity, tuned to exactly 1/2 of the mode spacing frequency. This modulator has the effect of periodically resetting the phase of all the modes to zero every transit time of the cavity. All the modes then add together coherently and the output power is initially increased by a factor of N, but, after a time of order $1/N\nu_{cavity}$, the different modes destructively interfere. The result is that the laser generates a pulse train of frequency ν_{cavity} and pulse duration $1/N\nu_{cavity}$. The peak power of the pulse is N times the power of the non-mode-locked pulse and the conversion efficency of the sum frequency conversion increased by a similar factor.

The FWHM spectral bandwidth of a transform-limited mode-locked laser is given by

$$\Delta \nu = 0.88/\Delta\tau \quad GHz \tag{20}$$

where $\Delta\tau$ is the width of the pulse in nanoseconds

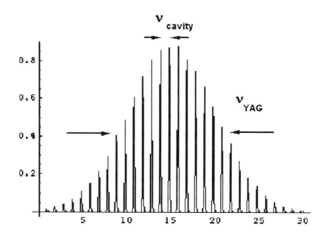

Figure 5. Mode structure of a typical laser

The laser spectrum consists of a set of modes spaced ν_{laser} GHz apart under a Gaussian envelope of FWHM $\Delta\nu$ GHz. If the mode locked pulse duration is too short, the linewidth of the laser light becomes broader than the Doppler width of the sodium atoms and the effective cross-section of the atoms to the radiation field is much reduced. The maximum FWHM gain envelope, ν_{YAG} is limited by the properties of the material but can be reduced by the addition of an etalon into the cavity so that, in practice, the value of N, and hence the mode locked pulse duration, can be controlled by the laser designer [5]. For a spectral bandwidth of 800 MHz, we require mode locked pulse widths of about 0.7 nanoseconds for each of the mode locked lasers.

Calculation of an optimum micropulse frequency is more complex. The interaction physics between a train of micropulses and the sodium atom requires the full solution of the Optical Block equations, briefly discussed in the previous Chapter. However it is possible to arrive at an approximate value of the optimum value of ν_{cavity}. If the micropulse frequency $\nu_{cavity} \gg 1/\tau$ (the lifetime of the sodium atom in the upper state), the atom will act as a low pass filter and will not respond to individual pulses but rather to the mean power level of the micropulses. However, the peak power in the micropulses is inversely proportional to their frequency (since the micropulse width has already been fixed by the requirement that it matches the dopplerwidth of the D_2 line). If $\nu_{cavity} \ll 1/\tau$ the atom will

---

[5] The bandwidth of yellow light is less than that of the two individual YAG lasers due to the non-linear nature of the mixing process. Because the power of the sum frequency light is proportional to the product of the two IR powers the spectral bandwidth of the yellow light is given by: $fwhm_{0,59} = [(fwhm_{1.06}^2 \times fwhm_{1.32}^2)/(fwhm_{1.06}^2 + fwhm_{1.32}^2)]^{1/2}$.

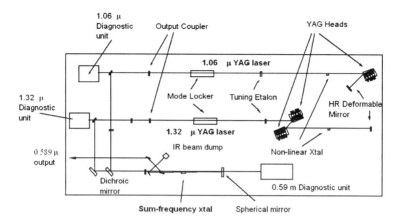

Figure 6. Optical layout of the sum frequency laser. Each YAG heads consist of double pass zigzag slabs pumped by 40 diode lasers. Infra-red light is passed twice through the LBO sum frequency crystal.

be activated by single micropulses and will tend to saturate, so we may posit that a good compromise would be for $\nu_{cavity} \approx 1/\tau$. In fact, although the physics is complex (it is, for instance, possible to achieve higher levels in the upper state than with a CW laser at some power levels but at still higher power levels the population is much reduced), detailed calculations show that the optimum cavity mode spacing is about 100 MHz, which is consistent with our hand waving value of $1/\tau$ Hz. At a 100MHz micropulse frequency, the sodium atom acts as though it sees a constant power level equal to the mean power of the micropulses. The sum frequency crystal, however, sees a factor of 14 times more peak power and (since the duty cycle is only now 1/14) should increase its conversion efficiency by the same factor. This however requires top-hat pulses of equal width and, in practice, more modest gains of about 5 are achieved, due mainly to the Gaussian beam profile, mismatch in the micropulse widths for the two lasers and errors in pointing of the two infra-red beams.

A drawing of the sum frequency built for our project by Lincoln Labs in shown in Fig. 6.

The laser is built on a 30" by 72" optical bench. The 1.06 micron laser is on the left, the 1.32 micron YAG in the center. The two infra-red beams are combined in the far right and focused onto the non-linear crystal which is operated in a double pass. A sodium cell is in the upper right. The laser is driven from a 1/2 height rack of electronics. The total power used by the laser and support electronics is 2300 watts.

A summary of the characteristics of the sum frequency laser are given in Table 2.

The sum-frequency laser has substantial advantages over other lasers

Figure 7. Sum frequency laser under test at Lincoln Labs tuned to the sodium D2 line. The laser is producing 9.5 watts with a duty cycle of 6% and linewidth of 0.9GHz.

Parameter	$1.06\mu m$	$1.32\mu m$	$0.59\mu m$
Micropulse length (nsec)	0.58	0.78	0.57
Calculated bandwidth (GHz)	1.52	1.13	0.90
Measured bandwidth (GHz)	1.55	1.1	0.90
Macropulse length (microsec)	180	150	150
Mode separation (MHz)	100.24	100.24	100.24
Mode locked power (watts)	13	10	9.5
Duty cycle	7%	6%	6%

TABLE 2. Characteristics of sum frequency laser

which have been developed for this work. These advantages include:

(1) Long pulse duration (150-200 μsecond > collision lifetime). This enables the laser to pump theoretically all sodium atoms into a two level state, increasing the backscattered return by a factor of up to 2.4.

(2) Pulse repetition period > light transit time to and from the sodium layer. This allows the light from Rayleigh backscattering from air molecules to be gated out. There is substantial amount of Rayleigh backscattered light even at high altitudes and it is necessary to use a defining aperture surrounding the laser beacon to mask out this light. Too small a defining aperture in wavefront sensor focal plane may introduce systematic errors into wavefront data. Too large an aperture may contaminate subapertures near to the laser beam and effect the noise properties of these subapertures.

(3) Wavefront data can be read from CCD immediately after the pulse has traversed the atmosphere. This effectively increases servo bandwidth

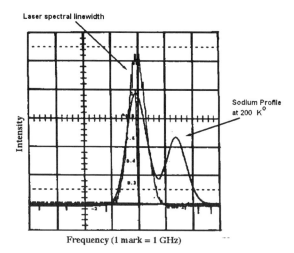

Figure 8. Lineprofile of the sum-frequency laser compared to the D2 absorption line in the mesosphere.

by factor of 1.5 for a given photon flux. Low bandwidth is major problem in high performance AO systems, especially since there is more power at high frequencies than predicted by simple theory.

(4) The mode locked pulse format enables the linewidth of the laser to be matched to the D_2 line and provides effective pumping of the atom .

(5) The beam quality of the laser is extremely good. Mode locked TEM_{00} YAG lasers already have very good beam quality; since the non-linear conversion process depends on the produce of the intensity of the beams,any halo round the infra-red beam at focus does not contribute to the formation of yellow light and reduces the conversion efficency . The yellow light which is formed has a higher quality wavefront than either of the two lasers ("beam clean-up").

(6) The solid state YAG lasers are efficient and easy to maintain. The current lifetime of the diode lasers that pump the YAG is over 10^{10} pulses or over 4000 hours. Replacement cost is <$100K. The entire laser has been rebuilt from scratch by the author and a student in an afternoon at 9200 feet altitude.

5. Laser field tests

5.1. LASER BEACON PROFILE AT THE LAYER

To measure the beacon profile and return flux generated by the sum frequency laser, the laser was installed at one of the Coude focal positions of the NOAO 75 cm Dunn Vacuum tower telescope at Sacramento Peak NM [8].

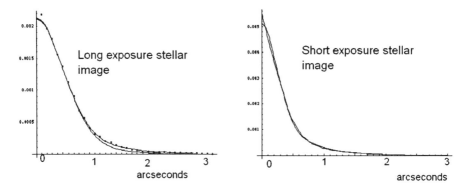

Figure 9. Radial profile of a star taken with the 75 cm NOAO Vacuum tower telescope

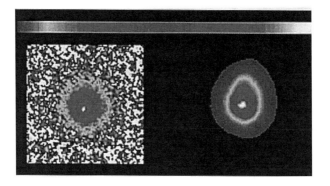

Figure 10. The image of a 9.5 V magnitude sodium beacon is shown on the left, a long exposure image of a m_V =2.6 star on the right. Laser power at the layer was 1 watt, r_o =0.105 m. The column density of the sodium atoms was $6 \times 10^9 / cm^2$.

This telescope was used both to focus the laser beam on the sodium layer and to observe the resulting beacon. Since imaging of the beacon through the atmosphere is tilt invariant in this configuration, the resolution of the telescope for imaging the beacon is increased and an accurate profile of the beacon can be measured. Long and short exposure point spread functions of the telescope are shown in Fig. 9.

The laser beam was expanded by the telescope to a $1/e^2$ diameter of 40 cm and focused onto the layer. A rotating shutter in front of the camera cut out Rayleigh backscattered light form the lower atmosphere and fluorescence within the optical beamsplitter was removed by suitable use of a filter. Direct images of the laser beacon and a bright star are shown in Fig. 10.

The radial profile of the beacon is shown in the left in Fig. 11, with the measured data displayed as points. The beacon profile at the mesosphere was obtained by deconvolution of this profile by the short exposure stellar

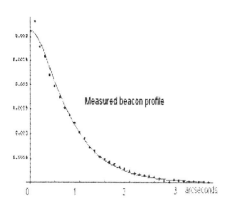 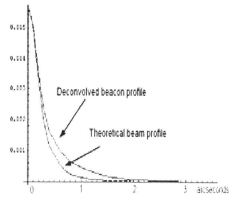

Figure 11. Measured and deconvolved sodium beacon profiles; $r_0 = 0.105$ m.

Gaussian	SF laser beacon		Ideal laser beacon	
	$1/e^{1/2}$ radius	Energy	$1/e^{1/2}$ radius	Energy
Core	0.17	13%	0.16	30%
Inner halo	0.54	47%	0.48	70%
Outer halo	0.90	40%		

TABLE 3. Parameters of Gaussians that give the best fit to the deconvolved laser profile.

psf shown in Fig. 9. The convolution of the beacon profile and the short exposure stellar image is shown as the filled line in Fig. 11 (left).

Also shown in Fig. 11 (right) is the theoretical profile calculated from Eq. 3 and already shown in Fig. 2. The deconvolved profile is well represented by three Gaussians with parameters shown in Table 3. The profile shape of theoretical profile is given for comparison, radii are given in arcseconds. Although the FWHM diameter of the beacon is 0.6 arcsecond, in good agreement with the theoretical value, the laser beacon has an extended halo caused by residual wavefront distortion in the laser beam, optics and launch telescope[6]. This halo significantly effects the accuracy with which the position of the beacon can be measured. However, once the profile is known we can compute directly the wavefront error of this laser as a function of laser power and system parameters. This exercise is carried out in the next section which discusses the system issues.

---

[6]Theoretical atmospheric mtf formulas also underestimate the effect of turbulence for $D/r_0 > 4$.

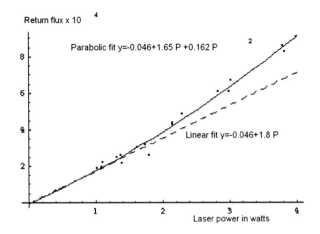

Figure 12. Return flux of the Sac Peak laser experiment as a function of laser power.

5.2. RETURN FLUX FOR THE SODIUM BEACON

The return flux was measured by calibrating the CCD detector with an absolute light source. Photometry of a standard star enabled the total transmission of the atmosphere, telescope and optics to be measured. The total transmission was 0.21, so that only 1 watt of yellow light was incident on the layer for 5 watts of power output at the laser. There were a large number of optical surfaces between the laser and the sky and this disappointing number illustrates the need to minimize the number of reflections in the optical train. The column density of the layer was estimated by defocusing the spot (to eliminate saturation effects) and measuring the total return in absolute flux units. The spectral linewidth of the laser was measured with a spectrum analyzer and an effective cross section calculated. This gave a column density of $6.\times10^9$ atoms/cm^2. The measured flux with 1 watt of laser power incident of the layer and a laser linewidth of 3 GHz was 122 T_a photons/cm^2/sec which is equivalent to a v magnitude of 9.6. The normal linewidth of the laser is 0.9 GHz. This would give a flux level of 66 T_a photons/cm^2/sec in a column density of 2×10^9 atoms/cm^2.

5.3. OPTICAL PUMPING EFFECTS

The return flux is shown in Fig. 12 as a function of laser power, which shows that the return flux efficiency increases with laser power.

This graph can be understood in terms of optical pumping of the sodium atom. From Table 4 we can derive an effective area of the laser beacon on the mesosphere of 3500 cm

From the previous Chapter, we know that the cycle time of a sodium atom is the time taken to absorb and reemit a photon.

$$\tau_{cycle} = 0.032(I + I_{sat})/I \quad \mu sec \quad (21)$$

where
- I = Intensity of the laser beam watts/m²/natural line width
- I_{sat} = Saturation Intensity of 64 watts/m²/natural line width

The cycle time is therefore about 6/ P μsec, where P is the laser power in watts. Significant optical pumping requires about 10 absorption/emission cycles of the sodium atom and since a collision with another molecule rethermalises pumping, we require (cycle time × 10) << Collision lifetime for significant optical pumping. For low powers (P<1 watt) there are not sufficient cycles/collision time to achieve effective pumping and do not achieve significantly pumping until we reach a power of about 5 watts. Saturation effects start to occur when the laser power is greater than 30 watts.

6. Estimation of the beacon position for the Gemini South telescope

We are now in the position to calculate the rms error due to photon noise and the finite beacon size for a sum frequency laser used with a typical large telescope such as the 8 meter Gemini telescope being constructed at Chile. We will assume that we are operating under adverse conditions, with a low column density of 2×10^9 atoms/cm² and a zenith angle of 45 degrees, under median seeing conditions of r_0=0.129 meters. We will assume the flux return calculated by Milloni of 0.63 $\times 10^6$ T_a^2/m²/sec/watt of power , which is in good agreement with the measured return flux from the Sac Peak laser scaled to a 0.9 GHz linewidth and 2×10^9 atoms/cm².

The RMS wavefront error is given by σ (0.2 + 0.09 ln[M])L meters, where M is the number of subapertures in the pupil used to measure the wavefront slopes and σ is the rms angle of arrival error in units of radians. To obtain an accurate value of σ we must first obtain the image profile of the sodium beacon in the focal plane of the WFS by convolving the beacon profile shown in Fig. 11 with the short exposure point spread function of the subaperture, given by:

$$i(\alpha) \approx \int_0^\infty r(\cos^{-1}(u) - u\sqrt{(1-u^2)})exp(-3.44 \ (d/r_0)^{5/3}) \ J_0(5.16 \ \alpha \ r) \ dr \quad (22)$$

The accuracy of determining the angle of arrival of the wavefront with an idealized quad cell is given by [7]:

Return flux (N) = $0.63 \times 10^6 \tau (1-0.15 \sec z)^2 \cos z\, (T_{laser} T_{8m} T_a \eta)\, L^2\, P$

Fried parameter at zenith	r_0	0.129 m at 589 nm
Zenith angle	z	45 degrees
Atmospheric transmission	T_a	(1 - 0.15 sec z)
Transmission of laser telescope	T_{laser}	0.8
Transmission of Gemini south	T_{8m}	0.8
Throughput of AO system to detector	T_{ao}	0.5
CCD Quantum efficiency	η	0.85
Read Noise	n_r	5 electrons
Servo bandwidth		100 Hz
Wavefront sensor: Shack Hartmann		
Centroid algorithm: Quad cell		
Subaperture size	L	meters
Laser power	P	watts

TABLE 4. Parameters used to calculate the wavefront error.

Subaperture	Laser power		
Size(m)	5 watts	10 watts	15 watts
0.25 m	156 nm	90 nm	68 nm
0.33 m	136 nm	83 nm	64 nm
0.67 m	107 nm	72 nm	60 nm

TABLE 5. RMS wavefront error for different subapertures and laser powers.

$$\sigma = 1/(4\ SNR) \int_0^\infty i(\alpha) \otimes i_b(\alpha)\, d\alpha) \qquad (23)$$

where $I(\alpha) \otimes I_b(\alpha)$ is the convolution of the subaperture and beacon angular psf

$$SNR = \eta N / (\eta N + 4 n_r^2)^{1/2} \qquad (24)$$

Putting in the numbers given in Table 5 we find the following rms wavefront errors for subapertures of different sizes and different laser powers

These are the minimum rms wavefront errors achievable under the conditions specified. For comparison, a 100 nm wavefront error is equivalent to a Strehl ratio of 0.7 at 1μm.

Although there may be some improvement in these numbers with a cleaner optical train for the Gemini telescope (there were 20 optical surfaces used in the Sac Peak experiment!), there will also be significant increase in the wavefront error due to:

(1)obliquity effects due to the finite thickness of the sodium layer. (2)saturation effects in the sodium layer. (3)misalignment of the system optics.

7. System issues

The finite height of the sodium layer sets a limit on the size of telescope that can be corrected by a single beacon via a parameter d_0, known as focal anisoplanatism [4], or the cone effect. The wavefront variance for this effect is:

$$< \phi^2 > = (D/d_0)^{5/3} \ rad^2 \qquad (25)$$

where
- D is the diameter of the telescope
- d_0 is a parameter which depends upon the $C_n^2(h)$ profile

For the Gemini south telescope $d_0 \approx 8$ meters for a zenith angle of 0 degrees, at a wavelength of $1\mu m$ under median conditions so that that at 1 micron the Strehl ratio of an otherwise perfect AO system is about 0.36 and its rms correction error is about 160 nm. A 10 watt sum frequency laser, even including obliquity, saturation and misalignment effects will achieve an rms error of about 100 nm at $1\mu m$ on the Gemini telescope at a zenith angle of 45° and low sodium column density(2×10^9 atoms/cm^2). We have shown that suitable lasers have already been developed which are capable of providing compact, bright beacons for astronomy. These lasers must be carefully integrated into the telescope and adaptive optics systems if their full potential is to be realized.

References

1. Methods of experimental physics - part a. volume 15, pages 325–359, 1979.
2. C.E.Max et al. *Science*, 277:1649–1652, 1997.
3. H.W. Friedman et al. *SPIE*, 3353:260–276, 1998.
4. G.A.Tyler. Rapid evaluation of d_0. *Report N0.TR-582 TOSC*, 1984.
5. T.H. Jeys, A.A. Brailove, and A. Mooradian. Sum frequency generation of sodium resonance radiation. *Appl Opt*, 28:2588, 1989.
6. E.J. Kibblewhite. The development of an infra-red adaptive optics system for large telescopes. *NSF proposal AST-89211756*, 1989.
7. E.J. Kibblewhite. Astronomical constraints of laser-beacon adaptive optics systems 1:the tracking problem. *SPIE*, 2201:265–271, 1994.
8. E.J. Kibblewhite and F. Shi. Design and field tests of an 8-w sum-frequency laser for adaptive optics. *SPIE*, 3353:300–319, 1998.
9. R.F. Lutomirski, W.L. Woodie, and R.G. Biser. Turbulence-degraded beam quality: improvement obtained with a tilt-correcting aperture. *Appl Opt*, 16:665–673, 1977.

What about lasers? (A. Quirrenbach - corner left - & N. Hubin)

CHAPTER 5
LASER GUIDE STAR OPERATIONAL ISSUES

C.E. MAX
Lawrence Livermore National Laboratory
7000 East Avenue, L-413
Livermore, CA 94550 - USA

In order to function well for astronomical observations, a sodium-layer laser guide star and its accompanying adaptive optics must work together as a system. This paper discusses the operational capabilities which such a system must have in order to take into account changes in guide star height and shape, to block or subtract unwanted Rayleigh-scattered light, to perform nodding (for infra-red observing), and to compensate for natural variations in the sodium-layer density. Safety issues such as eye and fire safety, aircraft avoidance, and spacecraft avoidance are also discussed.

1. Introduction

Laser guide star adaptive optics systems correct for the deleterious effects of atmospheric turbulence using a laser-generated reference source to accomplish turbulence measurements. These turbulence measurements are used as input to a real-time computer control system, which sends commands to a deformable optical element for wavefront correction.

Sodium-layer laser beacons [1][2] are generated by a laser tuned to 589 nm, the wavelength of the sodium D_2 line. Laser light excites resonance fluorescence in a layer of atomic sodium in the mesosphere, at altitudes between 95 and 105 kilometers. Several features of the naturally occurring sodium layer affect the design and performance of laser guide star adaptive optics systems:

- The sodium column density varies seasonally [3] and with latitude [4].
- The distribution of atomic sodium with altitude varies with a characteristic diurnal cycle [5], as well as with timescales of minutes to hours [6].
- The overall optical depth of the sodium layer to 589 nm radiation is only a few percent, because the sodium column density is modest. If

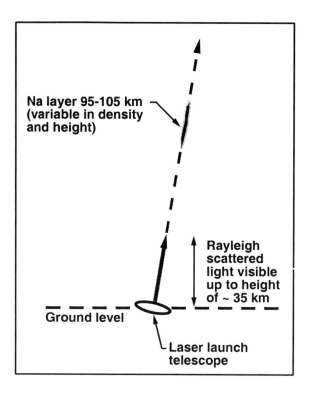

Figure 1. Schematic features of sodium laser guide star generation.

one attempts to obtain high guide star intensities by using high peak laser powers, saturation of the resonance transition can occur [1].

- Rayleigh scattering from altitudes up to about 35 km will be brighter than the desired high-altitude guide star, because the density of molecular scatterers is high at lower altitudes.

These features are summarized schematically in Figure 1.

2. Operational implications for the laser system

2.1. RAYLEIGH SCATTERING

When viewed through the guide camera of an astronomical telescope, the sodium-layer laser guide star "spot" appears atop a highly foreshortened "plume" due to Rayleigh scattering, as illustrated in Figure 2 for the laser guide star at Lick Observatory, which is launched from the side of the telescope.

The goal is to perform wavefront sensing on the small sodium-layer guide star spot, and to exclude the Rayleigh scattered light. Rayleigh scattered light in the wavefront sensor can be decreased via a focal plane stop or iris,

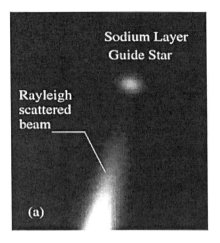 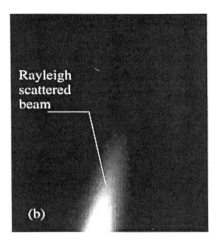

Figure 2. a) Sodium-layer laser guide star (top) and Rayleigh scattered light (bottom) as seen through guide camera on telescope. b) Rayleigh scattered light only (laser is tuned 3 GHz off central frequency of D_2 line).

with acceptance angle of a few arc sec centered on the sodium-layer guide star. It can also be decreased by careful consideration of the geometry for launching the laser beam from the telescope. If the laser beam is launched from behind the secondary mirror, it is sometimes possible to block much of the Rayleigh scattering via the secondary mirror itself. However with the type of under-sized infra-red secondary mirrors in common use, additional baffling around the secondary may be needed if the laser beam is launched from the back side of the secondary. Alternatively the laser can be launched from the side of the main telescope, at a sufficient stand-off distance that the Rayleigh scattered light does not intersect the telescope field of view until it has reached relatively high altitudes.

For tunable lasers it is helpful to have the ability to tune off of the D_2 central resonance-line frequency by about 3 GHz, so that one can measure the Rayleigh scattered contribution separately from the sodium-layer contribution (Figure 2b). Once this has been done, one can use the Rayleigh measurement as a "background" or "flat-field" which is subtracted from every wavefront sensor frame, in order to preserve only the sodium-layer contribution for use in the centroid measurement.

It is in principle possible to eliminate Rayleigh scatter by time-gating the wavefront sensor. This can be done if the laser has an appropriate pulse format. The simplest example is a laser whose pulse repetition interval is comparable to the round-trip travel time to and from the sodium layer, about 600 microsec, and whose pulse length is comparable to the round-trip time within the sodium layer, about 70 microsec. In this case the wavefront sensor could be gated to take a measurement only during the

70 microsec interval when the laser pulse is within the sodium layer, so that it would see no Rayleigh scattering at all from low altitudes. The sum-frequency solid-state laser developed by Lincoln Laboratories (as described in E. Kibblewhite's paper, this volume) appears to have the appropriate pulse format to accomplish this task.

2.2. FOCUS CHANGES

Focusing is a challenge for sodium laser guide star adaptive optics systems. The sodium layer is not located at infinity, and it varies in altitude during the night and as the zenith angle changes. The astronomical telescope and instruments must remain focused at infinity, while the wavefront sensor follows the altitude of the sodium layer.

Because the mean altitude of the sodium-layer is variable, the focus of the laser launch telescope and of the wavefront sensor must be able to respond to changes in the sodium-layer altitude. The laser launch telescope focus can be adjusted through movable mirror stages. The focus of the wavefront sensor is controlled by moving the whole wavefront sensor apparatus back and forth on a motor-controlled stage.

Currently with the sodium-layer laser guide star system at Lick Observatory, we focus and calibrate for Rayleigh scattered light in the following manner:

a) Focus the main telescope on a natural star (focus at infinity). b) Propagate the laser through its launch telescope, aimed at the natural star. c) Move the secondary mirror of the astronomical telescope to focus on the sodium layer as well as possible, and observe the sodium star on the telescope's guide camera at high resolution. Use the guide camera image to focus the laser launch telescope and obtain the smallest laser guide star image. d) Once the laser launch telescope has been focused on the sodium layer, re- focus the main telescope at infinity. Move the guider off-axis so that the laser guide star light now reaches the wavefront sensor. e) Stop down the iris in front of the wavefront sensor to block the Rayleigh scattered light as much as possible. Tune the laser on and off central wavelength for Rayleigh background subtraction. f) Measure the amount of focus on the laser guide star using the Hartmann wavefront sensor centroid positions. g) Feed back this focus information to move the wavefront sensor stage to the position of best focus.

Steps f) and g) can be replaced by use of a separate slow focus sensor, which continually measures the focus term for a natural star and feeds back this information to the wavefront sensor.

2.3. VARIATIONS IN SODIUM COLUMN DENSITY

Our observations at Lick Observatory, as well as those obtained from LIDAR characterization of the sodium layer [3][4][5], indicate that the total sodium column density can vary by at least a factor of five from summer to winter, with highest values in early winter (in the northern hemisphere). Hence we have judged it important to design a laser guide star system with roughly a factor of three margin, so that it can handle sodium column densities a factor of three above and below the annual mean.

This has practical implications that can affect the choice of which laser to select for use in a laser guide star system. For example, commercial CW dye lasers generally have average power less than 3 watts, and even custom versions are difficult to build with output powers above 4 - 5 watts. Figure 3 shows predicted Strehl performance for an adaptive optics system with 271 controlled degrees of freedom on a 10 meter telescope, at an observing wavelength of 1.1 micron. At 50% of the average sodium column density, a CW dye laser with 3 watts of average power is close to a "cliff" in predicted Strehl performance. In this case either pulsed dye lasers or solid-state lasers may be preferable, because of their ability to scale to the higher laser powers needed to stay away from the precipitous drop in performance shown on the far left side of Figure 3.

2.4. REQUIREMENT TO NOD THE TELESCOPE FOR INFRA-RED OBSERVING

Because of the higher sky and thermal backgrounds, it is common practice in infra-red astronomy to "nod" the telescope in course of observing. "Nodding" means that the telescope is pointed to different positions in the sky (e.g. separated by 5 - 30 arc sec) so as to position the astronomical object of interest on different regions of the infra-red array and so as to sample the sky and thermal backgrounds more thoroughly.

If this practice is to be possible with laser guide star adaptive optics, the laser control system must be able to "counter-nod" the laser so as to keep it pointed at a constant position relative to the astronomical object. In practice, a nodding request signal is sent from the adaptive optics supervisory controller to the laser controller, which then counter-nods the laser beam using the pointing and centering system or the uplink tip-tilt control (depending on the required angular size of the move).

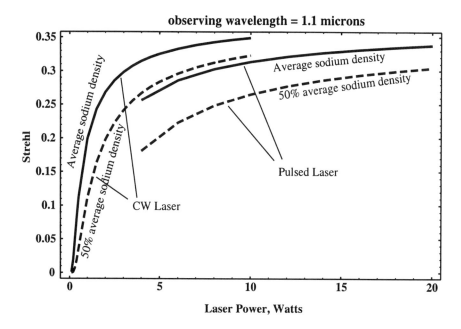

Figure 3. Predicted system Strehl for CW and pulsed dye lasers, as a function of laser power for 1.1 micron observation wavelength. A sodium column density of half the median $2.5 \times 10^9 \text{cm}^{-2}$ has been assumed. The laser was assumed to be projected from the side of a 10-m telescope; the adaptive optics system was assumed to have 271 controlled degrees of freedom.

3. Calibration of the adaptive optics system for sodium laser guide star operation

Adaptive optics system calibration with a laser beacon is more complex than that for a natural guide star, for several reasons.

The apparent laser spot shape is different on different subapertures for 8 - 10- m telescopes, because of the viewing geometry. The emitting region of the sodium layer is a long thin "pencil" of light, typically 10 km tall and only a meter or so in diameter, at a mean altitude of about 100 km. Regardless of whether the laser is projected from behind the secondary mirror of the main telescope, or from a separate aperture located on the side of the main telescope, the sodium spot will appear to be elliptical in some of the subapertures of the adaptive optics system. Figure 4 illustrates this effect.

In addition to the geometric effect described above, the apparent sodium spot shape will vary with time as the zenith angle changes, as the altitude structure of the sodium layer varies, and as the atmospheric "seeing" changes.

As a consequence of this variation in apparent sodium spot shape and size, the gain and accuracy of the centroid algorithm will be affected differ-

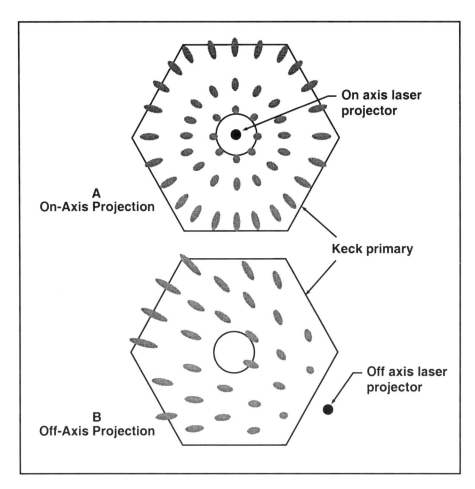

Figure 4. Predicted shape of sodium beacon spot as seen by the wavefront sensor of the 10-m Keck Telescope. Top frame shows sodium beacon shapes for projection behind the secondary mirror. Bottom frame shows predicted shapes for projection from the side of the main telescope.

ently in different subapertures, and the desired zero-point centroid positions for the control loop will be different from those which would be measured from a natural guide star of similar brightness. Calibration techniques are needed in order to determine appropriate centroid zero-points for use in the laser guide star adaptive optics control loop.

A variety of calibration techniques have been suggested and implemented in order to address these issues. We describe several such techniques in the following paragraphs. There is not yet sufficient experience with sodium-layer guide stars to understand which of the techniques will prove most useful in the long run.

3.1. TYPES OF INTERNAL CALIBRATION SOURCES

Adaptive optics systems generally carry several kinds of on-board calibration sources, for use in optical alignment and in static calibration of the adaptive optics system. It is common to use a HeNe laser for optical alignment. For natural guide star adaptive optics calibration, a diffraction-limited white-light source is frequently used so that it can be observed both on the wavefront sensor (for subaperture and centroid registration) and on the science instrument (e.g. an infra-red camera). In the laser guide star case, these need to be supplemented by an internal calibration source that has finite size (typical of the size of the laser spot as seen by the telescope), of a color near 589 nm, and at a focal position optically equivalent to the mean altitude of the sodium layer.

3.2. STATIC CALIBRATION

Static aberrations occur because of imperfect optical alignment of the adaptive optics components, because of non-common-path errors between the optical paths to the wavefront sensor and to the science instrument, and because of aberrations in the main telescope optics. Static calibration techniques can measure these aberrations, so that they can be cancelled out by the deformable mirror.

3.2.1. *Static calibration using internal calibration sources*

Because imperfect optical alignment and non-common-path errors occur within the adaptive optics package or after it, they can be addressed using light from a diffraction-limited on-board calibration source. Techniques such as image sharpening [7] and phase diversity [8] use images from the science instrument to diagnose imperfections in the detected wavefront. In the case of image sharpening, a suite of perturbations is applied to the deformable mirror and the resulting image changes are measured with the science instrument. A variety of algorithms can be used to select the deformable mirror shape which give the "sharpest" image on the science instrument (i.e. that image that is closest to the diffraction limit). This shape is then applied to the deformable mirror as a static offset for adaptive optics operation. In the case of phase diversity techniques, a phase map is constructed (via computer algorithms) from two images taken with the science instrument, usually one out-of-focus image and one in-focus image. The desired static corrections to the deformable mirror are deduced from this phase map.

3.2.2. *Static calibration using natural guide stars as references*

Residual aberrations due to the main telescope optics must be measured using starlight rather than an onboard calibration source. This can be done, for example, by running the natural guide star adaptive optics system for a long enough time to integrate over atmospheric turbulence effects, and then computing the average positions taken by the deformable mirror actuators during this time period. This average deformable mirror shape can be saved, and reflects the correction needed to overcome telescope aberrations. For telescopes with active primaries this information can be fed back to the primary mirror control system (some telescopes such as the Gemini Telescopes have dedicated on-board wavefront sensors for this purpose).

Natural guide stars can also be used to perform closed-loop laser guide star image sharpening. The idea is the following. After completing internal calibrations using internal reference sources as described above, the laser beam is pointed directly at a natural star known to be a "point source" (i.e. not a binary, not a red giant, and not a star undergoing accretion or mass-loss). The adaptive optics control loop is closed on the laser beacon, using the natural star as a tip-tilt reference. In order for this technique to work well, the natural star must be faint enough in the visible that its light does not contribute significantly to the wavefront sensor centroid measurements. The natural star is imaged on the science detector, and with the adaptive optics loop closed, image-sharpening is performed. The image-sharpened offsets of the deformable mirror are saved for later use as centroid offsets in the control system. This method directly addresses those wavefront errors which are introduced by the changing shape and size of the sodium guide star spot. However it can be cumbersome to use unless the natural star is much brighter at science-instrument wavelengths than at wavefront-sensor wavelengths, because otherwise long integration times are required to measure the natural star's point spread function.

3.3. AUXILIARY WAVEFRONT SENSORS

The use of two wavefront sensors, one for the natural star and one for the laser guide star, provides an alternative means of correcting for wavefront errors introduced by the changing shape and size of the sodium guide star spot. In this technique, first implemented by R. Fugate at the Starfire Optical Range (SOR), the laser beacon is pointed at a natural star. Dedicated fast wavefront sensors observe both the natural and laser guide stars, with a dichroic or notch-filter separating the light. At SOR the adaptive optics loop is closed on the natural star, and the difference in average centroid positions of the laser and natural stars is recorded. This determines the appropriate zero-point centroid offsets for laser guide star operation.

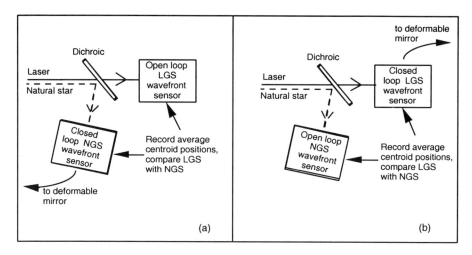

Figure 5. Two schemes for laser guide star calibration using an auxiliary wavefront sensor.

Figure 5 illustrates this concept. Figure 5a shows the scheme as implemented at SOR, in which the adaptive optics loop is closed around the natural guide star. Figure 5b shows the scheme as implemented at Lick Observatory, in which the adaptive optics loop is closed around the laser guide star. In both cases the second wavefront sensor is used to measure time-averaged centroid positions.

3.4. DYNAMIC CALIBRATION (REAL-TIME POINT-SPREAD-FUNCTION MEASUREMENTS)

None of the above calibration techniques can be performed during a long astronomical exposure. Yet atmospheric turbulence characteristics change on timescales of minutes, while astronomical exposures can last for hours or more. To obtain quantitative astronomical results it is crucial to understand what the actual point-spread-function (PSF) was during the whole time-integration by the science instrument.

One way to accomplish this would be to interrupt the astronomical exposure periodically, to nod the telescope away to a reference star, point the laser beam at the reference star, measure its point-spread-function with the adaptive optics loop closed, and then return to the astronomical object of interest. However for laser guide star adaptive optics this is difficult as well as time-consuming.

It is difficult because the star whose point spread function is to be measured must have a tip-tilt reference star closely resembling the tip-tilt star used for the astronomical observation: similar spectrum, magnitude, and offset from the point-spread-function star. At the faint magnitudes in question for tip-tilt stars (e.g. 15th magnitude for a 3-m telescope and 19th magnitude for a 10-m telescope) there are not yet complete enough star catalogues specifying color, stellar magnitude, multiplicity, etc. The technique is time-consuming because half of all faint reference stars will turn out to be binaries which may not be useful for true PSF measurements, and because of the need for many telescope moves in the course of an exposure.

Jean-Pierre Véran has implemented a technique for estimating the PSF for natural guide star adaptive optics, which does not require frequent pointing away from the astronomical target [9]. In this technique the adaptive optics wavefront sensor error signals and the deformable mirror voltages provide on-line information to reconstruct what the adaptive optics PSF was during the course of a long astronomical observation. The wavefront sensor directly measures the wavefront errors at spatial frequencies less than the inverse of the subaperture spacing, or for angles less than the observing wavelength divided by the subaperture spacing. The deformable mirror voltages provide the corrected wavefronts, which at low spatial frequencies are approximately equal to the actual wavefronts incident on the telescope. From the deformable mirror voltages one can get a good estimate of r_0, the atmospheric coherence parameter. Knowing r_0, we can statistically characterize the higher spatial frequencies (angles larger than the observing wavelength divided by the subaperture spacing). Data from CFHT [9] show that this technique works well for reference stars brighter than about 14th magnitude. An analogous technique could be developed for use with laser guide stars. Vérans' analysis applies for laser guide stars as well as natural guide stars, with the exception that tip-tilt information can no longer be obtained from the wavefront sensor. It will be necessary to keep additional records concerning tip-tilt control loop operation: voltages sent to the tip-tilt mirror, and tip-tilt sensor centroid signals. With this minor addition to the data stream, the technique should work well.

4. Safety considerations regarding laser guide star systems

Independent of specific technical requirements, each observatory should appoint a "safety officer" who will be the person responsible for all safety systems related to laser guide star operation. In addition whenever the laser is operated (whether during the night for astronomy or during the day for maintenance purposes), a specific individual who is present at the site should be designated as the safety officer on duty. These personnel should be well- informed about all potential safety issues of the equipment, and should have training in the appropriate actions should a safety incident occur.

4.1. LASER EYE SAFETY

Direct viewing of most laser guide star laser beams can cause instantaneous eye damage. The maximum permissible exposure limit is 2.3 milliwatts per cm^2 for continuous wave (CW) lasers, and about 1 milliwatt per cm^2 for most pulsed lasers. Damage to the eye's retina occurs before the blink reflex has time to take place.

Viewing astronomical laser guide star laser beams through the atmosphere (from the side) is safe. The safety of viewing the laser via its directly scattered light is variable. For example, should a white bird (e.g. a duck) fly through the laser beam, observers more than a few meters away would not be harmed. But viewing full-power laser light reflected from a white card at very close range could cause eye damage.

The most important safety measure is strong training of the technical staff concerning laser eye safety. In addition it is crucial to have an observatory management which emphasizes the importance of good employee safety practices.

Physical measures can be taken as well. Laser enclosures and the doors to laser laboratories should be interlocked, so that the laser is automatically shuttered if the enclosure or door is opened. (From time to time trained technical staff will have to bypass the interlock in order to perform repairs or alignment procedures.) If the laser is within the telescope dome, when laser enclosures are open and the laser is on, dome doors should be interlocked. The usual laser danger signs are brightly illuminated, and may cause unacceptable levels of scattered light on guider cameras and visible- wavelength science instruments. Acceptable alternatives will have to be developed.

4.2. FIRE SAFETY

Dye lasers generally use flammable solvents. Pulsed dye lasers use ethanol based dyes, while CW dye lasers use ethylene glycol. These solvents require fire-detection and fire-suppression systems. YAG lasers and other solid-state systems use no flammable solvents and require no special fire safety systems, apart from usual electrical safety and facility-wide fire suppression systems.

Fire detection for the type of alcohol fires which dye lasers can cause is best done using ultra-violet detectors rather than thermal detectors, because alcohol fires are not hot enough to trigger many thermal detectors. If additional surveillance is needed for remote locations, a UV fire-detection system can be supplemented by an ordinary television camera. In closed areas such as dye solvent cabinets, an alcohol fume detector is used.

CO_2 fire extinguishers are most appropriate for areas containing delicate optical components. Regular industrial fire suppression systems can be used for dye solvent cabinets.

Fire-detection and fire-suppression for the laser guide star should be integrated into existing observatory-wide alarm systems.

4.3. AIRCRAFT AVOIDANCE

4.3.1. *General considerations*

The lasers used for astronomical laser guide stars are not intense enough to cause physical damage to an airplane. However if an airplane pilot were to look directly down a laser guide star beam, eye damage could potentially result. Alternatively if the laser beam struck the airplane's windscreen and the windscreen lit up brightly, the pilot could be "flash-blinded" for a few seconds or minutes. This could ruin dark-adaptation and cause a safety hazard.

The general principle governing aircraft safety for laser guide star systems is that the laser beam will be shuttered whenever an airplane comes close to the observatory (e.g. within a few kilometers). In the US, the Federal Aviation Administration (FAA) enforces the relevant safety standards. It is the responsibility of the observatory to implement an aircraft-detection system which guarantees that the laser will be shuttered in time to avoid damage to pilots or aircraft. During laser guide star use one person should be designated as the "safety officer" on duty at the observatory, and is responsible for safe operations on the part of all the staff.

In the US the FAA has typically required a multi-level aircraft safety system. This consists of a wide-angle detection system which can see approaching aircraft while they are still at distances of four or more kilometers from the observatory, and a narrow-angle detection system which automatically shutters the laser if an aircraft is within seconds of flying into the beam.

4.3.2. *Wide-angle aircraft detection*

Wide-angle detection systems traditionally consist of visual observers (people) who hold "cut-out switches" in their hands which can shutter the laser beam on command. These visual observers are in audio communication with the safety officer on duty. Because most of today's large observatories are located in relatively hostile environments (e.g. at high altitudes or cold night- time conditions), methods for wide-angle aircraft detection are being developed which use wide-field CCD or infra-red cameras to detect incoming aircraft. The displays of these cameras can then be monitored from inside the warm telescope control room, avoiding the need to place personnel outside. Clearly such a wide-field camera must have a good wide-angle lens and be appropriately sited at a location having a clear 360 degree field of view. Summits with multiple telescopes could plausibly share the output from one such wide-angle system.

Observatories usually try to avoid deploying radars or other rf transmitters because of the potential for radio-frequency interference. But it is sometimes possible to obtain radar feeds from surrounding airport radars or from nearby military facilities, to supplement wide-angle camera systems located at the observatory itself. In the US a software package has been developed [10] which uses position information from such radar feeds to form a video display which can be monitored by the safety officer in the telescope control room.

It is hoped that the combination of wide-angle cameras and radar feeds from nearby facilities will be acceptable to the US Federal Aviation Administration as a substitute for the more traditional deployment of visual observers outside the observatory dome.

4.3.3. *Narrow-angle aircraft detection*

Two concepts have been implemented to date for narrow-angle or "boresight" aircraft detection at observatories. Both use detectors located on the telescope itself. The detectors are interlocked with the laser safety system; if an aircraft is detected, the laser shutter is automatically closed in a very short time interval (about 1/20 of a second).

The first concept uses a small phased-array radar, which covers a conical region surrounding the laser projection direction and moves with the laser. The beam half-width is typically a few degrees. The second concept uses a commercial 3 - 5 micron infra-red camera [11], in conjunction with electronics [12] which compares each frame with a reference frame taken about once a second. The electronics triggers the laser shutters upon detection of a signal difference between the frame and the reference. The fact that a new reference frame is taken frequently allows the safety system to take into account slowly changing backgrounds such as clouds.

4.4. SPACECRAFT DAMAGE AVOIDANCE

Lasers currently used for astronomical laser guide stars are not powerful enough to cause structural damage to spacecraft. But there is a possibility that the lasers could do damage to delicate imaging focal-plane detectors. This possibility would appear to be remote based on published detector properties and plausible collecting apertures. But since many spacecraft are highly classified, damage to spacecraft sensors remains a concern. The problem has dimensions beyond just the technical, because astronomers would like to avoid causing an international security incident by unintentionally damaging another country's spy satellites.

In the US, the Space Command has the responsibility for spacecraft avoidance. Current procedures are cumbersome. If Space Command judges that an astronomical laser is sufficiently powerful to be of concern, the observatory must submit in advance to the Orbital Safety Officer a list of potential "targets" (astronomical coordinates) for each observing night, via fax. This is referred to as a "Predictive Avoidance Request". A day or two later, Space Command sends the observatory a fax in return, listing times when the laser must be turned off corresponding to each target on the list. With typical astronomical observing programs the lists of targets and turn-off times are lengthy. The fact that there is no electronic submission makes the procedure error-prone as well.

With laser guide stars becoming the province of several different countries, there is a serious question about how to best coordinate an international plan to avoid damage to spacecraft. Efforts are ongoing to simplify and automate the process of spacecraft avoidance, and to address the international issues.

4.5. LASER COORDINATION ON MULTI-TELESCOPE SUMMITS

At locations hosting several telescope domes, there is the potential that a laser launched from one dome may ruin astronomical observations at other telescopes. The severity of such an event depends on the details of the observations being made. Infra-red detectors at neighboring telescopes would not be sensitive to the laser light, but their visible-light guide cameras would be. However in principle it would be possible to use narrow-band blocking filters in front of guide cameras to eliminate the laser light. Similarly, spectrographs would not be bothered by the laser's rest frame sodium D_2 line unless the laser light scatters into higher orders; blocking filters would be of help here as well. At Lick Observatory we have frequently used the laser guide star at the same time that an independent observing program is under way on a Coude spectrograph whose collecting optics is less than 20 meters from the laser. Observers on this spectrograph have not been able to detect whether the laser beacon is "on" or "off" unless we told them so over the telephone.

Given the complexities of multi-telescope summits, however, it is useful to develop a mechanism which would guarantee that a laser guide star operated at one dome will not "blind" a neighboring telescope (i.e. have the laser beam pass directly into the telescope's field of view). Such a system is being implemented at Mauna Kea in Hawaii. All summit telescopes wishing to participate will send their pointing coordinates (suitably blurred if they so desire) to a central computer, via the summit ethernet. This central computer, currently located at the Keck Observatory, will calculate the

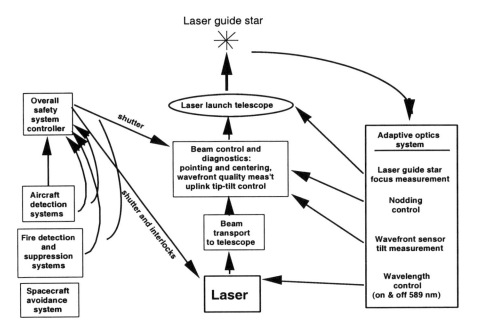

Figure 6. Schematic drawing of communications needed between the laser control system, the adaptive optics system, and the observatory safety systems.

geometry needed to predict when the Keck 2 Telescope's laser guide star is about to interfere with another telescope's field of view. It will then tell the laser beam to shutter itself until the potential conflict is over.

Simulations of typical observing programs on the main Mauna Kea telescopes indicate that the telescope most affected by the laser guide star on the Keck 2 Telescope will be the Keck 1 Telescope. Hence the Keck Observatory has a powerful incentive to make the laser avoidance system work effectively.

5. Conclusions

The operational issues described above have the effect of requiring tight integration between the laser control system, the adaptive optics system, and the various safety systems. A schematic of these relationships is shown in Figure 6. The laser itself is only a modest part of the whole laser guide star adaptive optics system. In practical terms this means that considerable manpower and expense will have to be devoted to the laser control system software and to observatory safety systems, as well as to the laser.

The main challenges for astronomical use of sodium-layer laser guide stars are the following:

- How to calibrate a laser guide star adaptive optics system, particularly for 8 - 10 meter telescopes.

- How to determine the point-spread-function which typifies a long time-integration.

- How to simplify the hardware, software, and observing procedures to achieve good observing efficiency with laser guide star adaptive optics.

Meeting these challenges will require careful systems integration and good engineering. In view of the potential for each observing step to be time-consuming, it will be important to automate as many of the functions as possible.

References

1. W. Happer, G. MacDonald, C. E. Max, and F. Dyson, J. Opt. Soc. Am. A 11, 263 (1994).
2. R. Foy and A. Labeyrie, Astron. & Astrophys. 152, L29 (1985).
3. G. Megie and J. E. Blamont, Planet. Space Sci. 25, 1093 (1977); A. J. Gibson and M. C. W. Sandford, J. Atm. Terrest. Phys. 33, 1675 (1971).
4. D. M. Simonich, B. R. Clemesha, and V. W. J. H. Kirchhoff, J. Geophys. Res. 84, A4, 1543 (1979).
5. C. S. Gardner, D. G. Voelz, C. F. Sechrist, Jr., and A. C. Segal, J. Geophys. Res. 91, A12, 13659 (1986).
6. C. S. Gardner et al. (1986), ibid.; A. J. Gibson and M. C. W. Sandford (1971), ibid.
7. R. K. Tyson, Principles of Adaptive Optics 2nd Edition (Academic Press, Boston, 1998), section 5.4.2 and references therein.
8. R. A. Gonsalves, Opt. Eng. 21, 829 (1982); M. G. Lofdahl and G. B. Scharmer, Astronomy and Astrophysics Supplement 107, 243 (1994).
9. J. P. Véran, F. J. Rigaut, D. Rouan, and H. Maitre, J. Opt. Soc. Am. A 14, 3057 (1997).
10. Remote Airspace Monitoring System, from Sandia National Laboratory.
11. Camera manufactured by Amber, using an indium antimonide array (256 x 256 pixels).
12. Frame-comparison electronics from Optical Electronics Inc., Tucson AZ.

Evening impressions: M. Chang (left), N. Jones & C. Dainty (right)

Left to right Foreground: S. Ragland, C. Holstenberg & M. Kasper
Background: K. Wilson, N. Iaitskova & C. O'Sullivan

CHAPTER 6
THE CONE EFFECT

R. FOY

Centre de Recherches Astronomiques de Lyon
Observatoire de Lyon
9 avenue Charles André
69561 Saint-Genis-Laval - France

1. Introduction

The laser guide star (hereinafter referred as LGS) is created by a physical process in the Earth atmosphere. Consequently it is located at a finite distance from the observer, which results in the so-called *cone effect*, or focus anisoplanatism. The volume of the atmosphere from the LGS to the telescope mirror is a cone, whereas it is a cylinder from any astrophysical source because it is located almost at infinity. This difference causes errors in the measurement of the phase disturbances of the incoming wavefronts [1].

In this chapter, I discuss in the following section why the cone effect degrades the images corrected with an adaptive optics device. In Section 3, parameters which characterize the cone effect are briefly described, and the consequences on astrophysical programmes and expected performances are discussed. Then I address two families of methods proposed to correct for this effect, which both are based on the projection of laser spot arrays in the sky: in Section 4, the stitching method, and in Section 5 the 3D mapping of phase disturbances in the atmosphere. The field of view and its improvement resulting from the correction of the cone effect are described in Section 5; this is a very important issue from the point of view of diffraction-limited ground-based astrophysics. The concluding section mentions ongoing or future actions required before we are able to fully correct the cone effect.

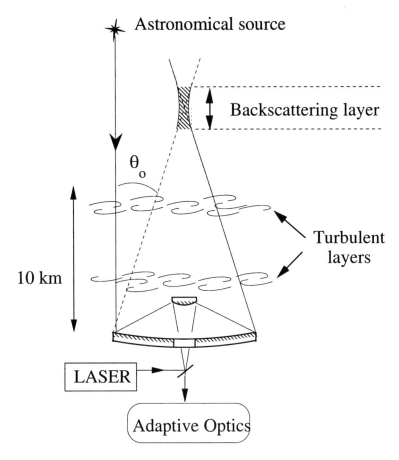

Figure 1. Beam propagation from the laser guide star at finite distance and from the astrophysical source at infinity to the telescope pupil through two turbulent layers. The conical beam from the artificial star does not lighten the whole surfaces of the cross section of the cylindrical beam at the level of the turbulent layers.

2. What is the cone effect?

The laser spot is created either in the $15-20\,km$ altitude range by Rayleigh scattering mainly by N_2 molecules, or in the $90-100\,km$ altitude range, in the mesosphere, by resonant scattering by sodium (Na) atoms. Figure 1 shows a not-to-scale description of the laser beam propagating from the LGS to the telescope pupil, and the beam from the programme object. For the purpose of clarity, the turbulence is assumed to be located within two thin layers, which is quite realistic for good astronomical sites (e.g. [2, 3]).

At the level of a turbulent layer, the annulus between the cross section of the conical beam from the LGS and the cylindrical one from the natural source is not sampled on the wavefront sensor. In addition, beams from the two sources hit the same area of the pupil after having travelled through different areas of the turbulent layers, which means that they have undergone different phase perturbations. Ignoring these differences leads to wrong corrections of the conjugated annulus on the AO deformable mirror, which do not balance the phase error. It has been shown that it makes the image to be corrected worse than if no correction at all is applied.

There are arguments to assert that the cone effect has definitely to be taken into account in the phase restoration algorithms.

3. Parameters and astrophysical implications

Fried and Belsher[4] have introduced the d_0 parameter to parameterize the error due to the cone effect; they defined it from the phase error on the telescope pupil due to the only cone effect $\sigma_\phi = (D/d_0)^{5/6}$. d_0 is the telescope diameter such $\sigma_\phi^2 = 1\,rad^2$.

It is rather straightforward from Fig. 1 that the strength of the cone effect varies with

- D, the telescope diameter
- H, the altitude of the backscattering layer,
- z, the zenith distance,
- λ, the wavelength of the observation, and
- $C_n^2(h)$, the vertical profile of the refractive index structure constant.

The phase error due to the cone effect is:

$$\sigma_\phi^2 = \frac{2\pi^2}{\lambda^2} \frac{1}{H \cos z} \int_0^H C_n^2(h)\, h^2 (h^{1/3} - 1)\, dh \qquad (1)$$

The cone effect strength increases with increasing telescope diameter, zenith distance and r_0, and with decreasing wavelength and altitude of the spot. The integral shows the role not only of the integrated C_n^2 but of its vertical distribution because of the weighting function $h^2(h^{1/3} - 1)$.

Figure 2 shows the variation of d_0 with wavelength for two altitudes of the LGS, 20 and 90 km, and in both cases for an optimistic and a pessimistic Hufnagel-Valley model of $C_n^2(h)$ [5].

It is worth noting that the cone effect can be considered as resulting from the anisoplanatism error between the LGS and the programme object as seen from off-axis points of the pupil: the smaller the anisoplanatism error, the smaller the cone effect.

Figure 3 shows the degradation of the Strehl ratio due to the cone effect solely. It is assumed that the deformable mirror is perfect and that there is no noise or bias; therefore the Strehl ratio should be 1 in all cases. Solid lines and dashed lines respectively refer to a 3.6 m and an 8 m telescope. Upper

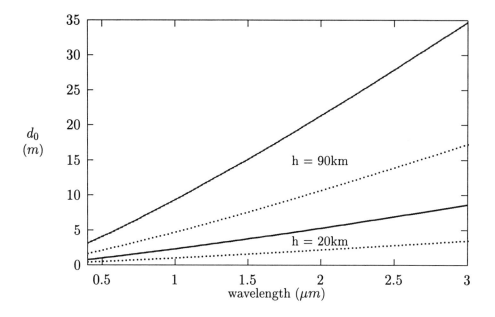

Figure 2. Variation of d_0, the diameter of a telescope for which the cone effect causes an rms error on the phase of $\lambda/6$, versus the wavelength of the observation. Solid lines: optimistic Hufnagel-Valley model for $C_n^2(h)$; dotted lines: pessimistic model; thick lines: 90 km high LGS; thin lines: 20 km high LGS [5].

and lower curves are computed respectively for an optimistic model and a pessimistic model: these curves have been computed using a standard model for Cerro Paranal [6]: a seeing of 0.6", two turbulent layers located 3 km and 8.5 km above the summit with 53% of the integrated $C_N^2(h)$, and a weighted wind velocity of 9.8 km/s. The figure emphasizes the importance of the cone effect; for average seeing conditions at Paranal, the cone effect should be corrected even in the K band (2.2 μm). At shorter wavelengths, the effect is still larger, as expected; even under the very best seeing conditions, the degradation factor increases from ≈ 0.8 in the J band to 0.5 in the V band.

Consequently, not correcting the cone effect leads to a decrease of the sky coverage of an AO at a given Strehl ratio, since the high values of the Strehl ratio cannot be obtained with a LGS. This is shown in Fig. 4 and 5. They have been computed with the optimistic seeing model at Paranal, in the galactic plane (galactic latitude $b = 0°$), in the galactic pole direction ($b = 0°$), and in the intermediate direction $b = 20°$. The abrupt decrease of the sky coverage toward high Strehl ratios means that in this domain, natural guide stars provide better performances, as shown by the dashed lines which are computed for a NGS-AO system. The frontier between the two domains occurs at lower Strehl ratios when moving toward shorter wavelengths.

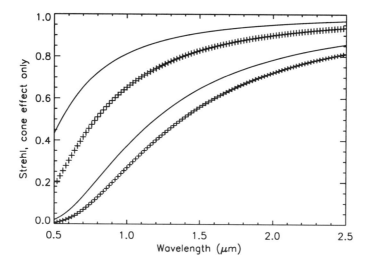

Figure 3. Strehl ratio degradation due to the only cone effect, as a function of the wavelength. Full and dashed lines: 8 m and 3.6 m telescopes respectively. Upper and lower curves: optimistic model and standard model for Cerro Paranal (see text).

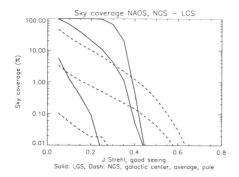

Figure 4. Sky coverage as a function of the Strehl ratio with a LGS (solid lines) and with a NGS (dashed lines), for galactic latitudes $b = 0, 20,$ and $90°$ from top to bottom. Optimistic seeing model; J band ($1.25\,\mu$m).

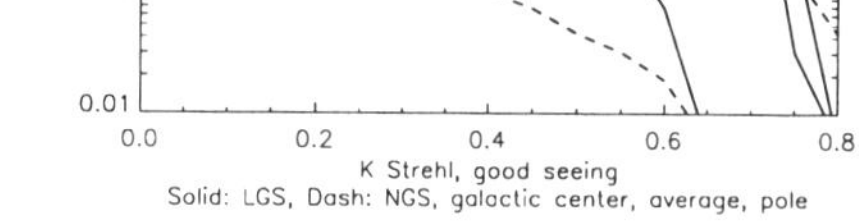

Figure 5. Same as Fig. 4 but for the K band ($2.2\,\mu$m).

The only way to overcome the cone effect is to reduce the projected distance of the LGS to the programme object as seen from the telescope mirror edge. Since one cannot accept a reduction in the size of the telescope, one has to make not a single LGS but a small array of LGSs. In the next two sections, I will describe the two families of methods which have been proposed to handle the signal from a multiple LGS(see Fig. 6).

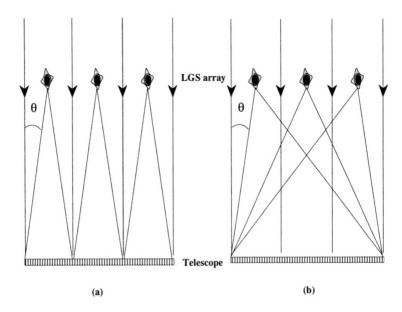

Figure 6. How to handle a multiple LGS. (a): the pupil is divided into sections. Each observes the LGS in the array in front of it, and the signal is processed independently of the other areas. (b): each of the pupil sections measures the whole array and the signal is processed globally [5].

4. Multiple LGS: stitching method

The principle of the stitching method is very simple: it divides the telescope pupil in areas small enough so that for each of them the cone effect is negligible for a LGS situated on the axis of this area (Fig. 6). These pupil areas may or may not overlap (respectively called the butting and stitching methods). It has been first experimented by the Lincoln Lab at Maui [7]. Figure 7 shows that the gain they got using 2 LGSs instead of one in order to correct for the cone effect at their 60 cm telescope is marginal.

This result is easily explained. For each of the LGSs facing a subaperture, the tilt is not corrected. Then when restoring the wavefront surface across the whole pupil, tilt errors combine to increase the phase error σ_Φ^2. Qualitatively, the benefit of the correction of phase errors due to the cone effect is lost because of the tilt errors induced by the stitching method. One could think of increasing the size of the spot array. But there is a mathematical limitation in the gain due to the wavefront discontinuities [8]. Indeed, with the so-called Hufnagel-Valley$_{5/7}$ turbulence model, $d_0(1\,\mu\mathrm{m}) \approx 6\,\mathrm{m}$ with a single spot and $d_0(1\,\mu\mathrm{m}) \approx 15\,\mathrm{m}$ with an infinite spot array. Then

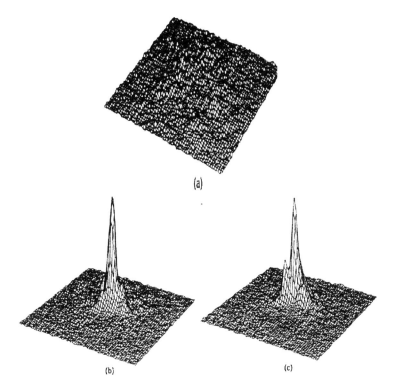

Figure 7. Observations of β Gem at the 60 cm telescope of the Lincoln Lab at Maui. a) uncompensated image, b) image compensated using a single LGS, and c) image compensated using two LGS to correct independently two halves of the pupil[7].

the ratio of two variances due to the cone effect solely is:

$$\frac{\sigma_\infty^2}{\sigma_1^2} = (\frac{6}{15})^{5/3} = 0.217 \qquad (2)$$

The maximum gain with an infinite number of spots is a factor of less than 5. But since one increases the volume of the turbulent atmosphere properly sampled due to the number of LGSs, then one decreases σ_ϕ^2. There is a trade-off [9] between these two errors, which is a function of the above mentioned parameters. Figure 8 shows the gain as a function of the LGS altitude using 4 LGSs instead of one at a 4 m telescope for a correction at 0.5 μm (respectively dashed line and solid line). The Hufnagel-Valley 21 turbulence model is assumed. At such a short wavelength, the cone effect is so large that the tilt stitching phase error is smaller than the cone effect phase error, even at the zenith distance $z = 45°$ (dotted line) and for the highest altitude of the LGSs. Of course the same calculations for an 8-10 m telescope would be less favorable.

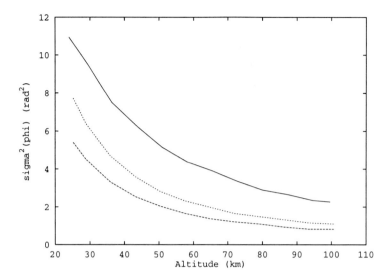

Figure 8. Piston and tilt removed phase variance at a 4 m telescope at 0.5 μm as a function of the altitude of a single LGS (solid line), or of an array of 4 LGSs at zenith (dashed line) or at 45° (dotted line)[9].

5. Multiple LGSs: the 3D-mapping method

The 3D-mapping method does not split the pupil into independent subapertures: each of the subpupils observe simultaneously the whole array of LGSs. Indeed, it aims at mapping the phase disturbances in the volume of the turbulent atmosphere crossed by the wavefront area intercepted by the telescope pupil [10]. These phase errors per atmospheric layer are then used to control a deformable mirror conjugated with the corresponding layer[1]. The set of these deformable mirrors is a multiconjugate adaptive optics device [11]. Extended studies of multiconjugate adaptive optics have been done in [12, 13].

Let us assume that we use a Shack-Hartmann wavefront sensor. Each of the N subpupils provides an image of the whole array of M LGS. Then we have $2 \times N \times M$ slope measurements. They are related to the $2 \times N \times m$ optical path deviations in the m turbulent layers through a matrix equation. The game is to invert this matrix, which is similar in some way to tomography problems. In the following we will describe two ways to write the equation array, considering the beam either propagates from the spot array to the telescope or backpropagates.

5.1. HYPOTHESES

To restore the three dimensional map of phase disturbances from the measurements of the spot array images provided by each of the wavefront sensor subapertures, several hypotheses are required, mostly about the properties of the turbulent atmosphere.

1. Turbulence is assumed to be concentrated in a small number of thin layers, typically 2 or 3. This is consistent with SCIDAR [14] campaigns to measure the vertical distribution of $C_n^2(h)$ [14] or balloon [15, 2], as shown in Fig. 9 and 10. Figure 9 shows a typical $C_N^2(h)$ profile, which is strong mostly within three thin layers. From Fig. 10, they contribute for $\int_{lower\ boundary}^{upper\ boundary} C_N^2(h)\,dh \,/\, \int_0^\infty C_N^2(h)\,dh = 33\%, 32\%$, and 13% respectively to the total seeing, i.e.: $\approx 80\%$ of the seeing is produced by only three thin layers.
2. The geometric propagation approximation works (scintillation is negligible). It implies: $\lambda/r_0 \leq r_0 \times \cos z \times h_n^{-1}\ \forall n$ where n is the number of the thin turbulent layer and h_n its altitude. At zenith, assuming that the altitude of the highest turbulent layer is 8 km above the telescope, at $\lambda = 550\,nm$ it yields the lower limit $r_0 > 7$ cm, which is a realistic condition for a good astronomical site.
3. The small perturbation approximation is valid; it means that a ray crossing the turbulent layer from a laser spot toward a point of the pupil does not depart significantly from a straight line, i.e. it crosses each turbulent layer in the same coherence area as the straight line from the laser spot to that point of the pupil. Then the optical path fluctuations caused by the turbulent layers simply add independently [16]. This approximation is considered as valid at zenith distance as far from zenith as $z \approx 60°$ [17].

In the following subsection, we will consider that the wavefront sensor is a Shack-Hartmann device.

5.2. TELESCOPE TO LASER SPOT ARRAY APPROACH

Let us consider that the telescope pupil is divided into N subpupils with central coordinates (X_{ij}, Y_{ij}). We will follow the beam backward from the pupil to the mesosphere, and using simple geometry we will compute the location of its impact onto each turbulent layer $\vec{P}_{ij}^{(k)}(h_k)$ [10]. At the level of the kth layer, the coherence length is $r^{(k)}$, and indices of the coherent area are rounded: $i = int(X/r^{(k)} + 0.5)$ and $j = int(Y/r^{(k)} + 0.5)$. Notations are summarized in Fig. 11.

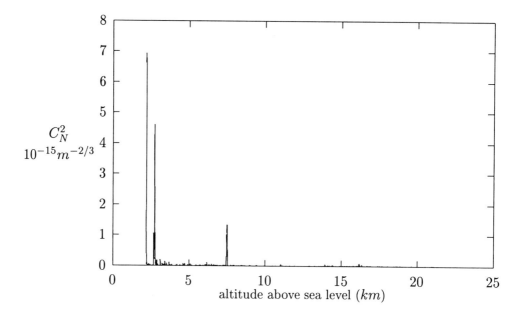

Figure 9. Distribution of C_N^2 with altitude on July 1991 above the Observatorio del Roque de los Muchachos, located at $2200\,m$ above sea level at the top of La Palma island (Canary Islands) [2].

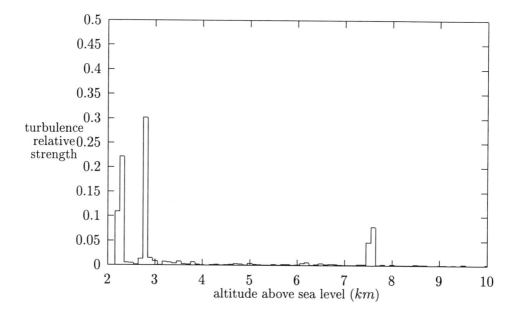

Figure 10. Same as Fig. 9, but for the relative strength of the turbulence [2].

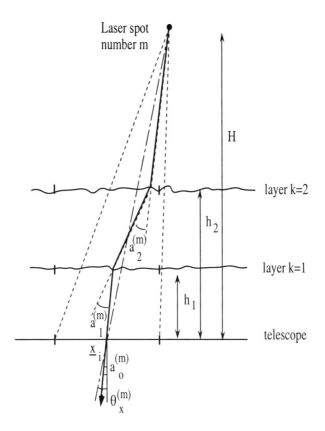

Figure 11. Notation used in the Eqs. 6 & 7 to compute the 3D-mapping of turbulence [5].

Between the pupil level (h_0) and the first layer, the beam location writes:

$$\vec{P}_{ij}^{(0)}(z) = a_{ij}^{(0)} z\vec{x} + b_{ij}^{(0)} z\vec{y} + \vec{P}_{ij}^{(0)}(0) \qquad (3)$$

where $\vec{P}_{ij}^{(0)}(0) = X_{ij}\vec{x} + Y_{ij}\vec{y}$ and the incident angles $(a_{ij}^{(0)}, b_{ij}^{(0)})$ are measured with the wavefront sensor. Above the first layer, where beams are deflected by the unknown angles $(a_{ij}^{(1)}, b_{ij}^{(1)})$, the beam (i,j) may be no longer above the subpupil (i,j), in particular for off-axis laser spots or for subpupils close to the edge of the mirror. Then indices of the coherent area crossed by the beam in the m_{th} layer are:

$$p = \text{int}\frac{X_{ij}(H - h_m) + X_{\mu\nu}}{r^{(m)}H} + \frac{1}{2} \qquad (4)$$

$$q = \text{int}\frac{Y_{ij}(H - h_m) + Y_{\mu\nu}}{r^{(m)}H} + \frac{1}{2} \qquad (5)$$

H is the distance to the backscattering layer and $(X_{\mu\nu}, Y_{\mu\nu})$ are the coordinates of the $\mu\nu$ laser spot. The (p, q) are known quantities, from the telescope diameter, the Shack-Hartmann sensor geometry, the altitude of the turbulent layers, the geometry of the spot array and the altitude of the backscattering layer. Then, above the first layer:

$$\vec{P}_{ij}^{(1)}(z) = (a_{pq}^{(0)} + a_{pq}^{(1)})(z - h_1)\vec{x} + (b_{pq}^{(0)} + b_{pq}^{(1)})(z - h_1)\vec{y} + \vec{P}_{ij}^{(0)}(h_1) \quad (6)$$

Equation 6 can easily be extrapolated up to the mth layer:

$$\vec{P}_{ij}^{(m)}(z) = (z - h_m)\sum_{k=0}^{k=m} a_{pq}^{(k)}\vec{x} + (z - h_m)\sum_{k=0}^{k=m} b_{pq}^{(k)}\vec{y} + \vec{P}_{ij}^{(m-1)}(h_m) \quad (7)$$

Replacing the last term of Eq. 7 by its expression written for the previous layer leads by iteration to:

$$\vec{P}_{ij}(z) =$$
$$(X_{ij} + (z - h_m)(\sum_{k=0}^{m} a_{pq}^{(k)} + \sum_{l=1}^{m}(h_l - h_{l-1})\sum_{k=0}^{l-1} a_{pq}^{(k)}))\vec{x}$$
$$+ (Y_{ij} + (z - h_m)(\sum_{k=0}^{m} b_{pq}^{(k)} + \sum_{l=1}^{m}(h_l - h_{l-1})\sum_{k=0}^{l-1} b_{pq}^{(k)}))\vec{y} \quad (8)$$

Writing Eq. 8 for all the subpupils yields to the system:

$$\begin{cases} X_{\mu\nu} - X_{ij} - Ha_{ij}^{(0)} = \sum_{k=1}^{m}(H - h_k)a_{pq}^{(k)} \\ Y_{\mu\nu} - Y_{ij} - Hb_{ij}^{(0)} = \sum_{k=1}^{m}(H - h_k)b_{pq}^{(k)} \end{cases}$$

5.3. LASER SPOT ARRAY TO TELESCOPE APPROACH

A more general formalism can be used considering the propagation from the spot array to the ground [18]. At the ground level, the corrugation of the laser spot at angular position $\theta_{\mu\nu}$ due to a turbulent layer at altitude h_k is magnified by a factor of $\gamma_k = H/(H - h_k)$. We can assume that in fact we have successively three operations:

- a parallel propagation along the path h_k/γ_k,
- a magnification by a factor of γ_k, and
- a shift of the image of $\gamma_k \theta_{\mu\nu} h_k$

Let be $\vec{\psi}_k$ the vector gathering the samples of the layer k. After its propagation to the ground along the path h_k/γ_k, its shape can be written as

$$\vec{\Phi}_k = \mathbf{P}_k(\vec{\psi}_k) \quad (9)$$

where $\mathbf{P}_k$ is a linear operator, due to the small perturbation approximation.

The magnification changes $\vec{\Phi}_k$ to $\Gamma_k \vec{\Phi}_k$, with $\Gamma_k = 1, \gamma_k^{-1}$, or γ_k^{-2} respectively if the $\vec{\Phi}_k$ are phases, slopes or curvatures. On the other hand, the magnified and shifted sampling grid projected on the ground does not coincide with the sampling of the wavefront sensor. One has to resample it, through a linear combination of the $\vec{\Phi}_k$ components: $\vec{\Phi}_k^{\mu\nu} = G_k^{\mu\nu} \Gamma_k \vec{\Phi}_k$.

Adding the contribution of each layer yields to the general equation system:

$$\vec{M}^{(\mu\nu)} = \sum_{k=1}^{m} G_k^{(\mu\nu)} \Gamma_k \vec{\Phi}_k \qquad (10)$$

where $\vec{M}^{(\mu\nu)}$ is the vector of measurements for the LGS $\mu\nu$.

There are as many Eqs 9 as turbulent layers, and as many Eqs 10 as LGSs in the spot array.

5.4. THE INVERSION

We get $2N$ equations per laser spot (μ, ν). The number of unknowns is $2Nm$. If the inversion is well conditioned, then one needs as many laser spots as turbulent layers. There is at least one case where the matrix cannot be inverted: if an array of 2 spots is observed with a 1-dimensional pupil through 2 turbulent layers [10]. In fact the rank of the matrix is too small by 1, so that an extra condition is required to constrain the inversion. This condition may be:

$$\sum_p \sum_q a_{pq}^{(k)} = 0 \text{ and } \sum_p \sum_q b_{pq}^{(k)} = 0 \qquad (11)$$

which results from that the global tilt is unknown, and that we should not attempt to measure it from the array of monochromatic LGSs.

5.5. MODAL APPROACH

After this paper had been prepared, a modal approach was proposed[19], which, in the opinion of the author should be easier to implement, to allow us filtering and handling of practical situation such as the telescope central obscuration.

6. Spot array geometry and field

In addition to the required number of spots to allow us to solve the equation system, the spots have to be distributed in such a way that each of the coherent area of the 3D-map is crossed by a beam from at least one LGS. In fact, this condition concerns the uppermost layer at altitude h_m. Figure 12 shows that if beams from the two spots cross close to the optical axis below the uppermost layer, a part of that layer is not sampled. In order to keep at the minimum the number of spots, they have to be distributed on the edge of the conical volume V defined by the telescope pupil and the d_m diameter sampled disc at altitude h_m. A spot at such a position provides a d_s diameter disc at h_m which is included and tangent to the sampled disc.

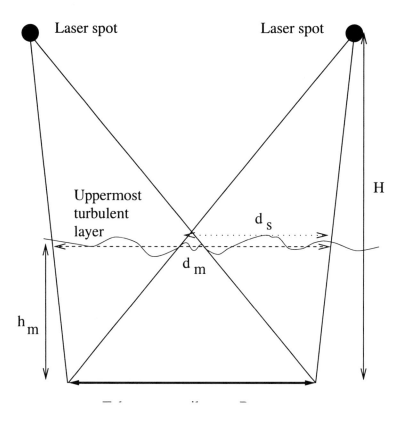

Figure 12. The volume sampled with a single LGS at altitude H lies between the discs d_s in the uppermost turbulent layer h_m and the telescope pupil D.

The geometry of the wavefront sensor has to be specified in such a way that the images of the individual LGSs of the array given by each of the subpupils (in the case of a Shack-Hartmann) never overlap, whatever the local slope of the wavefront. Then identification of the components of the array is derived from the array distribution on the sky and from the wavefront sensor geometry. Of course one can also use one wavefront sensor per LGS in the spot array. Finally let us note that the global wavefront sensing algorithm [20] could lead to less severe requirements for the wavefront sensor.

A major by-product of the 3D mapping of the phase shifts in the turbulent volume above the telescope is the widening of the corrected field. From the 3D-phase shifts, one can compute the total phase shift for each of the telescope subpupils in directions different from that of the reference source; therefore the size of the corrected field is independent of the anisoplanatic patch. For each sampled turbulent layer, restored phase disturbances have to be used to control an adaptive optics mirror conjugated of the layer: this is the multiconjugate adaptive optics mentioned above. Figure 13 illustrates this process, and show that one needs as many deformable mirrors as turbulent layers.

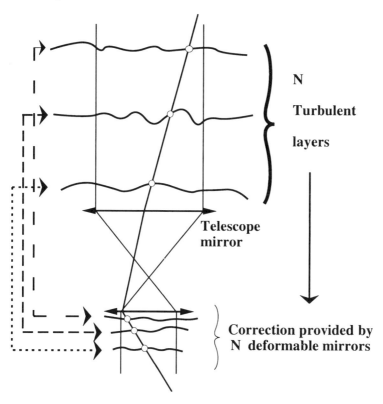

Figure 13. The principle of multiconjugate adaptive optics. In this case, there are three turbulent layers; an optics device forms conjugated images of these three layers onto three deformable mirrors. Each of these mirrors has to be fed with the phase disturbances in the conjugated layer.

The typical field in which all rays to an $8\,m$ telescope are corrected by a 3-spot and a 4-spot array are 7" and 50" respectively. This is significantly larger than the isoplanatic patch of $2-4''$ at short wavelengths. This wide field should increase the probability of finding a natural star to measure the tilt. Beyond this field, correction is not possible over the whole pupil: the corrected field may be much larger, but it is vignetted. Finally it is worthwhile noting that the spot array is distributed quite far away from the center of the corrected field, e.g. on a circle of radius of 24" and 41" in the 3-spot and the 4-spot arrays respectively. This will make it much easier to reject the laser light from the astrophysical field with a simple spatial filter. There are geometrical constraints which set the minimum number of spots to 3 for two turbulent layers when widening the fied of view beyond the isoplanatic patch [3].

Similar approaches have been proposed (e.g. [21]). As far as I know, no experimental test has been carried out of the 3D mapping of turbulence, either in the lab or at the telescope, and no multiconjugate adaptive optics device has been developed.

7. Conclusion

As of today, no solution to the problem of the cone effect has proven to be efficient on the sky. A lot of work remains to be done:

- the overall error budget of the different methods has to be estimated and compared;
- the 3D mapping of the turbulence method, or the merging method, requires an optimization of the inversion algorithm;
- the monitoring of the altitude of the turbulent layers requires more studies: is the low sky coverage of the SCIDAR technique sufficient?
- multiconjugate adaptive optics has to be modeled and built;
- laboratory experiments and then real observations have to be carried out.

These items remain to be addressed to allow adaptive optics to provide its contribution to a wide domain of astrophysical research.

References

1. Foy, R. and Labeyrie, A. *Astron. Astrophys.* **152**, L29–32 (1985).
2. Vernin, J. and Muũnoz-Tuñón, C. *Astron. Astrophys.* **284**, 311–318 (1994).
3. Tallon, M., Foy, R., and Vernin, J. In *Laser guide star adaptive optics workshop*, Fugate, R. Q., editor, 555–566 (SOR/Phillips Lab., Albuquerque (NM), 1992).
4. Fried, D. and Belsher, J. *J. Opt. Soc. Am. A* **11**, 277 (1994).
5. Tallon, M. private communication, (1997).
6. Le Louarn, M. private communication, (1997).
7. Zollars, B. *Lincoln Lab. J.* **5**, 67 (1992).
8. Fried, D. *J. Opt. Soc. Am. A* **12**, 939 (1995).
9. Sasiela, R. J. *J. Opt. Soc. Am. A* **11**, 379–393 (1994).
10. Tallon, M. and Foy, R. *Astron. Astrophys.* **235**, 549–557 (1990).

11. Beckers, J. M. In *Very Large Telescopes and their instrumentation*, M.-H., U., editor, number 30 in ESO conferences, 693–703 (ESO, Garching, Germany, 1988).
12. Johnston, D. C. and Welsh, B. M. *J. Opt. Soc. Am. A* **11**, 394–408 (1994).
13. Ellerbroek, B. L. *J. Opt. Soc. Am. A* **11**, 783–805 (1994).
14. Vernin, J. and Azouit, M. *J. Optics (Paris)* **14**, 131–142 (1983).
15. Bufton, J. L., Minott, P., Fitzmaurice, M., and Titterton, P. *J. Opt. Soc. Am.* **62**, 1068 (1972).
16. Roddier, F. *Progress in Optics* **XXI**, 281–376 (1981).
17. T.Young, A. *J. Opt. Soc. Am.* **60**, 248–250 (1970).
18. Tallon, M., Foy, R., and Vernin, J. In *Progress in telescope and instrumentation technologies*, M.-H., U., editor, number 42 in ESO conferences, 517–521 (ESO, Garching, Germany, 1992).
19. Ragazzoni, R., Marchetti, E., and Rigaut, F. *Astron. Astrophys.* **342**, L3–L56 (1999).
20. Cannon, R. C. *J. Opt. Soc. Am. A* (1995).
21. Jankevics, A. and Wirth, A. *SPIE* **1543**, 438–448 (1991).

Adriano Ghedina

R. Ragazzoni (left) & E. Gendron

CHAPTER 7
LASER GUIDE STAR ADVANCED CONCEPTS: TILT PROBLEM

ROBERTO RAGAZZONI
Astronomical Observatory of Padova
vicolo dell'Osservatorio 5
I-35122 Padova - Italy

The recovery of absolute tip–tilt information from a LGS–based adaptive optics system is (together with conical anisoplanatism) one of the open problems that the LGS community is faced with. Here a number of approaches are discussed in some detail. While there is no evidence that some specific technique is clearly better than the others, it is pointed out that more investigation is still required to assess the performance and to eventually point out some new uncovered schemes. The first experimental evidence of how some technique really work are also reported.

1. The framework

A number of optical ground–based telescopes with aperture larger than 6m are going to have first–light at the turn of this century. At the same time adaptive optics (Beckers, 1993) could escape from the realm of the experimental and hard–to–handle focal plane instruments and should be, in principle, able to boost resolution of these telescopes with the only constraint, in principle, of their diffraction capability. The latter is almost two orders of magnitude better than that imposed by the atmospheric seeing and still slightly less than one order of magnitude better than the Hubble Space Telescope.

The construction of 8m–class telescopes offers a unique opportunity to adaptive optics to exploit full–sky diffraction limited capabilities. Even without considering multi–conjugate adaptive optics, in fact, a pessimistic isoplanatic patch size estimation of a few arcsec still offers images with thousands of resolution elements on each side, when imaging is performed at 10 milli-arcsec (mas) resolution.

In other words, the *prize* for the realization of an 8m–class telescope with a LGS–based (Foy & Labeyrie, 1985; Hubin, 1997) adaptive optics system able to point toward any object in the sky is really very huge.

On the other side one could dream of a near future where some industrial applications of adaptive optics components (like optical communication or medical applications) will provide low–cost adaptive optics modules for the large number of 4m–class telescopes that, equipped with the proper LGS system, could routinely obtain HST–quality data in the visible and in the near infrared. The recent development of MEMS could be placed in this scenario.

While we are free to dream, there are still a number of engineering problems (that are treated elsewhere in this book) and a few *fundamental* problems. Among all the others are conical anisoplanatism and the absolute tip–tilt determination.

In this chapter we shall discuss a range of solutions to cover the absolute tip–tilt determination. The reader should be aware that this is just *one* of the conceptual problems involved in our dream and that engineering–related issues to both tilt and conical problems have not yet been investigated in detail and that a demonstration experiment, for instance, would require large dedicated efforts.

Finally, in this introductory section, I would like to point out to the attention of the reader the review by Esposito (1998).

2. Searching for new ideas

Regarding conical anisoplanatism, one would say that at least two solutions are already well known and there is some sort of general consensus that the major problems are now of technical nature. Of course noise propagation for these techniques is still unexplored but the adoption of multiple LGSs appears as an unavoidable need to correct the conical effects.

For the lack of tip–tilt information, the situation is, from a certain point of view, more intriguing. In 1992 Foy and coworkers introduced the concept of polychromatic LGS (Foy *et al.*, 1992, 1995) to solve the problem. For a few years no other solution had been figured out. The situation is now dramatically different (and, in a certain sense, more confusing: if you have 10 clocks in your house you will probably never know the exact time). In Table 1 I have tried to summarize the several techniques proposed. The subdivision and the proposed acronyms are very tentative and these can be subject of criticism. I like very much the tip–tilt problem for a specific reason: it is a very well defined problem. I think that the more difficult step in order to have good answers is to place a very specific question. In this chapter it is assumed that some basis of the problem of the lack of absolute

Technique	Reference	Acronym
Natural Guide Star		NGS
Optimum Sharing Aperture	Lukin (1996)	OSA
Sub Pupil Averaging	Riccardi, Ragazzoni & Esposito (1997)	SPA
Double Adaptive Optics	Rigaut & Gendron (1992)	DAO
Polychromatic LGS	Foy et al. (1992, 1995)	P-LGS
Inclination Perspective	Ragazzoni & Marchetti (1996)	IP
Propagation Delay	Ragazzoni (1996b)	PD
Near Term Prediction	Ragazzoni & Marchetti (1996)	NTP
Strip Averaging (Aux. Proj.)	Belen'kii (1995), Ragazzoni (1996a)	SA-AP
Strip Averaging (Aux. Tel.)	Belen'kii (1996)	SA-AT
NGS-based Perspective (Aux. Proj.)	Ragazzoni (1997)	NGS-P-AP
NGS-based Perspective (Aux. Tel.)	Ragazzoni, Esposito & Marchetti (1995)	NGS-P-AT
Tomographic Tilt Recovery	Ragazzoni (this work)	TTR
3D-Taylor	Ragazzoni (this work)	3DT
Inter-Zernike Correlation	Whiteley, Roggemann & Welsh, 1998	IZC

TABLE 1. The techniques proposed in the open literature to retrieve the absolute tip–tilt from a Laser Guide Star adaptive optics system.

tip–tilt reference are clear to the reader. A number of references are listed at the end and these can be consulted. The concept that the knowledge of the upward LGS tip–tilt is enough to solve the tip–tilt problem is also assumed.

Moreover there is not any firm indication that all the possible ways to have a suitable tip–tilt signal have been fully exploited. In fact we do not know if we are missing a simple or a very complicated way to retrieve the tilt signal without most of the practical problems, or technical and fundamental limitations affecting the various LGS tip–tilt retrieving techniques, including the polychromatic LGS.

Here I wish to recall a very simple *rule* to explore in depth the *space* of the various schemes and techniques to retrieve the absolute tip–tilt. The idea relies on the fact that sometimes a technique offers a certain degree of *symmetry*. For instance there are techniques that use auxiliary apertures. In some schemes these are used as auxiliary projectors to fire a further LGS into the sky; in some others they are used to collect the photons backscattered by the *main* LGS fired from the principal telescope where the adaptive optics observation is to be performed (see Fig. 1). For instance the NGS-P-AP is, from the described point of view, the *dual* of the NGS-P-AT. The same could be said between the SA-AP and SA-AT techniques. In these cases, the duality is evident, although more than one year has been required to get evidence of this. Using the concept of duality one can try to figure out new techniques starting from the proposed ones. In some cases

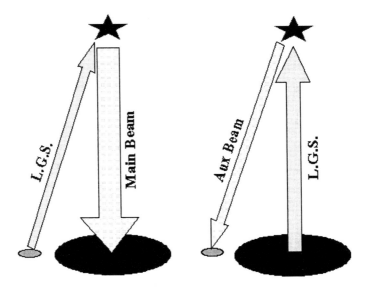

Figure 1. Duality between two techniques can help to figure out a new approach starting from its dual.

duality is not as evident as in the examples outlined a few lines before. Averaging techniques, for instance, can be performed on an *angular* basis (like in the SA–AT, where one averages the tilt from several sections of the same LGS) or on *pupil space* basis, like in the SAP technique. Duality can be very effective to find out new solutions but the nature of the dualism can be very hidden. A systematic attempt to find out all the duality inherent to these techniques is still missing.

3. Tuning the parameters

A few words should be said recalling the statistical nature of the lack of tip–tilt information from LGS. If one tries to use directly the LGS tip–tilt signal, in fact, there will certainly be an error. How large this error is will depend upon the atmospheric parameters characterizing the specific observation one is making. Moreover it will depend upon several other parameters like the diameter d_{proj} of the LGS projector, the diameter D_{tel} of the telescope aperture and their separation. For each specific amospheric situation (characterized, for instance, by r_0, L_0 or whatever is relevant) one can think of optimizing the separation and the aperture diameter of the LGS projector.

Lukin (1996) extended this concept to the case of sharing the main telescope optics with the LGS projector (Fig. 2). When the engineering problems of using the same aperture, both to fire the LGS and to collect

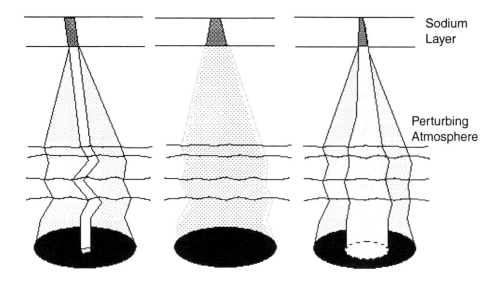

Figure 2. The optimization of the ratio between the shared aperture used to fire the LGS can lead to a moderate relaxation of the tip–tilt recovery problem.

the scientific data, are solved, it should be relatively easy to optically change the ratio β between the two apertures defined as:

$$\beta = \frac{d_{\text{proj}}}{D_{\text{tel}}} \quad (1)$$

In principle one can explore any point in this space. In practice $\beta > 1$ will translate into an unefficient use of the telescope aperture.

Lukin has shown that a maximum in the optimizing function can be found for some interesting values of β. Recent work is in progress to use also the separation of the LGS projector as a free parameter (Lukin & Fortes, 1997; Lukin, 1997, 1998b). While it is clear that this sort of optimization can hardly be used to obtain full–sky diffraction–limited capabilities, the results could be efficiently used to further optimize or relax the requirements of the several schemes that allow the direct detection of the absolute tip–tilt.

4. Looking for non–standard tiny effects

As was pointed out for the first time by Pilkington(1984), looking to the centroid motion of a LGS does not provide useful tip–tilt information. In order to get further information one should look for a second–order effect. Second order effects will, generally speaking, appear on scales much smaller than the usual ones, both in time and in angular space. For the second order techniques, in fact, one is faced to the typical problems of evaluating centroids with an accuracy that is one or two orders of magnitude smaller than the required accuracy one would obtain at the main telescope. Sometimes this is to be accomplished on a time–scale that is orders of magnitude shorter than what is usually required to close a simple tip–tilt closed loop. When the goal of diffraction limited capability for 8m–class telescopes is faced, this will translate into sub–milli–arcsecond centroiding in real–time, maybe at more than few kHz sampling rate. In principle this can be accomplished when enough photons are collected and this will translate into larger and larger laser power. Of course the tuning of the involved parameters can relax these requirements to acceptable levels (one of these cases is P–LGS that is treated elsewhere in this book). However, as anyone who has tried to perform fine tip–tilt correction is aware, there are several technical issues to be solved, like the stability of the sensor or the mechanical flexures within different parts of the module, and this will probably warn you about the technical complications that must be considered to overcome such limits. When this level of accuracy is required, one is not free to discard minor effects, like the variation of sensitivity across the pixels of the detector, the diffraction effects inside the wavefront sensor unit and the colour–related effects in the atmosphere, for example.

The Inclination Perspective (IP) technique (Ragazzoni & Marchetti, 1996) is a good example of second–order technique. This scheme is based upon the assumption that due to the upward LGS propagation tip–tilt, the LGS, as seen from the main aperture, will appear slightly elongated (see Fig. 3). Assuming that all the upward tilt is concentrated in a well defined perturbing layer, located at a given height h_l, the elongation between two extrema produced by resonant backscattering at the heights h_{Na1} and h_{Na2} will exhibit an angular displacement ε given by:

$$\varepsilon \approx \varphi \frac{2(h_{Na1} - h_{Na2})}{h_{Na1} + h_{Na2}} \qquad (2)$$

Provided that ε is much smaller than the isoplanatic patch size, one should assume that, while the absolute positions of the elongated spots is unknown, their relative distance is unaffected by the downward propagation. The measurement of such a tiny displacement (roughly one order of magnitude smaller than the searched tilt term) is hard to obtain on the elongated spot itself. It is much more realistic to have a pulsed LGS, to gate the same pulse at two different altitudes and to make a relative centroid measurement. It is to be pointed out that in looking for such a tiny effect, the weakest assumption here is the isoplanicity of the two spots. The difference in the downward propagation, in fact, will be amplified by the inverse of the factor in Eq.(2) and this could change dramatically the concept of isoplanicity as far as this scheme is concerned. A detailed analysis

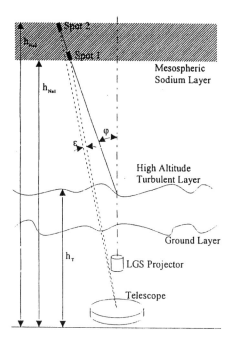

Figure 3. Tilt in the upward beam will translate into an elongation of the LGS as seen from the main aperture. A fine measurement of such elongation, for instance using a gated pulsed laser, should provide enough informations to retrieve φ.

stating the ultimate performance achievable is missing. Technical issues relate to the effects of mesospheric sodium density variation that could alter the measurements.

Another approach that uses a second order effect can be understood looking at the classical example of the LGS as seen from the same aperture used to *fire* the beam in the mesospheric layer. Nominally, the beam encounters exactly the same perturbing layers both in the up–going and down–going path. While the effective position of the excited sodium atoms will wander in the high altitude layer, the apparent position of the beacon will appear as frozen.

The typical travel time τ between the passage of the beam through the same perturbing layer, located at an altitude $\tilde{h}$, is given by:

$$\tau = \frac{2}{c}\left(H - \tilde{h}\right) \qquad (3)$$

where c is the speed of light and H is the sodium layer distance from the telescope. Looking at the zenith $H \approx 95km$ and typical values are $\tau \approx 0.5msec$. When observations are performed far from the zenith this

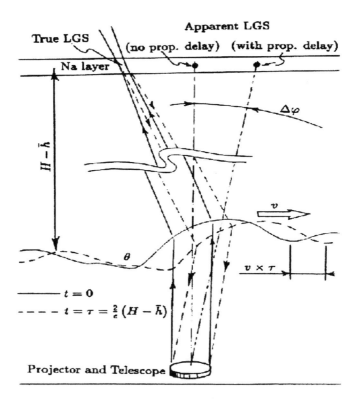

Figure 4. In the Propagation Delay technique a second order effect, due to the finite speed of light, is used to retrieve the tilt evolution with an error growing with time because of an integral operation involved. Again, a very faint NGS is required to overcome the overall resulting drift in the measurement.

figure scales up with a $1/\cos z$ law. The net result (Ragazzoni, 1996b) consists in the observation of a finite difference in the time evolution of the tip–tilt (see Fig. 4). The typical dispersion of the differences is, with a crude estimation by Lukin (1998a), on the order of 1/100 of the one-way tilt dispersion affecting the same aperture. Assuming that the finite difference is a good estimation of the derivative of the tilt evolution, one can integrate the absolute tilt, starting from an arbitrary starting point:

$$\theta(t_0 + \Delta t) \approx \theta(t_0) + \frac{t^*}{\tau}\sum \Delta\phi \qquad (4)$$

where t^* is the time step used to evaluate the finite differences $\Delta\phi$. Each of the $\Delta\phi$ measurements will be affected by a given error due, for instance, the photon shot noise involved in the centroiding of the LGS spot. The integration operation will accumulate these errors, leading to an overall drift of the estimated tilt evolution with respect to the true one. Assuming that these errors are fully uncorrelated, the final estimation errors will grow with the square root of the time. In detail:

$$\sigma_\theta \approx \sigma_{\Delta\phi} \frac{\sqrt{t^* \Delta t}}{\tau} \tag{5}$$

In order to have σ_θ smaller than λ/D for an 8m class telescope and using a laser power able to get back enough photons to close a high–order adaptive optics loop, one can rely on maximum Δt of the order of one second. In this time, however, one can acquire a very faint star and can use this very low frequency tip–tilt information to recover the overall drift.

While there are the technical drawbacks in using the same aperture both to fire and to sense the laser beacon (fluorescence, gating and other issues) it is to be pointed out that there are some fundamental limitations to the technique that deserve a detailed study. In fact the sodium layer extends for several kilometres and the finite difference will not be produced by a sharp time delay, because of an effective spreading of H. Moreover the effect of the propagation delay will depend upon the altitude $\tilde{h}$ of each layer so that there is an additional spreading on the time–base to be adopted in the scaling of the finite difference.

On the other hand one should mention that it is relatively easy, using some polarizer, to have back the LGS into a quadrant sensor, within a LGS projector. Because of the limited size of the typical projector it is hopeless to use this to solve at all the tip–tilt problem; however in this way one could get the high–bandwidth of the absolute tip–tilt, relaxing the requirements on all the other techniques described in this chapter.

The PD technique is a very good example of second–order technique, but it introduces also the concept of having some short–term evaluation of the tip–tilt while the final tip–tilt loop is locked on an extremely faint NGS located within the isoplanatic patch of the target (or using the target itself for this purpose).

This concept is fully developed in the so–called Near Term Prediction technique (Ragazzoni & Marchetti, 1996), where some predictive technique is applied to the LGS projector aperture. It is to be mentioned that the problem is, in a certain sense, much easier than the *usual* tilt prediction.

In fact, provided that the LGS is fired by a projector co–axial with the main telescope aperture, the portion of wavefront producing the upward tilt on the LGS beam is just the central portion of the whole wavefront sensed by the main telescope. This means that one needs to *predict* or to evaluate the absolute tilt of the central hole of an annular aperture using the information of the whole annulus as a support. Of course the global wavefront tilt of the main telescope is unknown but the time evolution of the central hole can be estimated by successive accumulations. The knowledge

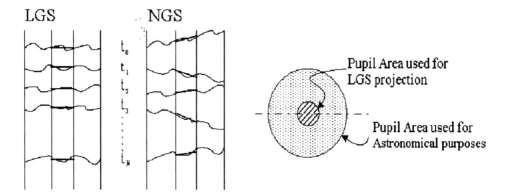

Figure 5. In the Near Term Prediction technique one predicts the behaviour of the up-ward tilt term within a certain limited time range. Using a very faint NGS located within the isoplanatic patch size of the target, and by integrating as long as possible, one should be able to correct the overall drift experienced by the prediction.

of the initial absolute tilt is meaningless because it can be taken arbitrarily and all the rest of the tilt evolution is rigidly shifted by the initial error.

5. LGSs in a new perspective

The LGS is, essentially, a column of continuously excited sodium atoms that emits photons. A tiny fraction of this light will be received by the main observatory: most of the light is, in a sense, wasted. Can this light carry useful information for the absolute tip–tilt retrieval? The answer is yes, and this can be accomplished in several ways.

First of all I would like to provide some useful numbers: the sodium layer is located at an altitude $H \approx 95km$ (Happer et al., 1994) and the approximate thickness of the layer is $\tau \approx 15km$ (the precise length depends upon the choice we made on the definition of the edges). These numbers are minimum figures for the distance of the sodium column and its extension: looking at 45deg of zenith distance the distance increases to $\approx 134km$ and the length of the column will grow to $\approx 21km$.

Observing the LGS from one side, the elongated nature will reveal itself dramatically: at $1km$ distant, a LGS will be seen with an apparent elongation of roughly one tenth of degree.

There are two known ways to use the light which is scattered from a LGS not exactly back from where it was fired. The first one relies on the possibility of using portions of the LGS that fall at distances much larger than the isoplanatic patch size; this technique relies just on the LGS and it provides good results because of some statistical cancellation of undesired tilt components. The second approach uses the LGS as a giant *ruler* that can link different positions in the sky; with the proper geometrical conditions one can relate in this way a bright NGS to any region.

While these two approaches will be described in the forthcoming sections, it is here recalled that techniques relying uniquely on LGSs will suffer the (technical) limitation of being vibration sensitive. In fact any vibration introduced in the observing telescope of an LGS is hard to be disentangled from a tilt term where the isoplanatic patch is much larger than the whole LGS extension. This problem applies to any other technique based solely on LGS, like P–LGS for instance. In principle, optical gyros are able to cancel out this effect but this would mean a further technical complication and stronger specification to the stiffness of the telescope.

5.1. A STATISTICAL APPROACH

If one looks at a LGS sideways and subdivides it into several small segments, a measurement of the tilt in a direction which is orthogonal to the LGS will exhibit the co–addition of two tilt terms (Belen'kii 1995, Ragazzoni 1996a). The first is common to all the segments and it has been introduced during the upward path of the beacon. The second tilt term is specific to any segment and it is due to the path from the segment to the observer. Because of the angular extension, in fact, the various segments will be interested by different optical path from the beacon to the telescope aperture. A few words should be given on the commonality of the first tip–tilt term. There are three basic concerns regarding this assumption:

1. Significant turbulence is zero at the altitudes where the sodium column backscatters useful photons;
2. The tilt introduced by the upward propagation will change the apparent position of the LGS beacon *and* its apparent position angle. The last term produces a differential shift between two points located at different position along the beacon;
3. Due to the finite speed of light there is a tiny (much smaller than the one described for the PD technique) perturbation to the straightness of the beacon as seen by one side, because it contains the integration of the tilt evolution during its propagation from the ground.

There is no strong evidence that item 1 could represent a serious problem; the error introduced by the item 2 is a second–order effect. When a LGS as wide as 1arcmin is seen, a 1 arcsec tilt will introduce a maximum 0.15mas displacement to the two extrema. Unless an extremely large strip is to be observed this cannot introduce any serious deterioration. The last item can be significant when very high performance is required, and it has been the subject of a detailed study (Ragazzoni *et al.*, 1998).

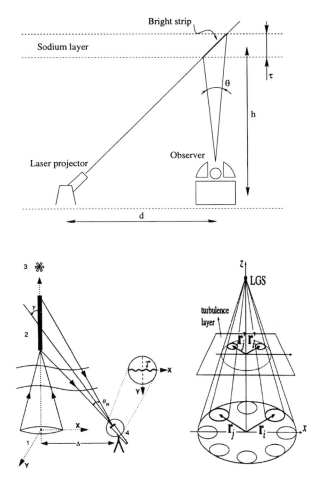

Figure 6. Three ways to average some tilt–related component of an LGS in the hope that the obtained estimation is enough close to zero: *i):* (top) portion of an auxiliary LGS as seen from the main telescope; *ii):* (bottom left) portion of the main LGS as seen from an auxiliary telescope and *iii):* (bottom right) sub–pupils of the same main aperture illuminated by the principal LGS.

The details of the performance that can be reached averaging out the non–common tilt terms are described in the following while, for the specific case mentioned above, we should remember that one can adopt the dual situation (that is to use the main LGS observing it from an auxiliary telescope, Belen'kii 1996) or (but this approach is much less efficient) averaging the same portion of LGS beacon from several subapertures (Riccardi, Esposito & Ragazzoni, 1997; Fig. 6).

A final technical note that deserves attention is the fact that one needs two *fixed* auxiliary stations, just to observe or fire the LGSs. There is not any specific need to move the auxiliary apertures around or radially with

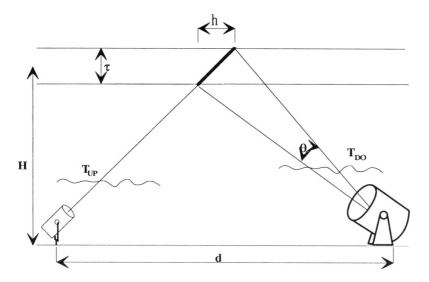

Figure 7. The basic concepts to be considered in this drawing in order to evaluate the effectiveness of the statistical approach to the absolute LGS tilt determination, can be traced from the discussion in the text.

respect to the main telescope. From a logistic point of view this can be a very positive fact because the implementation to existing telescopes can be relatively easy.

In order to carry out the calculations that form the basis of the statistical approach we shall consider the situation sketched in Fig. 7. Using the fact that the triangles defined by the baseline between the projector and the observer and the one defined by the LGS strip are similar, one can write:

$$\frac{\eta}{d} \approx \frac{\tau}{H} \tag{6}$$

and, as a result:

$$\eta \approx \frac{d\tau}{H} \tag{7}$$

Finally, in the small angle approximation, one obtains the apparent extension of the LGS strip as seen by the observer as:

$$\theta \approx \frac{d\tau}{H^2} \tag{8}$$

Dividing the strip in n small portions one can made several measurements of the tilt of each of these portions. Averaging all the measurements one obtains the tilt α defined as:

$$\alpha = \frac{1}{n}\sum_{i=1}^{n}\left(T_{\text{UP}} + T_{\text{DO}}^{(i)}\right) = T_{\text{UP}} + \frac{1}{n}\sum_{i=1}^{n} T_{\text{DO}}^{(i)} \qquad (9)$$

We introduce a coefficient Γ that gives the degree of decorrelation between the various portion of the LGS strip. Γ is defined in a way that $\Gamma = 1$ is for perfect decorrelation for each pair, and $\Gamma = 0$ means that the strip portions are perfectly correlated (in other words, in this last case, $\theta \ll \theta_0$). Given $C[p, q]$ is the correlation of the p and q variables, Γ is tuned in a way that:

$$\Gamma = 1 \Rightarrow C\left[T_{\text{DO}}^{(i)}, T_{\text{DO}}^{(j)}\right] = \delta_{i,j} \qquad (10)$$

Under these circumstances the sum on the right side of Eq.(9) is nearly zero with a certain error depending upon n and Γ. It is easy to see that the following holds:

$$\alpha = T_{\text{UP}} \pm \sigma_{T_{\text{DO}}}\left[1 + \Gamma\left(\frac{1}{\sqrt{n}} - 1\right)\right] \qquad (11)$$

The number of elements that one can retrieve from a strip of length θ is imposed by the isoplanatic patch size θ_0 and by a certain degree γ:

$$n \approx \frac{\theta}{\gamma \theta_0} \qquad (12)$$

A trade-off between γ and Γ is required in order to get the best performance (in fact, a small γ will translate into a larger n but with smaller degree of decorrelation Γ, and viceversa).

5.2. GAUGING WITH A NATURAL GUIDE STAR

There are good reasons to rely on a NGS rather than uniquely on a LGS. We mentioned the jitter in the observing telescope but, in addition, we could include some dome–seeing induced tip–tilt. The analysis of this and other similar effects depends upon a number of parameters. For instance, dome–seeing induced tilt does not deteriorate the performance of the P–LGS technique, while it will lower the performance of the statistical techniques described in the preceeding section.

A number of other issues, like the differential refraction between the LGS wavelength and the wavelength chosen for the scientific observation, can be controlled or taken into account in a variety of ways.

The effectiveness of these techniques could be lost due to potential sources of error or complications that were not considered at a preliminary investigation.

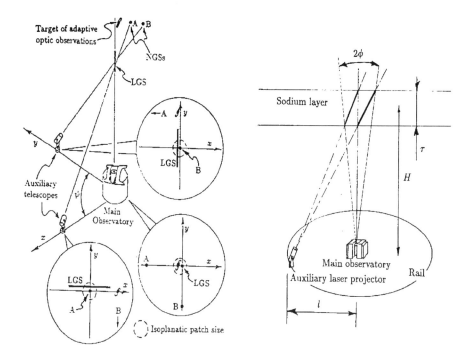

Figure 8. The two NGS–based techniques. On the left the one requiring two auxiliary telescopes; on the right the one requiring two auxiliary LGS projectors. With the proper handling of the data from a pulsed laser the number of auxiliary devices can be effectively reduced to one.

Most (but not all) of these problems are solved using LGS as a sort of giant *ruler* placed in the sky, linking the observed region, where a suitable NGS is not found, to another sky patch where there is a NGS.

This concept is the main driving idea of the NGS–based techniques with auxiliary telescopes (Ragazzoni, Esposito & Marchetti, 1995) or auxiliary LGS projectors (Ragazzoni, 1997; Fig. 8). In the first one the LGS is seen from a modest telescope placed at a certain distance from the LGS projector which instead is located within the main observatory. Provided that the geometry is the right one, in the field of view of the small auxiliary telescope the LGS and a suitable NGS will appear. Provided that these are located within the same isoplanatic patch the downward tilt of the LGS and the NGS from the auxiliary telescope is the same. As a result a differential measurement will produce the upward tilt of the LGS only (being the NGS a fixed star!).

In the second technique an auxiliary LGS is projected in a way that in the field of view of the main telescope, the beacon, appearing as a strip, falls within the isoplanatic patch of both the scientific target and a bright enough NGS. In both approaches there is a requirement to move the auxiliary telescope or the auxiliary LGS projector around the main observatory. However only the first technique will also strictly require some capability of radial displacement.

In the very first crude approach to these techniques two auxiliary apertures are required. In the second option, moreover, a total of three LGSs are fired in the sky (depending upon the involved power, it could be furthermore investigated to use the two additional LGSs to give the reference for the high-order adaptive optics loop in the main telescope). Using a pulsed laser beacon, provided that the pulse format falls within some specification imposed by the thickness of the sodium layer (essentially to avoid to have two or more pulses at the same time within the layer), one can *track* the pulse displacement clocking the pixels of the CCD used to centroid the beacon. A crude estimation shows that the clock frequencies involved and the timing accuracy required are within standard technology.

Simulations of real-time tracking of NGSs to randomly distributed targets show that some substantial movement of the auxiliary apertures should occour within several hundred meters from the main telescope.

A detailed calculation starts from the separation ρ between the LGS and the projected NGS. Denoting A the azimuth angle and h the elevation one the matrix formalism leads to:

$$\begin{pmatrix} \Delta x \\ \Delta y \end{pmatrix} = \begin{pmatrix} \cos A & \sin A \\ -\sin A & \cos A \end{pmatrix} \times \begin{pmatrix} \sin h & 0 \\ 0 & 1 \end{pmatrix} \times \begin{pmatrix} -\rho \cos m \\ \rho \sin m \end{pmatrix} \quad (13)$$

The calculations are carried out in the same way used to retrieve the parallactic angle, a classical derivation to retrieve the proper driving angle for the derotating mechanism of a modern altazimuthal telescope. In the case under consideration there is the additional complication of the evolution with time of the distance of the beacon from the observer (Marchetti & Ragazzoni, 1997). Due to the change in the elevation angle, in fact, the distance is to be constantly updated. This translates into a different scale change between the linear separation ρ at the level of the sodium layer and the corresponding angular one θ:

$$\begin{cases} x = \frac{H}{\sin h} \theta \left(\cos m \sin h \cos A - \sin m \sin A \right) \\ y = \frac{-H}{\sin h} \theta \left(\cos m \sin h \sin A - \sin m \cos A \right) \end{cases} \quad (14)$$

Sometimes an inverse approach can be much more interesting. Some 8m-class telescopes are built in order to perform interferometric capabilities in conjunction with fixed or movable auxiliary telescopes. This offers a unique opportunity to exploit these techniques using the same rail mechanism (or

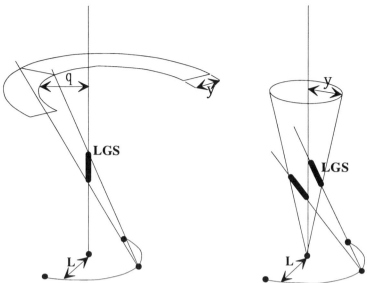

Figure 9. The two techniques using an NGS projected on the background of an LGS will exhibit different sky coverages that are roughly estimated in the text.

even the auxiliary telescope themselves). The rail pattern can be back–projected in the sky in the different observing conditions and sky coverage calculations can be performed in order to evaluate the probability to find out a suitable NGS star for tracking purposes.

A numerical simulation for the VLT case, using the rail system to be used for the auxiliary telescope of the VLTI system is described by Baruffolo, Farinato & Ragazzoni (1998).

While a detailed trade–off between the NGS–P–AP and NGS–P–AT techniques can be solely made case by case it is interesting to make an estimation of their relative capability of sky coverage for a given ground occupation around the main telescope (Ragazzoni & Esposito, 1998).

Looking at Fig. 9 one can derive the sky are for the AT case, provided that:

$$\psi \approx \frac{\tau l}{H^2} \qquad (15)$$

and $\theta \approx l/H$ where, as usual, H is the sodium layer altitude and τ its thickness.

With these assumptions the sky area for the AT case, assuming a possible rotation at a fixed distance from the main observatory, is given by:

$$S_{\text{AT}} \approx \pi \theta \psi \qquad (16)$$

while in the AP case:

$$S_{\rm AP} \approx \pi\psi^2 \qquad (17)$$

Their ratio will be roughly given by:

$$\frac{S_{\rm AT}}{S_{\rm AP}} \approx \frac{H}{\tau} \approx 8 \qquad (18)$$

where the numerical estimation will depend, as usual, from the adopted definition of τ. However, in the AP for the same sky area S can be found NGSs much fainter because one can use the whole large telescope aperture. Depending upon which star distribution model is taken into consideration the final results can slightly vary. Usually the AP technique is better. A detailed comparison for even a single case study is still missing.

6. A few remarks about thinking globally

Because of the complication inherent in adaptive optics and laser guide star systems, people involved in these issues tend to concentrate to a well-defined portion of the problem. A module able to routinely close a high-order loop on a NGS is still a formidable task for the typical efforts in terms of money and people that the astronomical community can usually involve in this type of project. Even from the theoretical side things are so complicated that one prefers to face a single problem at a time.

These simple considerations limit the possibilities to find out very simple solutions that take advantages from a synergy of concepts and technicalities. Conical anisokinetism is, in this field, a good example. Conical anisoplanatism of tilt only, concerning LGSs, has been pointed out as a potential limitation without any overcoming capability (Sasiela & Shelton, 1993; Esposito, Riccardi & Ragazzoni, 1996; Neimann, 1996). To solve the overall conical anisoplanatism it is easy to see that also conical anisokinetism can be removed to a great extent using the same information (Ragazzoni, Esposito & Riccardi, 1998).

The conceptual problem is to have the habit of thinking globally. In an 8m–class telescope facility with a LGS–based adaptive optics module, conical anisoplanatism *must* be solved, so it is not an additional requirement for any absolute tilt retrieval technique to have real–time Scidar and tomographic wavefront sensing capabilities.

Following this reasoning, other new solutions could be revealed as viable ones. We mention here, recalling to the NTP technique, that disentangling all the layers and assuming the Taylor hypothesis of frozen layers (Roggermann et al., 1995), the prediction of the tilt in each central portion of the wavefront for each of the several layers is a trivial task. Summing all

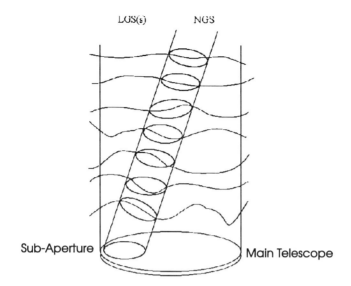

Figure 10. 3D–mapping of the wavefront allows to use a NGS far from the observed region to fix the absolute tilt of each layer.

these tilts one can easily get the overall tilt in the central portion of the wavefront.

When full 3D mapping of the wavefront is performed in real–time one can find out suitable NGSs at distances much larger than the typical isoplanatic patch size using a portion of the main large telescope (see also Fig. 10).

An hybrid approach is represented by the possibility of fixing the position of a single LGS in a multiple reference beacon system, allowing the constellation some degree of freedom in its position in the sky (Ragazzoni & Rigaut, 1998).

Global techniques involve a large number of concepts and systems and it is hard to predict their effectiveness. Noise propagation remains up to now unknown. If one includes in the concept of *global* also the crude cost of the instrumentation, the possibility of using Rayleigh beacons instead of mesospheric sodium ones should not be discarded (Ragazzoni & Marchetti, 1995; Esposito, Ragazzoni & Riccardi, 1997; Esposito, Riccardi & Ragazzoni, 1998).

Globality is to be intended in the widest sense as possible. In fact, recently Whiteley, Roggemann & Welsh (1998) pointed out, in their remarkable paper, a completely new class of possible solutions for the tilt indetermination problem. Their concept starts from the consideration that some non–vanishing correlation between tilt (affected by the upward beam propagation) and higher order terms (like coma, for instance) can be used to retrieve the first, measuring the second on a series of LGSs (the higher order terms are unaffected by the upward propagation beam). A concept that extends some study performed in the same direction by Molodij & Rousset (1997) and that could be suitable of significant new and interesting results in the near future.

7. Experimental verifications

In recent years non–conventional tilt–determination techniques are starting to be scrutinized by experimental tests in the sky. Although only two techniques have been the subject of dedicated, and very peculiar indeed, experiments, this means that in the following years we should have a more wise view of the tilt–indetermination problem based on much more solid basis.

Moreover, it is to be taken into account that tomography (Tallon & Foy, 1990; Ragazzoni, Marchetti & Rigaut, 1999) and MCAO (Ellerbroek, 1994; Johnston & Welsh, 1994) is also moving from the realm of numerical simulation to the one of sky experiments and the sketched link between these techniques and tilt indetermination could be easily fixed in the near future.

I mention here the attempts by Ghedina et al. (1998ab) to use the edge of the Moon to mimic a LGS. This concept could improve the feasibility of a LGS experiment, simply because some concept can be tested without really having a LGS facility.

To the best of my knowledge the very first experimental verification of a LGS tilt determination is by Belen'kii et al. (1999), where the SA–AT technique has been tested using a Rayleigh beacon and Polaris as reference star. The choice of this star avoids the tracking problem in the experiment and, in fact, the telescope has been bolted in place during the experiment. The LGS has been fired by a 1.5m telescope, used also to check the tilt measurements, while the absolute LGS measurement has been made by a 0.75m telescope over a $\approx$ 100arcsec strip length of the LGS. The best value for the correlation between the true and the sensed tilt appears to be 0.71 while an average value is close to 0.5 (in this case no amelioration is obtained to the tilt correction with respect to not providing any correction at all). The experiment has been performed at the SOR facility in Albuquerque.

The only other performed experiment (Esposito et al., 1999) refers to the NGS–P–AP technique, using two NGSs (one for the check of the goodness of the results). A mesospheric sodium LGS has been launched from a $\approx$ 0.4m projector while all the sensing has been performed on a 2.2m telescope. Correlations between desired and measured tilt ranges from 0.5 to 0.8 depending upon assumptions on the nature of the noise. The experiment has been performed at the ALFA facility in Calar Alto (Spain).

As one can deduce both the experiments are encouraging, because they are showing, although marginally, a potential improvement with respect to not performing any correction at all. Nevertheless a definitive experiment showing a correction up to a very fine detail is still missing. A further step is a demonstrative experiment where tilt is really corrected in real–time. Both the experiments described are somewhat an *a posteriori* check with comparison to a true NGS. Moreover they act along a single axis. It is to be pointed out that an *a posteriori* experiment requires some specific arrangement of NGSs (the Polaris, in the first case, and two NGSs aligned with the LGS strip in the second one) that makes these experiments, in a sense, more difficult to perform than a true tip–tilt sensing.

8. Future directions

I can see three different directions where one could work in the near future:

1. Working in depth for each of the proposed solutions one should be able to simplify the technique, to relax the tolerances involved, to lower its cost or to make easier the adaption to several telescopes. At this time the situation could become clearer and some techniques could be evidently preferable to others. An example of this approach is to use pulsed LGS in order to change from 2 to 1 the number of auxiliary apertures in the NGS–based perspective techniques (Ragazzoni & Esposito, 1999).
2. Combination of several techniques or pieces of them could substantially improve the feasibility of the absolute tip–tilt recovery. The number of schemes is becoming so large that it is hard to say if any possible combination have been investigated with enough detail to exclude such a solution. An example of this approach is the use of the Taylor hypothesis to predict the tilt in a portion of a single layer, using the data from 3D atmosphere tomography to cancel out the conical anisoplanatism. Moreover, in conjunction with MCAO, the probability to spot a suitable NGS within the corrected field of view could become so large that the tilt–indetermination problem will remain an historical issue linked with small telescopes where their ratio D/d_0 does not impose to solve the conical anisoplanatism problem.
3. Thinking deeply, or in a serendipitous fashion, one could, of course, find out a new technique much better than all the ones described in this book, and much easier to implement than any other described here. In order to have some chance to get an example of this last approach you could attend the next Cargese adaptive optics school.

References

1. Baruffolo A., Farinato J., Ragazzoni R. (1998) SPIE proc. **3353**, 340
2. Beckers J.M. (1993) ARA&A **31**, 13
3. Belen'kii M.S. (1995) SPIE proc. **2201**, 321
4. Belen'kii M.S. (1996) SPIE proc. **2471**, 289
5. Belen'kii M.S., Karis S.J., Brown J.M., Fugate R.Q. (1999) Opt. Lett. **24**, 637
6. Ellerbroek J.B.L. (1994) JOSA-A **11**, 783
7. Esposito S. (1998) SPIE proc. **3353**, 468
8. Esposito S., Riccardi A., Ragazzoni R. (1996) JOSA-A **13**, 1916
9. Esposito S., Ragazzoni R., Riccardi A. (1997) SPIE proc. **3126**, 476
10. Esposito S., Riccardi A., Ragazzoni R.(1998) MNRAS **294**, 489
11. Esposito S., Ragazzoni R., Riccardi A., O'Sullivan C., Ageorges N., Redfern M., Davies R. (1999) Exp. Astron. (*in press*)
12. Foy R., Labeyrie A. (1985) A&A **152**, L29
13. Foy R., Boucher Y., Fleury B., Grynberg G., McCullough P.R., Migus A., Tallon M. (1992) ESO conf. proc. **42**, 437
14. Foy R., Migus A., Biraben F., Grynberg G., McCullough P.R., Tallon M. (1995) A&AS **111**, 569
15. Ghedina A., Ragazzoni R., Baruffolo A., Farinato J. (1998a) SPIE proc. **3219**, 73
16. Ghedina A., Ragazzoni R., Baruffolo A. (1998b) A&AS **130**, 561
17. Happer W., MacDonald G.J., Max C.E., Dyson F.J. (1994) JOSA-A **11**, 263
18. Hubin N. (1997) ESO conf. proc. **55**
19. Johnston D.C., Welsh B.M. (1994) JOSA-A **11**, 394
20. Lukin V.P. (1996) OSA Tech. Digest **13**, AMB-35
21. Lukin V.P., Fortes B.V. (1997) Atm. and Ocean Opt. **10**, 34
22. Lukin V.P. (1997) Atm. and Ocean Opt. **10**, 609
23. Lukin V.P. (1998a) Pure and Applied Optics (*in press*)
24. Lukin V.P. (1998b) Appl. Opt. **37**, 4634
25. Marchetti E., Ragazzoni R. (1997) A&AS **125**, 551
26. Molodij G., Rousset G. (1997) JOSA-A **14**, 1949
27. Neimann (1996) Opt. Lett. **21**, 1806
28. Pilkington J.D.H. (1987) Nature **330**, 116
29. Ragazzoni R., Marchetti E. (1995) Atm. and Ocean Opt. **8**, 174
30. Ragazzoni R., Esposito S., Marchetti E. (1995) MNRAS **276**, L76
31. Ragazzoni R. (1996a) A&A **305**, L13
32. Ragazzoni R. (1996b) ApJ **465**, L73
33. Ragazzoni R., Marchetti E. (1996) SPIE proc. **2871**, 948
34. Ragazzoni R. (1997) A&A **319**, L9
35. Ragazzoni R., Esposito S. (1997) SPIE proc. **3126**, 94
36. Ragazzoni R., Esposito S., Riccardi A. (1998) A&AS **128**, 617
37. Ragazzoni R., Esposito S. (1999) MNRAS (*in press*)
38. Ragazzoni R., Farinato J., Ghedina A., Mallucci S., Marchetti E. (1998) Appl.Opt. **37**, 4645
39. Ragazzoni R., Rigaut F. (1998) A&A **338**, L100
40. Riccardi A., Esposito S., Ragazzoni R. (1997) SPIE proc. **3126**, 467
41. Rigaut F., Gendron E. (1992) A&A **261**, 677
42. Roggermann M.C., Welsh B.M., Montera D., Rhoadarmer T.A. (1995) Appl. Opt. **34**, 4037
43. Sasiela R.J., Shelton J.D. (1993) JOSA-A **10**, 646
44. Whiteley M.R., Roggemann M.C., Welsh B.M. (1998) JOSA-A **15**, 2097

CHAPTER 8
THE TILT PROBLEM - MULTIWAVELENGTH

R. FOY
Centre de Recherches Astronomiques de Lyon
Observatoire de Lyon
9 avenue Charles André
69561 Saint-Genis-Laval - France

1. Introduction: What is the tilt problem

The tilt is the phase gradient across the entire telescope pupil. At the telescope focus it results in image wandering. For the tilt, the atmosphere can be simulated as a two-dimensional prism whose angle continuously varies. The typical characteristic time of the tilt varies from a few milli-arcsecs (mas) to a few hundred mas depending on the wavelength and on the seeing. This is significantly longer than the round trip time of light to the mesosphere, which is between 0.6 and 1.2 ms, respectively, at the zenith and at 60° from the zenith. Consequently, the laser beam of a laser guide star (LGS) is equally deflected on its path to and from the mesosphere and the laser spot appears fixed with respect to the telescope's optical axis, independent of the true position of the spot in the mesosphere. This prevents us measuring the tilt[1].

The standard deviation of the angle of arrival, θ, in arcsecs, is:

$$\sigma_\theta = 0.062 D^{-1/6} r_0^{5/6} \cos z^{-2}, \qquad (1)$$

where D stands for the telescope pupil diameter and r_0 is the Fried parameter, both units of meters, and z is the zenith distance[2]. At an 8-m telescope, $\sigma_\theta = 0.21$ and $0.43''$ respectively at zenith and at 45° zenith distance for $r_0 = 0.15$ m. These values are much larger than the telescope diffraction limit, which is 0.055 and $0.014''$, respectively, in the K band ($2.2\,\mu$m) and in the V band (550 nm). Integrating on a detector instantaneous diffraction limited images formed with a LGS-fed adaptive optics device would hugely degrade the spatial resolution: one would pay for an adaptive optics device and for a LGS for a marginal gain in the resolution. Note that σ_θ varies with D because the phase gradient varies with its baseline according to the structure function of the air refraction index[2].

In this chapter, I will briefly review the methods proposed to overcome the tilt problem (Section 2) and I will present the principle of the polychromatic artificial star (PAS) (Section 3). Then I will explain the excitation process of the mesospheric sodium atom and the laser requirements (Section 4). In Section 5, I will report on the observing run we had with the AVLIS laser at the Lawrence Livermore National Laboratory (LLNL) to check in terms of photometry the feasibility of the PAS. I will conclude with a discussion of the difficulties of the PAS and of the possibilities to improve the entire PAS process.

2. Methods to measure the tilt

A few methods have been proposed to measure the tilt in the case that an adaptive optics device is used with an LGS. I will briefly review them here, except for the so-called "perspective" methods whom R. Ragazzoni describes in Chapter 7.

2.1. USE OF A NATURAL GUIDE STAR

At the time the tilt problem had been identified[1, 3], the first idea to overcome it has been to use a natural guide star (NGS). It was expected that the NGS could be much fainter and farther apart from the programme object than required for high-order mode measurements; indeed for the tip and tilt Zernike modes, both the entire pupil can be used and the mode isoplanatic patch is larger than that of higher orders of the wavefront. Then the probability to find such a NGS was thought to be high. Unfortunately this is false as pointed out by Rigaut and Gendron[4]. Indeed, typically 90% of the phase variance lies in the tilt. Therefore the required flux for the NGS is such that the contribution of the photon noise is negligible with respect to the phase error. It severely constrains the limiting magnitude of the tilt NGS. In addition, the relative anisoplanatism error has to be significantly smaller than for the higher modes which precludes using a NGS rather far apart from the programme object. Figure 1 shows the degradation of the Strehl ratio as a function of the distance of the tilt reference to the programme object. It shows why observations in the near infrared are considered as the decorrelation in the visible is very steep.

Two turbulence models are considered, corresponding to very good conditions (15% of the time) and to medium conditions[5] at Cerro Paranal; these models are defined in Chapter 9 of this book. It turns out that the probability of finding a reference star for the tilt is not very high. Figures 2

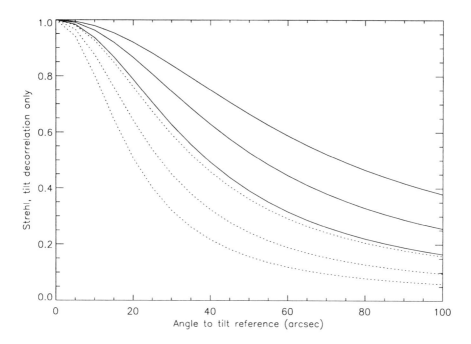

Figure 1. Degradation of the Strehl ratio versus the distance to the reference object for the tilt. Solid lines: good model at Cerro-Paranal. Dotted lines: medium seeing (see Chapter 9 in this book). For each set of curves, top: K band (2.2 μm), medium: H band (1.65 μm) and bottom: J band (1.25 μm). [5]

and 3 compare the Strehl ratio versus the magnitude for an adaptive optics device which is either fed with a NGS solely or with an adaptive optics device fed with a LGS for orders higher than the tilt and with an NGS for the tilt.

For bright objects, the Strehl ratio is lower with a LGS because the cone effect reduces it. Then, as soon as the photon noise affects the NGS-fed device, the LGS-fed device leads to better performances for quite a small magnitude interval: the drop in the Strehl ratio with the LGS-fed device is due to the decorrelation of the tilt mode between the wavefronts of the programme objects and of the tilt reference.

This method is the only one really ready to be implemented on a telescope. It is in use or it will be in use for the LGS systems in operation at Calar-Alto[6, 7, 8], at Lick[9, 10], and soon at the Keck[11] and Gemini telescopes[12].

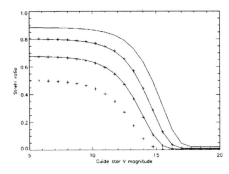
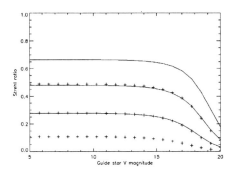

Figure 2. Strehl ratio versus magnitude in the V band ($\lambda = 0.55\mu$m) of the reference star for the adaptive optics device. There is no laser guide star. Solid lines: good seeing conditions at Cerro Paranal. Crosses: medium seeing (see Le Louarn lecture - Chapter 9 - in this book). For each set of curves, top: K band, medium: H band and bottom: J band. [5].

Figure 3. Same as Fig. 2 *with* a laser guide star for correction of the Zernike polynomials higher than the tilt order.

2.2. THE DUAL ADAPTIVE OPTICS CONCEPT

The basic idea of the Dual Adaptive Optics concept[4] relies again on an NGS. But in this case a second LGS is used, and a second adaptive optics device fed with that LGS is used to correct the NGS image. Then the accuracy in the center of gravity measurement of the corrected image is much better than that of the raw image. Hence it allows us to use effectively fainter NGSs which considerably improves the sky coverage.

TABLE 1. Probability to get a reference source as a function of wavelength for a seeing of 0.8" at a 6 m telescope with a perfect deformable mirror[4].

Wavelength	r_0 (cm)	No LGS	Single LGS-AO	Dual LGS-AO
3 μm	138	30%	100%	100%
2 μm	85	5%	70%	100%
1 μm	37	0.2%	5%	90%
600 nm	20	0.01%	1%	25%

Table 1 shows the maximum gain in sky coverage we can get with this tilt measurement method. Lower sky coverages are expected taking into account for instance photon noise and errors due to the finite sampling on the deformable mirror and aliasing errors.

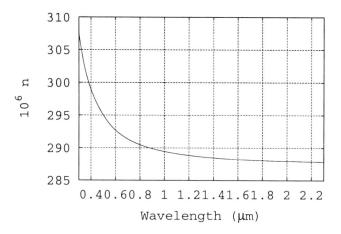

Figure 4. Variation of air refraction index with wavelength

It turns out that the dual adaptive optics concept is worthwhile to be considered for infrared observations. It is clearly not satisfying for astrophysical programmes at shorter wavelengths.

3. Principle of the polychromatic artificial star

The basic concept of the polychromatic laser guide star is to overcome the principle of the inverse return of light by exciting in the upper atmosphere a process which radiates at several wavelengths: a polychromatic process. One assumes generally that the wavefront is achromatic, for example, for the wavefront sensors. This assumption is valid to first order only. In fact, the refractive index of the air, n, varies with the wavelength (Fig. 4).

In particular, in the ultraviolet this variation is rather abrupt. Figure 5 shows that beams emitted at the mesosphere follow different optical paths to the telescope depending on their wavelength; this results in a differential tilt for the observer whereas the tilt itself remains not observable. It is straightforward to show that the spectrum of the polychromatic process has to span the largest spectrum possible, including the shortest wavelengths possible. The tilt, θ, can be derived from the differential tilt, $\delta\theta$, thanks to the principle of separability of n[13]; it allows us to write:

$$n(\lambda, P, T) - 1 = f(\lambda) \times g(P, T) \qquad (2)$$

where λ is the wavelength, and P and T are respectively the temperature and the pressure of the atmosphere. Taking the derivative of Eq. 2 with

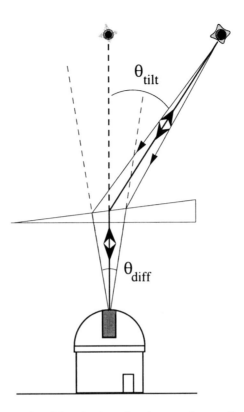

Figure 5. The beam emitted by the launch telescope is equally deflected on its way to and back from the mesosphere, but it produces a spectrum which radiates at wavelengths different from the emitted beam. For these wavelengths, the backscattered light follows different optical paths in the turbulent atmosphere. The resulting differential tilt is proportional to the tilt.

respect to the wavelength λ_i leads to :

$$\Delta n/(n-1) = \Delta F(\lambda)/\lambda_i \qquad (3)$$

for any wavelength interval and λ_i as long as the optical paths at the different wavelengths are not significantly different in terms of phase errors, that is as long as they stay within the isoplanatic patch. The dependency on P and T has disappeared in the derivative. Then, for a wavefront function corrected for the piston at a given point x in the telescope pupil we get :

$$\Gamma(\lambda_i) = ((n-1)/\Delta n)\Delta\Gamma(\lambda, x) \qquad (4)$$

This shows that the wavefront at any wavelength is proportional to its variation between two arbitrary wavelengths. The proportionality factor is the inverse ratio of the variation of the air refraction index between these wavelengths with respect to its value at the considered wavelength, λ_i. This property applies to any quantity derived from the wavefront using a linear operator, for example: the tilt, θ. Then, one gets the fundamental relation of the polychromatic LGS

$$\theta = \Delta\theta(n-1)/\Delta n. \tag{5}$$

Equation 5 means that, from the differential tilt measured between two wavelengths, one can derive the tilt itself without any natural guide star.

3.1. REQUIRED PHOTON FLUX

Since the polychromatic LGS relies on a differential measurement, it requires significantly more photons than a measurement of the tilt itself. If the required accuracy for the tilt determination is a fraction, f, of the Airy disc:

$$\sigma_\theta = \frac{1.22}{f}\frac{\lambda_c}{D}, \tag{6}$$

where λ_c is the wavelength for which the tilt has to be corrected, that is the wavelength of the image observation. Let us recall that:

$$r_{0r} = r_0(\lambda_r/500)^{6/5} \tag{7}$$

where λ_r is in units of nm [2]. From Eq.5 it follows that the error of the differential tilt:

$$\sigma_{\Delta\theta} = \frac{1.22}{f}\frac{\lambda_c}{D}\frac{\Delta n}{n-1}. \tag{8}$$

Typically for an 8-meter telescope, $\sigma_{\Delta\theta} \approx 1\,\text{mas}$. A large number, N, of photons is required to get such a high accuracy. Let us consider only the photon noise and let us assume that the detector for the differential tilt is perfect: that is the ideal case. Let us also assume that the LGS image is Gaussian; this is not a critical assumption. To get the required value of $\sigma_{\Delta\theta}$ needs:

$$N = p^2/\sigma_{\Delta\theta}^2, \text{ where } p = \frac{1.22}{2\sqrt{2}}\frac{\lambda_r}{r_{0r}}, \tag{9}$$

is the spot image size at the wavelength λ_r and r_{0r} is the coherence length of the atmosphere at that wavelength. With Eq. 8 and 7 we get:

$$N = 9.4\ 10^{-17} f^2 \frac{\lambda_r^{-2/5}}{\lambda_c^2} (\frac{D}{r_0})^2 (\frac{n-1}{\Delta n})^2. \tag{10}$$

It is worthwhile noting that N has to increase as D^2 for a given accuracy in Airy disc units. And of course, the number of photons collected by the telescope mirror to measure the tilt also increases as D^2. As a result, the laser power required to produce the polychromatic LGS is independent of the telescope diameter.

3.2. RAYLEIGH AND RAMAN SCATTERING

The Rayleigh scattering, mostly by N_2 molecules, and the associated Raman scattering may satisfy the condition that the polychromatic process has to emit in the UV. If the wavelength of the Rayleigh is 350 nm, the wavelength of the Raman process one is 380 nm. The flux ratio is $N_{380}/N_{350} \approx 9 \times 10^{-4}$. Pros and cons of this process are the following:

- pros :— speckle patterns at 350 nm and 380 nm are still a little correlated, since the spectral coherence length of the atmosphere is $\Delta \lambda = 9 \times 10^{-4} \lambda^2 (r_0/D)^{5/6}$[14], where r_0 is the Fried parameter at 500 nm in the direction of the LGS. Then, the accuracy in the measurements is dictated no longer by the blurred image diameter solely, but also by the size of a single speckle.

  — light pollution is limited to the ultraviolet domain

  — powerful excimer lasers are commercially available at comparatively low costs.

- cons :— the magnitude difference between the two spots at 350 nm and 380 nm is $\Delta m = 7.6$, which means that the photon noise limited accuracy in the relative motions of the spots is given by the faint Raman component.

  — the relative variation of the air refraction index is small, due to the small wavelength interval: $\Delta n_0/(n-1) = 1/125$. Thus, the number of returned photons has to be large (it depends on the N_2 density at the selected altitude).

  — significant Rayleigh scattering is produced at altitudes $\lesssim$ 20-30 km. The cone effect (see Chapter 6 in this book) is huge for such a close LGS for 8-10 m class telescopes. Then, more power is required to produce an array of LGS large enough to correct for it.

Therefore the Rayleigh-Raman LGS does not sound attractive.

3.3. THE EXCITATION OF THE $4P_{3/2}$ ENERGY LEVEL OF MESOSPHERIC SODIUM ATOMS

Currently, this process is the most attractive one to create a polychromatic LGS. It relies on the sodium atom, a simplified energy level diagram of which is shown in Fig. 6. The concept is to raise the Na valence electron from the ground energy level $3S_{1/2}$ to the high energy level $4P_{3/2}$.

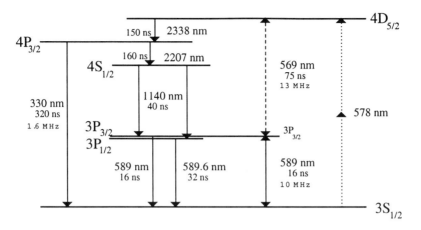

Figure 6. Simplified diagram of energy levels of the valence electron of sodium. Wavelenths (in nm) and lifetime (in ns) of the transitions are given, as well as the homogeneous widths (in MHz) of the 589, 569, and 330 nm transitions.

From the $4P_{3/2}$ level, valence electrons decay to the ground state following two paths. They either fall down directly through the transition $4P_{3/2} \to 3S_{1/2}$ which emits a line in the ultraviolet close to the atmosphere transparency cutoff due to ozone absoption, at 330 nm. Or they cascade through the $4S_{1/2}$ level, emitting in the infrared at 2.207 μm, and the two $3P$ levels, emitting at 1.14 μm and then emitting the two D lines. The radiative cascade has a line spectrum spanning the range from the ultraviolet, at 330 nm to the near infrared at 2.21 μm, which make this process remarkably suited for the polychromatic laser guide star. In the following, we discuss

TABLE 2. Branching ratios of the radiative cascade from the sodium energy level $4D_{5/2}$. Numbers of photons are relative to a population of 27 atoms at the $4D_{5/2}$ level.

Wavelength (μm)	0.330	0.569	0.5890	0.5896	1.14	2.21	2.34
Photons	3	18	22	2	6	6	9

possible mechanisms to excite the $4P_{3/2}$ level.

4. Excitation of the $4P_{3/2}$ atomic level of sodium

At least three mechanisms can be envisaged to populate the $4P_{3/2}$ level of mesospheric sodium from the ground state.

4.1. THE $3S_{1/2} \to 4P_{3/2}$ ABSORPTION

The simplest process is to excite the $4P_{3/2}$ level directly with a laser beam at 330 nm. The absorption coefficient A is

$$A \leq \rho\sigma(\Delta\nu_H/\Delta\nu_T) \approx 1.6 \times 10^{-3}, \qquad (11)$$

where $\rho = 5 \times 10^9$ atoms.cm^{-2} is the sodium column density in the mesosphere, $\Delta\nu_H = 1.6$ MHz is the homogeneous width of the transition, $\Delta\nu_T = 3$ GHz is the total (mostly Doppler width and hyperfine structure) width, and σ is the effective cross section for the transition from the most probable hyperfine component of the ground state. This absorption coefficient is low. In addition:

- the strong absorption of the atmosphere at 330 nm affects both the beam to and from the mesosphere;
- fluorescence and scattering on the optical surfaces and in the lower atmosphere may raise operational problems;
- possibly heating of the innermost atmosphere could occur, resulting in a degradation of the seeing.

This direct excitation is therefore not seriously considered.

4.2. THE $3S_{1/2} \to 4D_{5/2}$: NON-RESONANT TWO-PHOTON ABSORPTION

In the following, the $4P_{3/2}$ level of sodium is populated from the decay of the $4D_{5/2}$ energy level. This decay has two branches: the $4D_{5/2} \to 3P_{3/2}$ transition emitting the line at 569 nm and the $4D_{5/2} \to 4P_{3/2}$ transition emitting the line at 2.338 μm. Relative probabilities are 1/3 and 2/3, respectively (see Table 2), in the inverse ratio of the lifetimes as indicated in Fig. 6. From the $4P_{3/2}$ level, the radiative decay is the same as described in Section 3.3.

The direct non-resonant two-photon absorption involves a virtual transition at 578 nm. This second-order non-linear process has a very low efficiency and is not suitable for the polychromatic laser guide star.

The resonant two-photon absorption is, of course, much more efficient. In this case, the Na valence electron is raised from the ground energy level $3S_{1/2}$ to the first energy level $3P_{3/2}$ with a laser beam at 589 nm as in the case of the "classical" monochromatic LGS. Then, before the valence electron returns to the $3P_{3/2}$ level, a second laser beam, at 569 nm, raises it to the highly excited $4D_{5/2}$ level. Note that the first beam at 589 nm produces the classical D_2, whereas the radiative decay from the $3P$ levels produces both the D_2 and the D_1 lines.

Of course, the two beams at 589 and 569 nm have to be superimposed in the mesosphere. At a zenith distance $z = 45°$, the differential atmospheric refraction is 0.055″. It has to be corrected in case the spot in the mesosphere is formed at the diffraction limit (with an adaptive optics system which pre-compensates the wavefront distorsions before the laser beam crosses the turbulent atmosphere).

4.3. SATURATIONS

Let us consider a monomode laser: the line width is narrower than the natural bandwidth $\Delta\nu_H \approx 10\,\text{MHz}$ of the transition. Ideally, the maximum absorption in a column occurs when the cross sections of Na atoms perfectly cover the column section. In this case, σ^{-1} atoms are involved, and equally as many photons. The power required to excite these atoms has to take into account both the lifetime of the $3P_{3/2}$ level, 16 ns, and the photon energy. Then, one derives that the saturation occurs at

$$P_{sat_1} = h\nu_1/(\sigma_1\tau_1) \approx 18.5\,\text{mW.cm}^{-2}, \tag{12}$$

where h is the Plank constant, $\sigma = 1.14 \times 10^{-9}\,\text{cm}^2$ is the cross section of the first transition, and τ_1 is its lifetime. This is the saturation power for a single mode. The total number of modes is approximately

$$N_m \approx \Delta\nu_T/\Delta\nu_H \approx 300 \tag{13}$$

where $\Delta\nu_T \approx 3\,\text{GHz}$ is the total width of the line, including both the Doppler broadening ($\Delta\nu_D \approx 1\,\text{GHz}$) and the hyperfine structure ($\Delta\nu = 1.77\,\text{GHz}$). Then the total power at saturation is $\approx 5.5\,\text{W.cm}^{-2}$.

Note that, in spite of the high cross section, the efficiency of the absorption is rather low

$$A_1 = \rho\sigma_1\Delta\nu_H/\Delta\nu_T \approx 0.028. \tag{14}$$

The cross section of the $3P_{3/2} \to 4D_{5/2}$ transition is $\sigma_2 = 1.47 \times 10^{-10}\,\text{cm}^2$. The saturation of a single mode occurs at

$$P_{sat_2} = h\nu_2/(\sigma_2\tau_2) \approx 47.5\,\text{mW.cm}^{-2}. \tag{15}$$

The variation with time of the level populations as a function of the laser powers can be computed in the simplest way by using kinetics equations. It is more efficient if the derivative of the populations with respect to the laser intensities is still steep. The optimal tradeoff is obtained when 10% of the atoms are in the $4D_{5/2}$ level; this requires that the 589 nm beam intensity is at 45% from saturation and that the 569 nm one is at 18% from saturation. Having 14% of the sodium atoms in the $4D_{5/2}$ level would require to double the laser powers! The ratio of the saturation intensities of the two beams is 185 W/475 W. It is very close to the inverse ratio of the optimal intensities with respect to saturation 18%/45% such that the total laser power has to be shared equally between the two beams at 589 and 569 nm.

More sophisticated calculations based on density matrix formalism to compute the population of the energy levels lead to the conclusion that the flux ratio should be $F_{569}/F_{589} \approx 2 - 3$ (Petit, private communication).

4.4. INTEGRATION TIME

The radiative lifetime of the energy level $4P_{3/2}$ sets up an upper limit to the power of the laser beams and thus to the minimum time of integration T_i. Indeed, valence electrons have to go down to the ground level to emit the light we need. This lifetime is $\tau_{4P_{3/2}} = (\tau^{-1}_{4P_{3/2} \to 4S_{1/2}} + \tau^{-1}_{4P_{3/2} \to 3S_{1/2}})^{-1} =$ 108 ns. The power emitted by the mesosphere in the 330 nm transition, in units of photons/second, is proportional to the volume where the emission occurs in the mesosphere and to the number of sodium atoms in the $4P_{3/2}$ state and inversely porportional to the lifetime of that energy level

$$N_{4P_{3/2}} Sh / \tau_{4P_{3/2}} = a\rho_h S / \tau_{4P_{3/2}}, \tag{16}$$

where S is the beam cross section in the mesosphere (assumed to be a cylinder and not a bicone), h is the thickness of the sodium layer, ρ_h is the sodium column density, and a is the fraction of sodium atoms in the $4P_{3/2}$ state. The telescope mirror intercepts only a fraction of that power: $a\rho_h S / \tau_{4P_{3/2}} \times D^2 / 16H^2$, under the (somewhat) pessimistic assumption that backscattering is isotropic and that the detected signal is still a lower fraction, due to the global optical efficiency η of the optics and of the detector. The energy detected during T_i has to equal or exceed the required number of photons which we have derived in Eq. 10, which leads to the constraint

$$T_i \geq 1.3 \times 10^{-15} \frac{\tau_{4P_{3/2} \to 3S_{1/2}}}{a\eta} f^2 \rho_h \frac{\lambda_r^{-12/5}}{\lambda_c^2} \left(\frac{n-1}{\Delta n}\right)^2. \tag{17}$$

This integration time is dependent neither on the distance to the sodium layer, that is on the zenith distance, nor on the telescope diameter, nor on r_0. These parameters cancel through the spot area $S = \frac{\pi}{4}(\frac{1.22\lambda_r H}{r_0})^2$. For example, if one measures the differential tilt between 330 nm and 2.2 µm for an observation at 900 nm, then $2800/\eta$ photons are required per tilt coherence time τ_c.

Another fundamental constraint of the polychromatic LGS is that τ_c is larger than the minimum integration time at the image observation wavelength, taking into account that τ_c varies as r_0 with the wavelength. Figure 7 shows how these two times vary with the wavelength of observation. The short wavelength frontier of the polychromatic LGS strongly depends on the typical coherence time of the site.

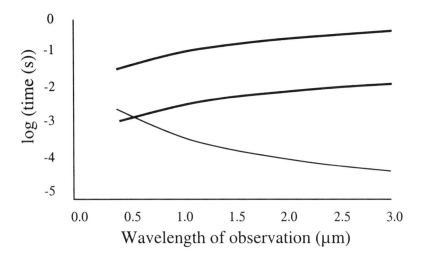

Figure 7. Logarithm of the minimum integration time (in seconds) of LGS images to get a correction for the tilt with an accuracy of half the Airy disc versus the wavelength using the baselines 330 nm-2.2µm (thin line). Upper curves are the tilt coherence time at 500 nm, respectively 2 ms (bottom) and 60 ms (top). In the useful wavelength domain, the coherence time curve is above the integration one.

4.5. LASER POWER

From the discussion in Section 4.3, the laser peak power should be $P_{589} \approx P_{569} \approx 85$ W.m^{-2} for one homogeneous width. Then, the necessary total peak power is

$$P_{peak} = 170 S \Delta \nu_t / \Delta \nu \, \text{W}. \tag{18}$$

With $r_0 = 0.15$ m, the required total *peak* power is close to 7 kW. The total *average* power is

$$P_{average} = P_{peak} T_i / \tau_{\lambda_c}. \tag{19}$$

It ranges from 250 W to, at worst, 7 kW. It should be emphasized that these powers are strongly dependent on the spot size. Using adaptive optics devices both for the launching beam and for the differential tilt measurement may decrease this requirement by more than a magnitude.

5. From the theory to the experiments : PASS

A first experiment had been set up at CEA (French Agency for Atomic Energy) in the laboratory[15]. They used two dye lasers each being pumped with an excimer laser with a laser pulse of 30ns. The double beam irradiated a sodium cell. They compared successfully their relative measurements with both rate equation and density-matrix models.

5.1. DESCRIPTION OF THE PASS EXPERIMENT

Another experiment was done on the sky in January 1996 using the AVLIS laser at the Lawrence Livermore National Lab (LLNL)[16]. The purpose was to check that we are able to populate the $4D_{5/2}$ level of mesospheric sodium atoms and that the returned flux at 330 nm is sufficient to measure and therefore to correct for the tilt.

The AVLIS laser has been used by the LLNL staff to demonstrate the feasibility of the (monochromatic) LGS for the Shane telescope at Lick Observatory and then for the Keck[17, 18, 19]. Because we used a single dye laser for both wavelengths at 589 and 569 nm, the total power was ≈ 350 W. Both beams were linearly polarized (parallel directions).

The $5 \times 8 \, \text{cm}^2$ beam was emitted from an underground vault through a siderostat, allowing us to form the LGS at any direction, for example close to a natural star for the purpose of photometric calibration and to track the diurnal motion. The 50 cm telescope to observe the LGS was at 5 meters distance from the launch telescope. We measured the seeing with a double hole mask installed on the telescope pupil from time to time . We found $r_0 \approx 6$ cm, typical for a low altitude site in winter time.

A dedicated instrument, PASS, was built at Lyon Observatory to be able to observe simultaneously on a single CCD camera the spot images at 330, 589 (D_2 line) and 589.6 (D_1 line) nm. Figure 8 shows the optical layout of this instrument of which the challenges were:

- to filter out most of the natural guide star at 330 nm
- to separate the D_1 line image from the D_2 one, in spite of the fact that they are only 0.6 nm apart
- to be a low-cost and light instrument

An example of the images provided by AVLIS and the PASS experiment is given in Fig 9, showing both the images of the polychromatic LGS and the dispersed images of the natural star SAO 60855. Some laser parameters have been varied. The pulse repetition rate was set either to 12.9 kHz or to 4.3 kHz, the bandwidth to 1 or 3 GHz, the total power to 350 W and the power balance between the two beams to 0.3 , 0.5, or 0.7 through 0.5. Additionally, the central frequencies have been scanned to maximize the returned flux.

5.2. RESULTS AND DISCUSSION

5.2.1. *The Saturation of the Sodium Layer*

Due to the seeing and the beam size, the spot energy distribution in the mesosphere was dominated by a single speckle, with a full-width-half-maximum ≈ 1.06 m. The laser peak power at 589 nm was $\approx 4.2 \times 10^5$ W at pulse

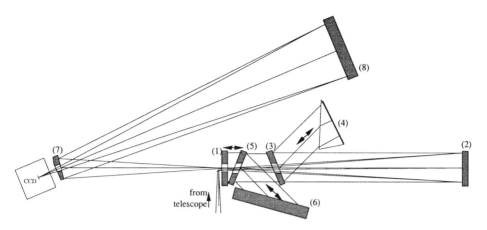

Figure 8. Optical layout of the PASS instrument. From the telescope Cassegrain focus, the beam crosses through holes matching the telescope central obscuration the flat (1) and the two tilted dichroic beam splitters (5) and (3) to hit the spherical mirror(2), Then the beam is collimated. The ultraviolet wavelengths are reflected by the beam splitter (3) and cross the quartz prism aluminium coated on its back face; it is reflected again by (3) and focused by (2). The beam then is folded with the flat (7), and the UV spectrum of the field, dispersed by the prism (4), is projected onto the CCD camera by the spherical mirror (8). The yellow-orange range follows the same scheme, with the exception that it crosses the UV dichroic beam splitter (3) and it is reflected by the yellow-orange dichroic beam splitter (5), and dispersed by the échelle grating (6); both wavelengths D_1 and D_2 are enough separated so that a spatial filter rejects the D_2 component, which is very bright because of the direct fluorescence from the $3P_{3/2}$ level. Finally the light of the collimated beam which is neither ultraviolet nor yellow-orange crosses the two beam splitters, is reflected by the flat (1) and is focused by the spherical mirror (2); the non-dispersed image of the field in these complementary colours is focused again onto the CCD camera.

repetition rate 12.9 kHz. Then, the peak intensity was 4.8×10^5 W/m^2, approximately 10 times the saturation intensity of the D_2 line (which is $\approx 4.5 \times 10^4$ W/m^2). This assumes that i/ there was no energy spread outside of the brightest speckle, ii/ the spectral profile of the laser perfectly matched the Doppler line profile, and iii/ the sodium column density was 5.10^9 atoms/cm^2. Hence this factor of 10 above saturation is overestimated.

5.2.2. *The calibrated returned fluxes*

We pointed the laser beams toward the natural star SAO 60855, whose ultraviolet magnitude is $m_U \approx 8.2$. At 365 nm, the flux expected from outside the atmosphere is ≈ 800 photons/ Å /s. The optical efficiency of the telescope, the PASS instrument and the CCD camera was ≈ 0.065, in particular due to the 25% quantum efficiency of the CCD camera in the ultraviolet. The measured flux was 26 photons/ Å /s. One deduces that the atmospheric transmission was ≈ 0.5. Extrapolating this transmission

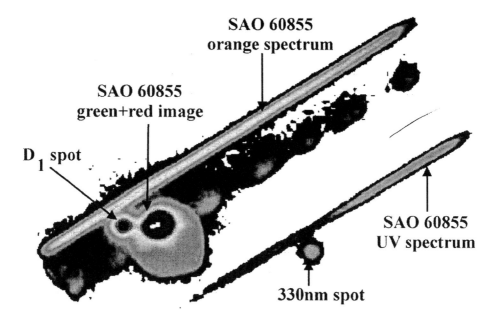

Figure 9. An image of the polychromatic laser guide star observed at the Lawrence Livermore National laboratory with the AVLIS laser with the PASS instrument shown in Fig. 8. The two stripes are the spectrum of the natural star SAO 60855 in the yellow (top) and the ultraviolet (bottom). The two components of the polychromatic LGS are visible (D_1 at the top left and 330 nm at the bottom right). The saturated image of the field in other colours is in the central area. The spots along and under the yellow spectrum are grating ghosts lit by the very bright Rayleigh beam at the wavelength of the D_2 line, of which the very top is barely visible at the bottom right of the field image. The total average laser power was ≈ 350 W.

to 330 nm leads to 0.3. At this wavelength the optical efficiency was only ≈ 0.035 due to the UV dichroic beam splitter being out of specifications.

The number of photons detected in the UV component of the polychromatic LGS was ≈ 800 photons/s. Above the atmosphere and with a perfect efficiency of 1, this number would be $\approx 7.6 \times 10^4$ photons/s. With a reasonably good optical efficiency of 0.4 and an atmosphere transmission at 330 nm of 0.6 (at an excellent astronomical site), one can expect to detect $\approx 2.3 \times 10^5$ photons/m²/s. This should allow us to measure the tilt at an accuracy of half the Airy disk down to the red without any adaptive optics.

5.2.3. *The Pulse Repetition Rate*

The pulse repetition rate was changed from 12.9 kHz to 4.3 kHz at constant average power. Since we were at saturation, the returned flux at 330 nm should have approximately constant (it should increase as the peak intensity, i.e. by a factor of 3, and it should decrease accordingly to the repetition rate, i.e. by a factor 3). At contrary, it increased by a factor of ≈ 1.9. A possible interpretation is that at 4.3 kHz, the field is 3 times higher, and then the Rabi frequency, or the power broadening, was 1.7 times higher (it varies as the square root of the intensity). This broadening may have at least partly filled the gaps in the modulator spectrum[20]: it is crucial for the process efficiency that the two modulators are perfectly phased. If this is true, it proves that, to get a given returned flux at 330 nm, one needs to improve the matching of the phase modulation of the two beams to fully benefit from the two-photon coherent absorption, rather than to rely on the Rabi frequency which implies an increase of the laser power.

Another interpretation of this behaviour of the 330 nm flux is of course that another process occurs in the mesosphere which is not taken into account in the above reasoning.

5.2.4. *The Modulation*

We switched the modulation width of the 589 nnm beam from 1 GHz (centered on the stronger line from the $F = 2$ ground level hyperfine level) to 3 GHz (spanning both hyperfine components). Then the returned flux at 330 nm increased by a factor of ≈ 1.55. This is expected since at 3 GHz 60% more sodium atoms from the lowest hyperfine structure component are excited.

5.3. THE POWER BALANCE BETWEEN THE TWO BEAMS

When the ratio of the average power at 569 nm to the total average power varied from 0.3 to 0.7, we found that the returned flux at 330 nm is maximum close to 0.5, as expected at saturation. We note, however, that the measurements are scattered.

6. From the experiment to a full demonstration: ELP-OA

A second PASS experiment is necessary to determine better the role of the experimental parameters in the excitation process. It should allow us in particular to optimize the laser specifications in order to be able to decrease the pump average power. Parameters such as the average output power, the modulation and the polarization will be varied. In particular a gain by a factor of 1.5 is expected with circularly polarized laser beams. This experiment will be carried out at the Pierrelatte site of CEA in October 1999. The SILVA laser chain will be used to shoot at the sky[21]. An oscillator free from spectral jitter will be also tested on this chain. In addition to six laboratories of the Centre National de la Recherche Scientifique (CNRS), partners are the CILAS company, the CEA, and the Office National d'Études et de Réalisations Aérospatiales (ONERA). r_0 will be measured simultaneoulsy with the LGS observations using a GSM device.

The column density of sodium in the mesosphere will also be monitored during all the nights of the run, thanks to a collaboration with the National University of Ireland at Galway (NUIG).

This will provide strong constraints for the French programme ELP-OA (Étoile Laser Polychromatique & Optique Adaptative) which aims at proving with an experiment at the 1.52 m telescope of the Observatoire de Haute-Provence (OHP) that the polychromatic LGS coupled with an adaptive optics device allows us to get long exposure images close to the diffraction limit in the red or the near infrared.

The ELP-OA programme is divided into the six following workpackages.

6.1. ATOMIC PHYSICS

This workpackage will provide a density matrix model of a 48 energy level sodium atom. An experiment will be set up to populate the $4D_{5/2}$ energy level with two lasers at 589 and 569 nm. The lab experiment will be used to check the sodium model. This work will be done at CEA. Simpler models, either with less energy levels or based on kinetics equations will be developped at CRAL and at ONERA and compared with the CEA model in order to be able to run numerical simulations with a large number of parameter configurations.

6.2. LASERS

The average laser power sent to the sky will be ≈ 25 W per beam. This value is thought to be the best tradeoff between the need for high laser power to excite the largest number of sodium atoms, and the lowest laser power to prevent thermal and light pollution of astronomical sites.

The pump lasers will be two CuHBr lasers currently developed at Laboratoire de Spectrométrie Physique (LSP, CNRS)[22, 23]. Each of them delivers ≈ 100 W. The pulse repetition rate can vary from 10 to 20 kHz. The pulse width is ≈ 50 ns. The beam quality is close to diffraction limit. The beam profile is Gaussian. The plug efficiency is remarkably high: $\approx 2\%$ and the temperature is quite low: $\approx 600°C$; both these last features are important in view of a routine operation at an astronomical site.

Oscillators for both transitions will be either dye lasers similar to those developed for the Uranium isotopic separation programme at CEA or OPOs currently developed at LSP.

6.3. OPTICS

This workpackage includes the following items:

Imaging through adaptive optics The adaptive optics system BOA developed at ONERA[24] will be attached to the Nasmyth focus of the telescope to observe the exposure images. BOA will be modified in order to measure the tilt independently, no longer with a natural guide star, but from the differential tilt. The focusing of the wavefront sensor onto the sodium layer has to be studied.

Emitting through adaptive optics In order to improve the accuracy in the differential tilt measurement, the laser spot images have to be as small as possible. Thus, they will be formed through another adaptive optics system with an auxiliary telescope. The tradeoff between the size of both the emission adaptive optics bench of the auxiliary telescope versus the gain in differential tilt $\delta\theta$ accuracy in terms of costs will be evaluated, taking into account the gain in returned flux due to non-linear excitation when the intensity in the mesosphere is increased.

Laser diagnostic tools The beam quality and the average power at the time of the launch will be monitored, as it was done, for example at the Lick Observatory[19].

Focal test instrumentation At least two instruments will be installed at the focus of the BOA:

- a spectral analyser to measure the spectral profile of the spot and
- an imaging camera, possibly with two channels in the near infrared and in the red. It will sample the Airy disk of the telescope at the wavelength of the observation.

Launch telescope Its diameter has to be defined. It will be at $\approx 10\,\mathrm{m}$ distance from the main telescope so that a field stop will reject the Rayleigh beam top. Additionally, telescope and atmosphere defocusings have to be disentangled, using the property that the time average of atmosphere defocusing is zero.

Laser beam transport Whether the laser beams will be transported to the launch telescope by free-space optics or by fiber optics remains to be decided.

6.4. INFRASTRUCTURE

This workpackage addresses handling, size of the rooms, cooling capabilities, power supply, as well as all safety requirements (for the operators, observers, detectors, air traffic, and satellites).

6.5. INTEGRATION

The Centre de Recherches Astronomiques de Lyon (CRAL) will manage the integration of subsystems and the final integration at the telescope.

This workpackage also includes the hardware and software to control the whole experiment.

6.6. MEASUREMENTS AND ANALYSIS

The strategy for the measurements campaigns at the telescope will be defined including the objectives which will contribute to the definition of the specifications of any component of the system. The data reduction and their interpretation in terms of the performance of the system is an essential part of this workpackage. Another kind of measurement is that of the mesosphere properties at the time of the observations. Indeed, we do need to know the altitude of the sodium layer for proper focusing and the sodium density, which is an input to the models to interpret the returned fluxes quantitatively, in particular at 330 nm. This mesosphere monitoring will likely be carried out by NUIG.

7. Conclusion

Within four years, the ELP-OA demonstrator should be operating. Then, one will definitely know whether the tilt can be measured at large telescopes with an LGS alone and without any natural guide star. The tilt measurement and the cone effect are the most important problems to be solved to get efficient performances with LGS adaptive optics system at 8-10 meter class telescopes. They are a major goal of the R&D programmes of several teams, in particular within the frame of the LGS network of the "Training and Mobility" programme of Researchers of the European Union.

Acknowledgements

I thank M. Schöck for a critical reading of this manuscript.

References

1. Pilkington, J. *Nature* **330**, 116 (1987).
2. Roddier, F. *Progress in Optics* **XXI**, 281–376 (1981).
3. Séchaud, M., Hubin, N., Brixon, L., Jalin, R., Foy, R., and Tallon, M. In *Very Large Telescopes and their Instrumentation,* M.-H., U., editor, ESO/NOAO Conferences, 705–714 (ESO, Garching, Germany, 1988).
4. Rigaut, F. and Gendron, E. *Astron. Astrophys.* **261**, 677–684 (1992).
5. Le Louarn, M., Foy, R., Hubin, N., and Tallon, M. *Mon. Not. R. astron. Soc* **295**, 756–768 (1998).
6. Davies, R., Hackenberg, Ott, Holstenberg, A. E. C., Rabien, S., Quirrenbach, A., and Kasper. In *Adaptive Optical System Technologies,* Bonaccini, D. and Tyson, R. K., editors, volume 3353 of *Proc. SPIE,* 116–124, (1998).
7. Hippler, Glindemann, A., Kasper, Kalas, Rohloff, Wagner, Looze, and Hackenberg. In *Adaptive Optical System Technologies,* volume 3353 of *Proc. SPIE,* (1998).

8. Davies, R., Hackenberg, W., Ott, T., Eckart, A., Rabien, S., Anders, S., Hippler, S., Kasper, M., Kalas, P., Quirrenbach, A., and Glindemann, A. *Astron. Astrophys. Suppl. Ser.* **138**, 345–353 (1999).
9. Olivier, S. S., Gavel, D. T., Friedman, H. W., Max, C. E., An, J. R., Avicola, K., Bauman, B., Brase, J. M., Campbell, E., Carrano, C., Cooke, J., Freeze, G., Gates, E., Kanz, V., T, C, K., Macintosh, B., Newman, M., Pierce, E., Waltjen, K., and Watson, J. In *Adaptive optics systems and technology*, Tyson, R. K. and Fugate, R. Q., editors, volume 3762 of *Proc. SPIE*, (1999).
10. Gavel, D. T. and Friedman, H. W. In *Adaptive optics systems and technology*, Tyson, R. K. and Fugate, R. Q., editors, volume 3762 of *Proc. SPIE*, (1999).
11. Wizinowich, P., Acton, D. S., Gregory, T., Stomski, P., An, J., Avicola, K., Brase, J. M., Friedman, H. W., Gavel, D. T., and Max, C. E. In *Adaptive optical system technologies*, Domenico and Tyson, R. K., editors, volume 3353 of *Proc. SPIE*, 568–578, (1998).
12. Sebag, J., d'Orgeville, C., Chun, M., Filhaber, J., Oschmann, J., Rigaut, F., and Simons, D. In *Adaptive optics systems and technology*, Tyson, R. K. and Fugate, R. Q., editors, volume 3762 of *Proc. SPIE*, (1999).
13. Filippenko, A. V. *Pub. Astron. Soc. Pacific* **94**, 715 (1982).
14. Roddier, F. In *Scientific Importance of High Angular Resolution at Infrared and Optical Wavelengths*, M.-H., U., editor, ESOConferences, 5–23 (ESO, Garching, Germany, 1981).
15. Aussel, H., Petit, A., Sjöström, S., and Weulersse, J.-M. In *ICO*, volume 16, (1993).
16. Foy, R., Tallon, M., Friedman, H., Baranne, A., Biraben, F., Grynberg, G., Louarn, M. L., Petit, A., Weulersse, J.-M., Migus, A., and Gex, J.-P. In *High-Power Laser Ablation 1998*, Phipps, C., editor, volume 3343 of *Proc. SPIE*, 194–204, (1998).
17. Max, C. E., Avicola, K., Brase, J., Friedman, H. W., Bissinger, H. D., Duff, J., Gavel, D. T., Horton, J., Kiefer, R., Morris, J., Olivier, S. S., Presta, R., Rapp, D., Salmon, J., and Waltjen, K. *J. Opt. Soc. Am. A* **11**, 813–824 (1994).
18. Avicola, K., Brase, J. M., Morris, J. R., Bissinger, H. D., Duff, J. M., Friedman, H. W., Gavel, D. T., Max, C. E., Olivier, S. S., Presta, R. W., Rapp, D., Salmon, J. T., and Waltjen, K. E. *J. Opt. Soc. Am. A* **11**, 825–831 (1994).
19. Friedman, H. W., Erbert, G., Kuklo, T., Salmon, T., Smauley, D., Thompson, G., Malik, J., Wong, N., Kanz, K., and Neeb, K. In *Adaptive optics*, Cullum, M., editor, volume 54 of *ESO Conferences*, 207 (ESO, Garching, Germany, 1996).
20. Morris, J. *J. Opt. Soc. Am. A* **11**, 832–845 (1994).
21. Scho"ck, M. and Foy, R. In *Adaptive optics systems and technology*, Tyson, R. K. and Fugate, R. Q., editors, volume 3762 of *Proc. SPIE*, (1999).
22. Coutance, P., Naylor, G., and Pique, J.-P. *IEEE J. Quantum Electron.* **31**, 1747–1752 (1995).
23. Coutance, P. and Pique, J.-P. *IEEE J. Quantum Electron.* (1998).
24. Madec, P.-Y., Rabaud, D., Fleury, B., Conan, J.-M., L.Rousset-Rouvire, Mendez, F., Montri, J., Michau, V., Rousset, G., and Schaud, M. *La Lettre de l'OHP* **16**, 2–3 (1997).

T. Roberts

Special 'lecture' for P. Doel (left), P. Lucas, C. O'Sullivan & J. Baker (right)

CHAPTER 9
SKY COVERAGE WITH LASER GUIDE STAR SYSTEMS ON 8M TELESCOPES

M. LE LOUARN
ESO
Karl Schwarzschild Straße 2
D-85748 Garching - Germany

1. Introduction

In order to work, an adaptive optics (AO) system needs a reference object in order to measure the wavefront perturbations created by atmospheric turbulence. This object has to be bright enough to provide a good signal to noise ratio on the wavefront sensor, so that a good correction can be achieved. If the science object is bright enough, it can be used as a reference. Otherwise, another reference (usually a star) has to be found. This star has to be close to the astronomical object (within the isoplanatic patch), so that the measurements made on the star also apply to the science object. These two constraints severely limit the use of AO systems, because a suitable star cannot usually be found. As will be seen, Natural Guide Star (NGS) systems sky coverage is usually only of order a few per-cent, at best.

The purpose of a laser guide star (LGS) is to increase the sky coverage ([3]) i.e. the number of objects that can be observed. This artificial star can be placed anywhere in the sky. One could think that this would completely solve the problem and allow to make corrections on the whole sky. This is unfortunately not the case. The tilt of the wavefront cannot be measured from a LGS, because of the inverse return of light. Several solutions have been proposed to solve this problem (see Chapters 7 & 8 in this book). However, for the first generation of LGS systems, the solution is to use a nearby reference star to measure the tilt. This solution increases sky coverage, because only the tilt has to be measured, a fainter or more remote star can be used. However full sky coverage cannot be achieved (see eg. [14], [16], [8]). If the tilt problem was solved, a full sky coverage would be achieved.

The goal of this lecture is to show how sky coverage calculations are done and what the results are. First, the model used to simulate the AO system and the laser system will be described. Then, two different approaches to the sky coverage calculations will be shown : a statistical method, and a cross correlation method. A comparison will be made between the performance (in terms of sky coverage) of an AO system with and without a LGS.

2. Sky coverage computation

This section shows what components are needed to compute the sky coverage and how one can compute it from the models.

2.1. AO SYSTEM MODEL

In order to compute the sky coverage, a model of the AO system has to be used. Indeed, the fainter reference star the system can use, the more stars will be available to perform wavefront sensing, and thus the higher the sky coverage will be.

2.1.1. *The AO system*

This system is made of several elements. Each element induces a source of error that can be modeled. Summing the individual contributions yields a total error budget. This can then be converted to a final Strehl ratio of the corrected image. The modeled system consists of a Shack-Hartmann Wavefront sensor (WFS) and piezo-stack mirror system. A separate tip-tilt sensing system (based on a quad-cell avalanche photo-diode) is included.

The error sources associated with AO that are considered are:
— The time delay error. It occurs both on the tilt sub-system and on the higher order correction loop. It is due to the delay between the measurement of the wavefront and the correction that is applied on the deformable mirror.
— Photon noise. This error which is also common to both sub-systems, comes from the quantum nature of light and is the fundamental limit to the sensitivity of AO systems.
— Aliasing and fitting errors come from the finite size of the inter-actuator spacing and of the finite size of the Shack-Hartmann WFS sub-apertures.
— Detector noise. This is mainly due to the read-out noise of the CCD which is used on the WFS. Even if this noise is very low (3 electrons rms per pixel and per frame), it contributes to the reduction of the limiting

magnitude of the system. A very small noise was also included for the APDs, to allow for imperfections.
- Sky background. The magnitude of the sky background was set to 21.5 at the wavelength of the wavefront sensor, to allow for the noise coming from airglow. This parameter is mostly relevant for the LGS case, where the limiting magnitude of the tip-tilt sensor is close to the sky background.
- Spot size of the LGS. This reduces the performance of the WFS, by decreasing the efficiency of spot centroid measurement on the SH, and is due to the limitations of the laser and transmitting optics. This size was set to 1.5 arcsec, which is compatible with measurements of spot sizes made on the sky ([7]).
- Isoplanatism. This effect arises when the reference star is not aligned with the science object. For NGS systems, all-order (piston removed) anisoplanatism is used. For LGS systems, only tilt anisoplanatism is computed, since the LGS provides all other information.
- Cone effect. This effect affects only LGS systems. It results from the finite altitude of the laser star. Its effect is to reduce the Strehl ratio of the system. The cone effect and possible solutions are discussed in more details in Chapter 8.

We have neglected the sodium layer height and column density variations, and suppose focus is measured on the laser star. If the sodium layer height varies rapidly, spurious focus measurements will be made causing a bluring of the corrected image. Not much is known about high temporal frequency variations of the height of the sodium layer. However, it seems that they will not seriously affect the correction quality. More measurements are however needed (see [11], [9]).

Using analytical formulae, one can compute the wavefront variance due to each error source. Then, the variances are added assuming the errors and independent. This yields a total variance. Analytical formulae are then used to derive the Strehl (see eg. [12]).

2.1.2. *Atmospheric model*
In order to see the effect of atmospheric conditions on the results, two model atmospheres were used. The "good" model, represents the 20% best seeing and turbulence distribution at Cerro Paranal, the location of ESO's Very Large Telescope. A median model is also used. A histogram of the seeing in Paranal can be seen on Fig. 1. It shows that the 20 % best seeing is approximately 0.5″ and half of the time one can expect to get a seeing better than 0.7″. Turbulence distribution statistics were taken from [5]. These measurements were made during the site testing campaign, where

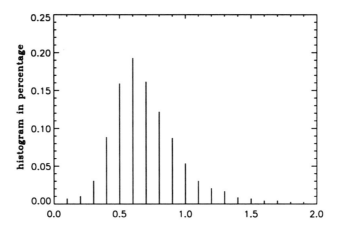

Figure 1. Histogram of seeing (in arcsec) at Paranal

the turbulence distribution was measured using a SCIDAR and balloons.

Flexible scheduling, which selects the nights having the best seeing conditions for high angular resolution observations, should permit the best use of the site and hence 20 % best seeing can be representative of the conditions when using the AO instrument.

In order to assess the temporal behavior of turbulence, one has to use a wind model. In the Taylor frozen turbulence hypothesis, one assumes that thin turbulent layers move at the same speed as the wind. The chosen wind model is a modified Bufton profile ([2],[1]). The profile can be seen in Fig. 2.

Turbulence distribution, integrated seeing and a wind profile are used to define some parameters used in the AO simulations. Table 1 summarizes the main characteristics of the atmospheric model.

In this table: $h_{ao} = (\mu_{5/3}/\mu_0)^{3/5}$ is the weighted altitude of turbulence for adaptive optics [15], corresponding to a characteristic height of the turbulence, $\theta_0 = (2.91k^2 \sec(\zeta)^{8/3} \mu_{5/3})^{-3/5}$ is the isoplanatic angle (in arcseconds) ([4]), and τ_0 is the correlation time of the atmosphere ([6]):

$$\tau_0 = (2.91k^2 \sec(\zeta) v_{5/3})^{-3/5} \qquad (1)$$

where $v_n = \int dh C_n^2(h) v(h)^n$ is the nth wind moment.

For more details about these simulations, such as hardware parameters and details of the formulae, see [8].

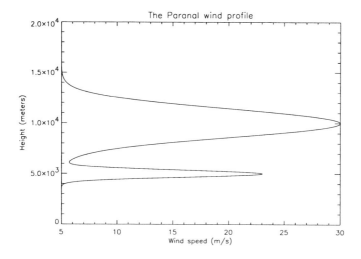

Figure 2. Wind profile for Paranal

TABLE 1. Atmospheric parameters

	Paranal (good)	Paranal (med.)
Seeing[1] ($''$)	0.5	0.7
% 1st layer [2]	0.89	0.7
% 2nd layer [2]	0.11	0.3
H_1 (km)	2.5	2.5
H_2 (km)	10.0	10.0
θ_0[1] ($''$)	3.5	1.7
h_{ao} (km)	3.8	5.5
τ_0[1] (ms)	6.6	3.0

[1]: at 0.5 μm, zenith
[2]: in % of the total atmospheric seeing

2.1.3. *Performance*

In this section, some results from the simulation described in the previous section are presented.

When the science object is used as a reference, one gets the Strehl performance shown in Fig. 3. The effect of the LGS on the performance can be seen when comparing the plots on the right and left. Two main effects occur. The first one is to increase the limiting magnitude of the system: one loses 50 % of the Strehl near magnitude 15 for the NGS (K

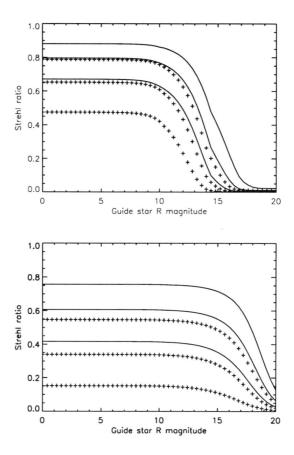

Figure 3. Predicted Strehl vs guide star magnitude (for an NGS system (top) and LGS system (bottom)) at 2.2 μm, 1.65 μm, 1.25 μm, good atmosphere model (solid) and median model(crosses). The guide star is on-axis.

band, good seeing), and 18.5 for the LGS system. This is due to the fact that with a LGS, one measures only the tilt mode on the faint reference star and that one can use noiseless avalanche photodiodes to perform this sensing. The second effect is to lower the maximum achievable Strehl ratio. It drops from 86 % to 75 % in K (good seeing). However, at shorter wavelengths and median seeing distribution, the loss is more severe (loss of 60 % in J band, median seeing). This loss reflects the fact that the cone effect is heavily turbulence height and wavelength dependent.

Astronomers are usually more interested in the FWHM of the image rather than in the Strehl ratio. In order to estimate this parameter, one can use an analytical formula giving the FWHM as a function of the wavefront error ([10]). This formula decomposes the point spread function into

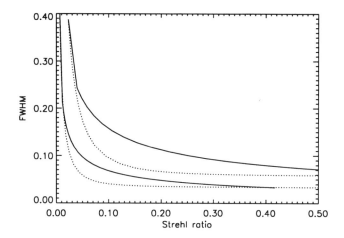

Figure 4. FWHM (arcsec) vs Strehl at 2.2 μm (top curve) and 1.25 μm (bottom), good seeing model. NGS AO (dotted) and LGS AO (solid).

two parts, a diffraction limited core and a seeing limited halo. At high Strehl ratios, the results are in accordance with measurements. However, at low Strehls, the results are less reliable, because the real structure of the image is more complicated than just a two component model ([12]). The FWHM (in arcsec) of the image versus the Strehl is plotted in Fig. 4. Two regimes appear: above a Strehl of about 0.2, the image is nearly diffraction limited. Below, the image FWHM increases rapidly. However, even though the FWHM is large, a diffraction limited core can still be present, indicating that high spatial frequency information is still present.

2.2. STATISTICAL SKY COVERAGE

Statistical sky coverage is obtained by combining two types of information: the on-axis system performance, and the anisoplanatism effects. Indeed, the on-axis performance is the maximum performance the system can achieve for a given reference magnitude. If an off-axis reference is needed, anisoplanatism will lower the Strehl.

Statistically, one will be able to find a faint star closer to the object than a bright star. However, the brighter the star is, the better the performance (see Fig. 3). So there is an optimum between the distance of the star and its brightness. To find this optimum, one can plot iso-Strehl curves which show the Strehl as a function of guide star magnitude and the distance between the guide star and the science object. The optimum is found at the knee of the curve (see dotted curves in Fig. 5).

In order to see how far, on average, one has to go in order to find a suitable reference star, a model of our Galaxy has to be used. This model predicts, for a galactic latitude and longitude, the density of stars brighter than a given magnitude. Density variations are large, since the regions of galactic poles have low star densities, and regions in the disk and towards the galactic center are rich in stars.

Given these densities, one can compute the probability P of finding at least one reference star within a given radius r around the science object, by assuming Poisson statistics:

$$P_{N_{\text{stars}}>0}(m,r) = 1 - e^{-\frac{\pi r^2 \eta(m)}{3600^2}} \qquad (2)$$

where $\eta(m)$ is the density of stars brighter than magnitude m (per square degree) in the considered galactic region. The model described by Robin and Crézé ([13]) was used to get the values for $\eta(m)$.

This allows us to plot the curve of probability as a function of guide star R magnitude for given galactic coordinates (solid lines on Fig. 5). When one overlays the iso-Strehl curve with the probability curves, one can read the sky coverage at the point where the knee of the iso-Strehl intersects the probability curve. This represents the probability to find a reference star yielding the given Strehl.

A summary of the results obtained can be found in Tables 2 and 3. They give sky coverage for two different observation wavelengths, for several positions in the sky, and for different target Strehls. The tables compare the performances of NGS and LGS systems. The first number is for good seeing conditions, the second one for median seeing conditions. It is clear that the LGS significantly improves the sky coverage that can be achieved with the AO system. The LGS insures full coverage with a Strehl of 0.5 near the galactic center, when NGS system offers only 20 per-cent (good seeing, K band). Even in the worst conditions of the K band (looking towards the galactic pole, with median seeing), the LGS improves the performances, even if the coverage remains very small. (from 0.02 % to 0.08 %). In the J band, the situation becomes more complicated. The cone effect significantly reduces the advantage of using a LGS. In good seeing, the LGS improves performances. In median seeing, the NGS can be better (see Strehl of 0.2, average galactic coordinates), but not always (e.g. galactic center).

A plot of sky coverage versus Strehl (Fig. 6), shows that the LGS does not always give the best sky coverage, even in K band, if one is interested in very high Strehl ratios. This could be for example the case for the search of faint and close stellar companions like brown dwarfs, where light concentration is important. If the target Strehl is near 0.8 (in K), the NGS system gives better coverage than the LGS. This is because the cone effect

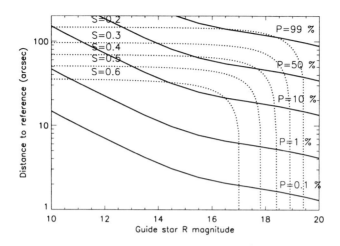

Figure 5. Iso-Strehl curves (dotted). Probability to find a reference star (solid) at galactic longitude (l) = 180, latitude (b) = 20. Notice logarithmic scale for distances.

TABLE 2. Statistical sky coverage, K band

K Strehl	Center	Average[1]	Pole
$S_{\rm NGS} = 0.2$	80 - 15	10 - 1	0.5 - 0.05
$S_{\rm LGS} = 0.2$	99 - 99	99 - 33	10 - 0.8
$S_{\rm NGS} = 0.3$	60 - 8	8 - 0.7	0.2 - 0.02
$S_{\rm LGS} = 0.3$	99 - 99	75 - 12	5 - 0.08
$S_{\rm NGS} = 0.5$	20 - 1	2 - 0.1	0.07 - < 0.01
$S_{\rm LGS} = 0.5$	99 - 8	30 - 0.5	0.7 - <0.01

[1]: l=180, b=20.

prevents one from obtaining these highest Strehls. In J band, where the cone effect is more severe, this effect is better seen, since the crossing point is between a Strehl of 0.2 and 0.4.

2.3. SKY COVERAGE: OBJECT-COUNTS

In the previous section, the percentage of the sky that can be observed was estimated. Another approach to sky coverage is to compute how many objects one can observe with a NGS or a LGS system. This question is driven by astrophysical considerations. Indeed, it is well known that some objects are not distributed uniformly in the sky (e.g. pre-main sequence (PMS)

TABLE 3. Statistical sky coverage, J band

J Strehl	Center	Average[1]	Pole
$S_{\rm NGS} = 0.1$	30 - 2	2 - 0.2	0.07 - <0.01
$S_{\rm LGS} = 0.1$	99 - 20	70 - 1	1 - <0.01
$S_{\rm NGS} = 0.2$	10 - 0.5	0.8 - 0.07	0.03 - <0.01
$S_{\rm LGS} = 0.2$	99 - <0.01	25 - <0.01	0.08 - <0.01
$S_{\rm NGS} = 0.3$	5 - 0.08	0.5 - 0.02	<0.01
$S_{\rm LGS} = 0.3$	70 - <0.01	6 - <0.01	<0.01

[1]: l=180, b=20.

stars are concentrated in star-forming regions), and hence the statistical sky coverage does not easily translate into a number of PMS stars one can observe.

In order to answer to this question, one can use cross correlations of catalogues: first, one has to find a catalogue containing a significant number of "interesting" objects. For example, the Veron-Cetty 96 (Veron-Cetty & Veron [17]) contains 8609 quasars and 2833 active galactic nuclei (AGN). In the stellar domain, one can use the SIMBAD database, which contains 4279 Miras, 2182 Semi Regular variables (SRs), and 928 Pre-Main Sequence stars (PMS). Once the coordinates of these objects are found, one can look in another catalogue (for example the US Naval Observatory A-V1.0 catalogue (USNOC)) if stars can be found around these science objects. If stars are found, one can compute, using the AO system model, the Strehl ratio that can be achieved using the reference star or the science object itself for tilt sensing (LGS) or for all modes (NGS). This study leads to Tables 4 and 5. The first four columns are related to NGS and the last four to the LGS. For each system, statistics are gathered on the mean Strehl achieved with the system (in per-cent), the dispersion on this Strehl and the number of objects one can observe with a Strehl greater than 0.1 and 0.2. GS stands for good seeing and MS for median seeing. For stellar objects, two numbers are represented. The first one is for the object at maximum brightness, the second one at minimum brightness. For many stellar objects, the minimum brightness is not quoted in catalogues. To cope with this, it was decided to take, for these stars, a brightness at minimum 4 magnitudes (SRs: 2 magnitudes) fainter than at maximum (for which a value was found in the catalogue).

Some results in this table can seem surprising at first glance. For example, the mean Strehl on Miras (at maximum intensity) is greater with a

TABLE 4. 8m - astrophysical targets, K band. GS: Good seeing, MS: Median seeing. Strehls are in percent.

Object	$< S_{NGS} >$	$\sigma_{S_{NGS}}$	$N_{S>0.1}$	$N_{S>0.2}$
Quasar - GS	4.4	7.2	819	357
Quasar - MS	1.2	2.0	63	21
AGN - GS	15.4	23.1	895	698
AGN - MS	6.9	15.2	426	315
PMS - GS	45.2 - 19.8	27.1 - 23.1	771 - 439	710 - 317
PMS - MS	25.4 - 7.6	23.7 - 14.2	568 - 180	443 - 127
SR - GS	61.4 - 49.5	23.0 - 26.6	2020 - 1896	1961 - 1768
SR - MS	42.2 - 29.4	23.9 - 25.5	1835 - 1426	1663 - 1159
Miras - GS	63.5 - 35.4	21.3 - 27.2	4084 - 3189	3961 - 2660
Miras - MS	44.2 - 17.4	22.4 - 22.0	3772 - 1811	3530 - 1342

Object	$< S_{LGS} >$	$\sigma_{S_{LGS}}$	$N_{S>0.1}$	$N_{S>0.2}$
Quasar - GS	37.2	17.0	7651	6803
Quasar - MS	16.7	10.6	5953	2893
AGN - GS	43.0	23.3	2329	2077
AGN - MS	22.6	17.0	1850	1393
PMS - GS	63.0 - 50.0	14.0 - 19.2	918 - 868	892 - 822
PMS - MS	38.9 - 26.5	12.4 - 14.0	870 - 777	842 - 624
SR - GS	57.8 - 56.9	18.3 - 18.0	2050 - 2055	1984 - 1990
SR - MS	33.8 - 32.6	15.5 - 15.1	1877 - 1884	1721 - 1696
Miras - GS	57.9 - 54.1	18.1 - 18.4	4096 - 4041	3960 - 3906
Miras - MS	33.8 - 29.7	15.4 - 14.9	3748 - 3679	3450 - 3148

TABLE 5. 8m - astrophysical targets, J band. GS: Good seeing, MS: Median seeing. Strehls are in percent.

Object	$< S_{NGS} >$	$\sigma_{S_{NGS}}$	$N_{S>0.1}$	$N_{S>0.2}$
Quasar - GS	0.7	1.5	43	12
Quasar - MS	0.2	0.2	1	0
AGN - GS	4.9	12.2	350	241
AGN - MS	1.6	5.3	148	68
PMS - GS	17.9 - 5.4	18.7 - 10.9	476 - 147	346 - 104
PMS - MS	6.5 - 1.3	10.6 - 3.9	203 - 35	124 - 14
SR - GS	30.4 - 21.1	20.2 - 20.6	1670 - 1213	1411 - 898
SR - MS	13.7 - 8.7	13.9 - 12.9	1017 - 619	632 - 418
Miras - GS	31.7 - 12.4	19.2 - 0.17	3516 - 1497	2975 - 1045
Miras - MS	14.1 - 4.2	13.3 - 9.0	2156 - 616	1294 - 371

Object	$< S_{LGS} >$	$\sigma_{S_{LGS}}$	$N_{S>0.1}$	$N_{S>0.2}$
Quasar - GS	11.0	7.6	4214	1119
Quasar - MS	2.1	1.9	40	0
AGN - GS	15.4	12.5	1568	965
AGN - MS	3.6	3.9	287	0
PMS - GS	27.3 - 18.2	9.5 - 10.2	854 - 702	760 - 422
PMS - MS	7.1 - 3.9	3.5 - 2.99	196 - 46	0 - 0
SR - GS	22.9 - 22.1	12.1 - 11.7	1736 - 1727	1406 - 1360
SR - MS	5.5 - 5.2	4.3 - 4.0	387 - 296	0 - 0
Miras - GS	23.0 - 19.8	12.2 - 11.5	3481 - 3231	2701 - 2271
Miras - MS	5.6 - 4.4	4.4 - 3.8	868 - 428	0 - 0

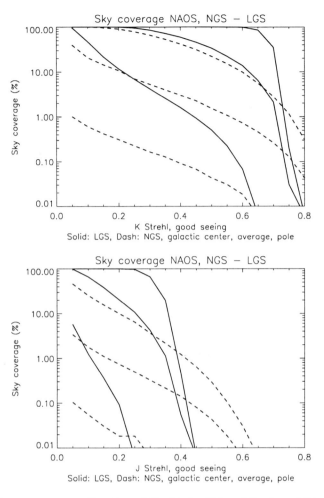

Figure 6. Sky coverage as a function of Strehl, for K band (top) and J band (bottom). Three galactic latitude and longitude positions are represented: pole, average (l=180, b=20) and center. Solid is LGS, dashed is NGS.

NGS than with a LGS, in K band. This can be explained if one considers the mean magnitude of this kind of object at maximum, 12.6. The situation is reversed if one considers the numbers for minimum brightness. This is phenomenon is explained by Fig. 3. Indeed, for bright objects, it is better to use the NGS since there is no cone effect which reduces the on-axis Strehl. However, for fainter objects, the LGS performs better, because its limiting magnitude is considerably fainter. This trend can be seen also with SRs (magnitude 12.3 at maximum). On the other hand, for extragalactic objects (quasars, mean magnitude 18.2 or AGN, 16.7), the LGS considerably improves the performance.

Another characteristic is that the dispersion on the mean Strehl is considerable. Several factors contribute to this dispersion: the intrinsic magnitude dispersion of the science objects and the reference stars. Another effect is due to the different conditions in which the objects are observed. Indeed, the zenith angle of the object was chosen so that it corresponds to the time of the year when the object is highest in the sky as seen from the Paranal observatory. Some objects can barely be observed from Paranal, and they therefore contribute to lower the mean Strehl and they also increase the dispersion.

The number of objects with Strehls above a given value indicates if a sufficient sample of this class of objects can be observed to infer statistical properties on them. For example, the LGS allows one to considerably increase the statistics on extragalactic objects, like AGN and quasars. Indeed, one goes from a sample of 357 Quasars (S > 0.2, in K band) to 6803 and from 21 AGN to 2893. In the stellar domain the improvements are less spectacular, but one can see that the domain of variable stars at minimum brightness is increased by a factor up to 2.6 (for PMS).

The two Strehl values of 10% and 20% were chosen because with a Strehl of 0.2, one can expect diffraction limited images. With a Strehl of 10%, diffraction limit can be achieved with the aid of deconvolution, provided the signal to noise ratio on the image is good and that the PSF is known or can be reconstructed.

The situation in the J band is, as with the statistical approach, more contrasted. In some situations, improvement is possible with a LGS, and sometimes this improvement is tremendous (like for the quasars, one goes from 12 objects with Strehl greater than 0.2, to 1119, in good seeing). However, in median seeing the situation is not favorable to the laser star. Improvements are small (e.g. AGN, one goes from one object to 40, but high Strehl are not achieved). Sometimes, the LGS degrades the performances compared to NGS (Stellar cases, like SRs and Miras, which do not seem to benefit much from LGS, except for the faint Miras).

Both tables, and especially J band data, show the importance of having favourable turbulence distribution conditions. In good seeing, in J band, one benefits tremendously from the LGS for Quasars, whereas in median seeing, this gain disappears. Therefore, a device measuring, on-line, the seeing distribution would allow one to diagnose the gain a LGS can bring, in real time.

This statistical approach, even if it brings an astrophysical insight to sky coverage, has biases. The completeness of the catalogues is one issue. For example, the USNOC has objects to magnitudes near 20, which is compatible with the performance curves of the LGS AO system. However, some faint stars (but bright enough for wavefront sensing) may not appear in the catalogue. There can also be mis-identifications (defects on the plates identified as stars) in the catalogue and voids which have not been digitized. Therefore, it cannot be excluded that not all possible reference stars have been accounted for. Another incompletenss is that of the science catalogues, which sometimes do not list the faintest sources.

Other biases exist. For example, the magnitude for AGN is an integrated magnitude. The magnitude of the nucleus might be different from the quoted one and hence, if one is interested in imaging the central regions, one might not get the predicted performance. This bias is small for quasars, which are less extended. For the on-axis performance calculations, it was assumed that the science objects are point-like. This is not true for AGN, in which the central region can be extended and the signal to noise ratio on the WFS is not as good as with unresolved sources, and so performances will be degraded compared to this study. The spectrum of the object is also important, since the number of photons seen by the WFS depends on it. This effect was taken into account by calculating mean V-R correction factors for the stellar sources. Such a correction was not done on extragalactic objects because of the variability of the spectra from an object to another.

Despite these limitations, the numbers given in the Tables should show the astrophysical performances of LGS systems.

3. Conclusion

As we have seen, the sky coverage remains an issue for first generation laser guide stars on 8m telescopes. The sky coverage is increased by a large factor in the K band (for example from 8% to 75% at average galactic coordinates, for a Strehl of 0.3), but in J band the gain is less obvious (gain can go from a coverage of 2% to 70 % at average coordinates for a Strehl of 0.2, but can be non-existent for a Strehl of 0.3, average coordinates, median seeing). The shorter the wavelength, the larger the variations of the performances as a function of the turbulence conditions are. The sky coverage limitations are due to the two limitations of the LGS, namely the tilt determination problem and the cone effect. The tilt problem limits the sky coverage and the cone effect limits the spectral coverage to the near infrared. If the tilt can be measured from the LGS, one can expect to get full sky coverage. However, only a solution to the cone effect will allow one to extend this 100% sky coverage to the visible.

References

1. D. Bonaccini. Apds for optimal tip-tilt on the vlti. Technical Report VLT-TRE-ESO-11670-1121, ESO, June 1996.
2. J.L. Bufton. Comparison of vertical profile turbulence structure with stellar observations. *Appl. Opt.*, 12(8):1785–1793, August 1973.
3. R. Foy and A. Labeyrie. Feasability of adaptive optics telescope with laser probe. *A&A*, 152:L29–L31, 1985.
4. D. L. Fried. Anisoplanatism in adaptive optics. *J. Opt. Soc. Am.*, 72(1):52–61, January 1982.
5. A. Fuchs and J. Vernin. Final report on parsca 1992 and 1993 campaigns. Technical Report VLT - TRE - UNI - 17400 - 0001, ESO, 1993.
6. D. P. Greenwood. Bandwidth specification for adaptive optics systems. *JOSA*, 67(3):390–392, 1977.
7. B. Jacobsen, T. Martinez, R. Angel, M. Lloyd-Hart, S. Benda, D. Middleton, H. Friedman, and G. Erbert. Field evaluation of two new continuous-wave dye laser systems optimized for sodium beacon excitation. In *Adaptive optics for astronomy Proc SPIE 2201*, pages 342–351. SPIE, 1994.
8. M. Le Louarn, R. Foy, N. Hubin, and M. Tallon. Laser guide star for 3.6m and 8m telescopes: performances and astrophysical implications. *MNRAS*, 295(4):756, 1998.
9. M. Le Louarn, N. Hubin, and R. Foy. Performances of natural and laser guide staradaptive optics for 8m class telescopes. In R. K. Tyson and R. Q. Fugate, editors, *SPIE, Adaptive Optics and applications*, volume 3126, pages 8–17. SPIE, SPIE, 1997.
10. R. R. Parenti. Adaptive optics for astronomy. *Lincoln Laboratory Journal*, 5(1):93–114, 1992.
11. G. C. Papen, C. S Gardner, and J. Yu. Characterization of the mesospheric sodium layer. In *OSA conf. on Adaptive Optics*, volume 13 of *OSA*, pages 96–99. OSA, OSA, 1996.
12. R. R. Parenti and R. J. Sasiela. Laser guide star systems for astronomical applications. *J. Opt. Soc. Am. A*, 11(1):288–309, February 1994.
13. A. Robin and Créze. Stellar population in the milky way: a synthetic model. *A&A*, 157:71–90, 1986.
14. F. Rigaut and E. Gendron. Laser guide star in adaptive optics: the tilt determination problem. *A&A*, 261:677–684, 1992.
15. F. Roddier, J. M. Gilli, and G. Lund. On the origin of speckle boiling and its effects in stellar speckle interferometry. *J. Optics*, 13(5):263–271, 1982.
16. D. G. Sandler, S. Stahl, J. R. P. Angel, M. Lloyd-Hart, and D. McCarthy. Adaptive optics for diffraction limited infrared imaging with 8m telescopes. *J. Opt. Soc. Am. A*, 11(1):925–945, February 1994.
17. M. Veron-Cetty and P. Veron. A catalogue of quasars and active nuclei. ESO scientific report 17, ESO, 1996.

CHAPTER 10
GROUND BASED ASTRONOMY WITH ADAPTIVE OPTICS

STEPHEN T. RIDGWAY
Kitt Peak National Observatory
National Optical Astronomy Observatories[†]
P.O. Box 26732
Tucson, AZ 85718 - U.S.A.

1. Introduction

Adaptive optics for astronomy has been under intensive development for a decade now. Numerous facilities are in or near operation. The performance of ground based natural guide star systems is fairly well understood, and laser beacon development has made enormous progress also. For the first time, a substantial body of scientific results from adaptive optics is in the astronomical literature. This is an excellent opportunity to review the progress of adaptive optics from the user's point of view, highlighting the successes, describing and understanding the limitations, and identifying areas where significant progress may be imminent. It is also timely for observatories which are not yet implementing adaptive optics to decide if it is time to start.

This contribution summarizes and expands on lectures presented at the institute. The search for published science based on adaptive optics extends through May 1999.

2. The Investment in Adaptive Optics

To some extent the development cost of adaptive optics for astronomy has been shared between the astronomy and military communities. Due to very different objectives and R&D styles, however, the opportunity to share development effort is limited. In recent years, it appears that the civilian

---

[†]Operated by the Association of Universities for research in Astronomy, under cooperative agreement with the National Science Foundation

TABLE 1. Identified U.S. spending on adaptive optics for astronomy 1990–1997 plus known future commitments.

Organization	Spending
National Science Foundation	$12,300,000
Department of Defense	17,525,000
NASA	5,800,000
Private and State	9,650,000

market for adaptive optics, for example in opthalmology, may prove to be an increasingly important driver for commercial product development.

In the U.S., it has been possible to prepare a reasonably accurate accounting of spending specifically for adaptive optics in astronomy. The total spent or committed for the interval 1990–1999 is approximately $45M dollars. The breakdown by source is shown in Table 1.

To a considerable extent the spending figures are fairly realistic estimates of actual costs, including personnel, with the possible exception of the salaries of some scientist participants. Perhaps not coincidentally, this total is close to the the spending level recommended for adaptive optics in the 1990's by the Astronomy and Astrophysics Survey Committee (NRC, 1991). A histogram of this spending with time shows a steady increase from 1990 to 1997, now perhaps leveling off.

For an estimate of activity in Europe, the participation of approximately 50 persons from the European community in this summer school, as well as the number of independent groups working on adaptive optics systems, lasers or laser beacons, suggests that current European spending in excess of several $M per year is likely.

These numbers testify to a remarkable level of support, reflecting the great confidence and salesmanship of the developers, and the eagerness of major observatories to augment their facilities with the expected benefits of adaptive optics. There is also, perhaps, the fear of being left behind by a development which by its very nature is expected to change the rules of the game. With such expectations, one is eager to ask the obvious question.

3. Does Adaptive Optics Work?

The very good news is that natural guide star adaptive optics works well, and the performance and limitations are reasonably understood. Information for two natural guide star systems is collected in Table 2. The sys-

TABLE 2. Demonstrated performance of operational natural guide star adaptive optics systems. BVRI data are from the ADOPT system (Shelton, 1998), and JHK are from the CFHT PUEO system (PUEO WEB homepage).

	Mt. Wilson				PUEO		
	B	V	R	I	J	H	K
Bright Star Strehl	0.02	0.07	0.15	0.30	0.27	0.41	0.56
R mag (50% loss)		10.0			14.3	15.0	15.7
Offset (50% loss)					20"	30"	40"

tems are the ADOPT[1] natural guide star system on Mt. Wilson, designed for visible operation, and PUEO[2], designed for infrared operation, at the Canada-France-Hawaii telescope on Mauna Kea. In this table, the Bright Star Strehl gives the on-axis Strehl achieved in median conditions for a bright wavefront reference star. (The higher Strehl at I than J may seem unexpected. Remember that the ADOPT system is optimized for visible operation, and that median conditions on Mt. Wilson are quite different from median conditions on Mauna Kea.)

The actual Strehl observed will be reduced for fainter reference stars, and for off-axis angle away from the reference star, and these reductions must be multiplied together when both apply. The 50% reduction values for these parameters are also given in Table 2.

To the extent that the table entries differ from model expectations, it is in a favorable direction. First, the limiting magnitudes are a little brighter than expected with foreseeable detector developments, so there is some room for improvement. Second, the size of the corrected field of view is significantly larger than most predictions, indicating that models were conservative and actual capability is greater in this regard than expected. Improving measurements of the turbulence profiles over observatories will probably bring models and performance into agreement.

Laser beacon adaptive optics systems are at least 5 years behind natural guide star systems. Tests so far show that both Rayleigh and sodium beacon systems can probably be built to function as conceived, though the ultimate Strehl is not yet clear. Laser beacon supported images have been reported, but it is not yet clear if they are better than could have been obtained with natural beacon systems on the same sources. The goal to image faint fields,

---

[1] Mount Wilson Institute
[2] *Probing the Universe with Enhanced Optics*

TABLE 3. The number of adaptive optics facilities in the world astronomy community.

Description/status	Number
Natural guide star systems in operation	8
Natural guide star systems nearing completion	4
Natural guide star systems in advanced planning	5
Laser beacon systems in operation	3
Laser beacon systems nearing completion	5
Laser beacon systems in advanced planning	2

using reasonably faint reference stars, still requires solution of numerous thorny problems.

4. Where are the Adaptive Optics Systems

The large expenditures on adaptive optics have in fact led to a large number of systems on telescopes at many different sites. Any list is rapidly obsolete, so just some statistics will suffice here. These are shown in Table 3.

The particularly remarkable number is the total of 11 operational adaptive optics systems. Most astronomers would not guess such a large number. There are several clear reasons for this. Only five have produced science publications, and of those only three are heavily utilized for astronomy. The number of publications prior to 1997 is fairly modest. Many of the publications tend to be outside the mainstream of astronomy, which currently emphasizes dark sky and multi-object observational programs.

5. What Will Adaptive Optics Do for Astronomy?

Perhaps it seems a little late to be asking this question. Actually, the expectations for adaptive optics have been immoderate in some cases, and some clarification is still needed. The basic answer is that adaptive optics will improve the angular resolution of our telescopes. There are a number of potential derivative gains, especially in sensitivity, but these must be examined with special care.

The angular resolution gain is relatively straight-forward in the infrared, where even simple shift-and-add imaging has shown great power to achieve high angular resolution with medium to large telescopes. In the infrared partial adaptive correction produces the well-known core-halo image, with the core approximating an Airy function (including diffraction rings), in which case angular resolution to or near the diffraction limit is immediate.

The terminology used to describe adaptive optics performance should be clarified here. As soon as a narrow Airy core is achieved, the angular resolution is effectively diffraction-limited, in the sense that the width of the Airy core (to the first zero) is $2.44\frac{\lambda}{D}$. This is true even if the Strehl is low – an Airy core can be obtained with Strehl of 0.1–0.2. This is the kind of diffraction limit that high resolution imaging attains. However, in optics, the term diffraction-limited is by convention reserved to describe an image quality with Strehl of greater than 0.81 (Schroeder, 1987) – a far more stringent criterion, and really not expected with most adaptive optics systems.

The obvious domain in which improved angular resolution will accrue benefits in increased information content is direct imaging, including imaging spectroscopic techniques such as long-slit and integral field spectroscopy. Improved resolution will produce an information gain which can be of great importance for observations of sources which have angular structure between the seeing limit and the diffraction limit. Depending on wavelength and telescope aperture, the diffraction limit is in the vicinity of a few hundredths to a few tenths of an arcsec. Not coincidentally, many sources in this angular size range are relatively bright, and hence are possible candidates for natural guide star adaptive optics.

6. Sensitivity and Speed of Observations with Adaptive Optics

In certain circumstances adaptive optics may improve sensitivity or speed of general astronomical observations. However, it is necessary to be fairly specific about the circumstances. Adaptive optics may fail to give a sensitivity gain due to details of the image correction process, or due to the nature of the measurement. Adaptive optics aficionados have a special obligation to thoroughly understand the basis for their claims in this area.

6.1. RAW SENSITIVITY

While relatively low Strehl ratios can give high angular resolution (thanks to the formation of the narrow image core), most improvements in sensitivity require overall concentration of flux. This concentration allows a higher ratio of signal to background (which may be sky or telescope emission, depending on wavelength). This will be called here, raw sensitivity. The ideal sensitivity gain should be approximately the ratio of seeing-limited FWHM to Airy width. This can be an order of magnitude. But including Strehl ratios, transmissions of adaptive optics systems, and where appropriate emissivity, more realistic expected gains may be in the vicinity of $2\times$.

TABLE 4. Estimating the gain in raw sensitivity with adaptive optics.

	V	I	H	K'	K
Seeing (arcsec)	0.50	0.44	0.40	0.37	0.37
$1.22\lambda/D$	0.016	0.028	0.050	0.069	0.069
Ideal gain	31	16	8	5	5
Bright Star Strehl	0.07	0.30	0.41	0.56	0.56
Strehl reduction	0.5	0.5	0.5	0.5	0.5
Transmission	0.80	0.80	0.80	0.80	0.80
Emissivity factor	1.00	1.00	1.00	1.00	0.41
Net gain	0.8	1.9	1.3	1.1	0.5

This subject appears to be a sensitive one, so let's take a closer look at an example. Consider an 8 meter telescope at an excellent site, where the natural seeing is 0.6 arcsec at 0.5 μm, with an adaptive optics throughput specification of 0.8 and a telescope emissivity specification of 0.04. Table 4 shows, as a function of wavelength, the expected seeing limited FWHM (using scaling from Kolmogorov turbulence) and the diffraction-limited resolution[3], $1.22\lambda/D$. A column has been added to show both K' and K filters. The center wavelengths differ by little, but the K' filter cuts off on the long wavelength side to reject the thermal background, while the K filter includes thermal background and sensitivity is limited by the background.

For ideal, lossless adaptive optics, and ignoring a factor of order unity depending on the detailed image shapes, the ratio of these numbers should give the relative sensitivity of an adaptive optics equipped telescope to a tilt correction only telescope. (We have implicitly squared the ratio to get the ratio of the number of background photons, and then taken the square root to get the ratio of background photon noise). This ideal gain in sensitivity is shown.

The next column gives an entry for the bright star Strehl ratio from Table 2. Of course, the bright star case will normally not apply. Increased raw sensitivity will be scarcely required for large telescope study of stars so bright. The interest in improved sensitivity will generally involve faint reference stars, often at the maximum useful angular distance. A survey of the published science confirms that most of the interesting sources are near the limit, and adaptive optics is employed at well below the bright star

---

[3]The factor 1.22 strictly applies to the clear aperture case (Schroeder, 1987), but the distinction scarcely matters here.

limit. Let's suppose that the reference star brightness and distance impose a Strehl reduction of 0.5 (as described in Table 2). This factor is shown as Strehl Reduction. This computation of the actual, useful Strehl, is clearly the crux of evaluating sensitivity gains. Preferably, an estimate should be motivated by consideration of the specific science program.

Next, a transmission factor for the adaptive optics system must be applied. Here, a value of 0.8 is used, reasonable for any general purpose system which must make compromises to function over a range of wavelengths.

The net gain in raw sensitivity is computed as the product of the Ideal Gain, the Bright Star Strehl ratio, the Strehl reduction, the transmission, and the emissivity factor. Gains much more than 2× greater are rather unlikely. This (partly empirical) calculation suggests that modest sensitivity gains may be most plausible in the near infrared near 1 μm.

There are exceptions. First, the table is constructed for demonstrated Strehl under median conditions. Higher Strehl will sometimes be achieved under better conditions, and with flexible operating scenarios it is possible to effectively exploit instrument performance which is only rarely available.

Second, in the magnitude range brighter than about 16, stars may be bright enough to serve as wavefront reference, but a little faint for high resolution infrared spectroscopy. In this particular case, the gain in sensitivity should be about 2× greater than shown in the table, since the full bright star Strehl can be employed.

A faint source very close to a bright source will profit from the bright star Strehl, but it is risky to ignore the scattered light and extended halo from the bright star, so gain in sensitivity in that case is problematic.

If the observation is in the thermal infrared (approximately 2.4 μm and beyond), there is also a loss due to emissivity. Assuming that the emissivity of the adaptive optics system is given by 1.0 minus the transmission (that is, 0.20), and assuming a telescope emissivity of 0.04, the emissivity after the adaptive optics will have increased by a factor 5.8, for an increase in background noise of 2.4×. This loss must be included in the thermal infrared. The result for the K band shows that even 0.8 throughput in a warm system destroys any hope of much sensitivity gain at wavelengths beyond 2.4 μm. In the thermal IR, there is also reason for concern with respect to modulation of the background by the AO system itself (Geoffray, 1999).

The use of laser beacons will change this calculation considerably. Tyler & Ellerbroek, (1999) describe detailed calculations of slit coupling power for laser beacon AO supported spectroscopy for the Gemini-North AO system, showing coupling values (AO "efficiencies") greater than 0.6 for 80-100% of the sky, at H and K bands. Performance like this would yield a factor 2× or so increase in Net gain over Table 4.

6.2. MULTIOBJECT OBSERVATIONS

A powerful observational technique in observational astronomy is the simultaneous study of large numbers of sources spread over an extended field. Wide field imaging and multi-object spectroscopy can easily cover fields 100X larger in solid angle than an adaptively corrected field, and this gives a gain in speed which far exceeds any promised by adaptive optics. Of course adaptive optics can be critical in the specific case of a confusion limited field, as witnessed by recent exciting observational results, but no concept currently on the drawing board will compete with wide-field instruments currently in operation.

For sources which are extended to near the seeing width or larger, there would in general be no sensitivity value of adaptive optics, either in imaging or spectroscopy. This should be clear. In the case of both imaging and integral field spectroscopy, where the goal of adaptive optics will be to provide increased spatial resolution, observations will obviously be slower when increased resolution produces lower photon flux per spatial resolution element.

7. The Isoplanatic Patch

Though defined precisely, the term "isoplanatic angle" or "isoplanatic patch" is most often used loosely to describe the size of the corrected field of view achievable with adaptive optics. The etymology suggests that the PSF (point spread function) might be approximately uniform over the isoplanatic angle, but this is not correct. The Strehl can vary by 2× between the on-axis position and an off-axis distance equal to the isoplanatic angle. Furthermore, the definition of isoplanatic angle does not map directly to Strehl loss.

There is some good news, however. The isoplanatic angle is found to be somewhat larger than expected. This seems to be due at least partly to actual observatory site turbulence profiles more benign than models used in many early performance predictions.

There is an interesting relationship between telescope aperture size and isoplanatic angle (Chun, 1997). Telescopes of order 1–3 meters, operating in the infrared on excellent sites, may have an isoplanatic field of view as large as several arcminutes.

8. The Point Spread function

For all its aggravations, a time-averaged, atmosphere-limited PSF usually has some nice characteristics, including a single peak, smoothness and monotonic decrease from the core. With adaptive optics, complex systematic effects are introduced. The PSF is expected to vary over the field of view, in a rather complex way which is difficult to predict even in principle. The "on-axis" PSF, that is toward the reference star or beacon, depends on many of the parameters that determine adaptive optics performance, including reference star magnitude, natural seeing and wind velocity in turbulent layers. The variation of the PSF with off-axis angle depends on the distribution of turbulence with altitude and the velocity of turbulent layers, in addition to angular decorrelation of the aberration which, ideally, can be estimated from a general Kolmogorov turbulence model subject to determination of the parameters which characterize it - including outer scale. And these are relatively well-behaved effects.

It is well known, though not so widely discussed, that AO can produce many erratic PSF artifacts, from the dreaded "waffle" pattern of Shack-Hartmann wavefront sensors, to radial smearing relative to the reference star. Non-common-path errors arise where the optical aberrations are not fully sampled by the wavefront sensor. Small temperature or orientation dependent drafts may easily change the AO-corrected PSF. Systematic effects of unknown nature may masquerade as genuine structure in images (eg, Currie et al, 1998; Chapman et al, 1999).

Imaging with no included point source provides a special challenge. Measuring the PSF in a separate observation of a reference star requires extreme care to ensure that the PSF is as similar as possible: for example, minimum elapsed time, staying at the same zenith distance and at the same reference star magnitude. We might speculate that selection of comparable data sets based on monitoring of some seeing parameters could help. For this reason, AO instruments for precision work should be augmented with such facilities as a full-time seeing monitor and SCIDAR (Klückers et al, 1997). Véran et al (1997) determined PSF's from the recorded residual wavefront sensor errors – this appears to be a powerful technique, but limited to reference stars significantly brighter than the AO faint reference limit. These issues do not go away with laser beacons, but in fact become even more severe (Max et al, 1998).

9. What about Photometry?

The utility of imagery is greatly increased if accurate surface brightnesses and other photometric quantities can be extracted. Depending on scientific objectives, accuracy of 1–10% may be needed. This accuracy range is fairly standard in conventional observations. In adaptive optics enhanced observations, questions arise because the PSF may be more difficult to determine. If the Strehl ratio can vary by 2× within the "corrected" field of view, as noted above, we might expect problems in extracting precise photometric information.

Ideally, a field richly populated with point sources provides enough internal information to determine the variation of the PSF over the field, using techniques developed to deal with HST images. Initial experience in extragalactic observations (Davidge et al, 1997b; Bedding et al, 1997) has produced photometry with errors of 0.1–0.2 magnitudes. Diolaiti et al (1999) present an algorithm for crowded field analysis with tests which show (magnitude dependent) residuals in the range 0.1-0.5 magnitudes.

A valuable theoretical and empirical study of photometry with adaptive optics is in preprint form as of this writing (Esslinger & Edmunds, 1998). This paper includes extensive modeling and examples, based on observed Strehl and PSF variations, for the dependence of photometric accuracy on such parameters as integration time and anisoplanatism. It is a valuable guide to selection of aperture sizes and observing strategy. While difficult to summarize briefly, the recommendations pertaining to K band observations with a 4 meter telescope note that accuracy of 0.01–0.02 magnitudes may be impossible, and that accuracy of 0.05 magnitudes may be impossible in crowded fields. PSF determination from separate images requires repeated repointings, and may achieve no better than errors of a few percent error (Rigaut and Sarazin, 1999).

A common requirement for study of binary stars will be precision relative photometry of two point sources in the visible, for stars which are separated by a significant fraction of the isoplanatic angle, but with PSF's which overlap. This may be a particularly difficult case, since it implicitly requires some information about the variation of the PSF with minimal internal diagnostic information. For examples, see the discussion of binary star science below.

10. Coronography, Nulling, Super AO and Speckle

A widely suggested application of adaptive optics is in support of coronographic observations, in order to detect faint sources near a bright point source. The benefit is three-fold. First, it concentrates the point source flux in a small area where it can be blocked with a small mask. This allows longer integrations without saturation. Beuzit et al (1999) have characterized a coronographic augmentation of the ESO ADONIS[4] system. They find a reduction in central star peak brightness as large as 10000×. This is a significant, even crucial, convenience, though not a direct gain in signal-to-noise.

The second coronographic gain is the reduction of illumination in the halo of the bright point source. This gain is much more difficult to obtain,

---

[4]ADaptive Optics Near Infrared System.

as the halo brightness is roughly proportional to 1-S, where S is the Strehl ratio. Typical achieved Strehls of 0.5 only produce a gain of order $\sqrt{2}$ in the photon noise limited case. Also, the AO halo reduction gain is limited to the region interior to the diffraction limit of the effective subaperture size (Ryan et al, 1998).

The third gain is of course the concentration of flux if the "faint" sources are unresolved, and the discussion of raw sensitivity applies.

The AO corrected wavefront produces speckles, which limit the detection of faint sources near bright ones. Ironically, AO also increases the speckle lifetime. This must be taken into account if the averaging process is to be understood correctly (Langlois et al, 1999).

ADONIS (Beuzit et al, 1999) shows a detection limit $\Delta m = 10$ below the central star at 1 arcsec radius, and $\Delta m = 12$ at 3 arcsec radius, for 300 seconds integration with the 3.6 m telescope. The Hokupa'a AO system has demonstrated point source detections with $\Delta m = 12.5$ at 3 arcsec radius at the 3.35 m CFHT (Close et al, 1999).

There are interesting alternatives to classical coronography. Baudoz et al (1998) have achieved achromatic nulling of a stellar source, employing AO in combination with a π-dephasing scheme, which has attractions such as a cleaner optical path and a tunable rejection ratio, but a disadvantage that it flips the pupil before nulling, hence nulls only the symmetric part. Guyon et al (1999) have demonstrated in the laboratory the use of a phase shifting mask, which has the advantage of a small mask and nulling of the Airy rings, but is more difficult to achromatize.

All of these gains together are impressive but may be less than the dynamic range differences that one wishes to defeat to detect host galaxies around quasars or planets around stars. Walker et al (1999) describe a multi-wavelength technique for additional gains, applicable when one member of a multiple system has broad and deep absorption bands. Even greater dynamic range may be achieved with special-purpose facilities. For the NASA interferometry augmentation of the Keck telescopes, expected to directly detect the indirectly discovered "hot Jupiters", the single Keck telescope adaptive optics is required to provide Strehl ratios of about 0.98 at 10 microns. Angel (1995) has proposed adaptive correction with a 10,000 element deformable mirror, in order to achieve high Strehl and detect extrasolar planets by a single aperture imaging technique.

Another interesting opportunity is the combination of adaptive optics with speckle imaging. Speckle is based on post processing of short exposure images. This may seem retrograde with adaptive optics, but with incomplete correction speckle may be an interesting alternative to deconvolving a residual PSF. Several groups have carried out preliminary experiments combining natural guide star and laser beacon adaptive optics and speckle data analysis (Koresko, 1997; Tessier & Perrier, 1996; Dayton et al, 1998) showing that adaptive optics and speckle can work effectively together. Tessier (1997) shows that speckle processing is most useful for Strehl ratios smaller than about 0.1.

11. Adaptive Optics and Spectroscopy

A frequently mentioned potential benefit of adaptive optics is the achievement of high spectral resolution with a small spectrograph. This is possible because high resolution astronomical spectrographs for large telescopes are slit limited – the slit size required to collect a large fraction of the source flux limits the spectral resolution. With a small adaptively corrected image, the slit could be reduced in size and the resolution increased. This attractive concept will be most interesting for very high spectral resolution, which will generally be possible for relatively bright sources, which can be useful as reference stars for the adaptive optics wavefront sensor. The concept has been demonstrated with a visible echelle spectrometer (Woolf et al, 1995) Of course, any efficiency loss must be much smaller than the slit throughput gain, if the strategy is to be generally useful – thus far there are no reports that this has been achieved. It is interesting to note that the Keck NIRSPEC infrared spectrometer, which is designed to be used with adaptive optics, will not initially exploit the narrow slit for higher spectral resolution – this option was judged too expensive.

The adaptively improved image size is a useful element of design flexibility, but it is only one of a number of factors facilitating high resolution spectroscopy, including new sites with higher quality seeing, tilt correction, immersion gratings, mosaic gratings, and slitless interference spectroscopy. Also, in the infrared, diffraction and grating resolution limits become important and the slit size is less significant.

Scientifically, however, AO in combination with spectral resolution will greatly enhance the diagnostic power over imaging or spectroscopy alone. The GraF instrument (Chalabaev et al, 1999a) offers integral field capability with spectral resolution of order 10,000 for the ADONIS AO facility.

12. How Adaptive Optics will Make the Rich Richer

Since reaching the sky background limit, astronomers have only gained sensitivity roughly as the aperture diameter of their telescopes rather than as the aperture area. Adaptive optics attacks that barrier in two respects. Since the resolution with adaptive optics improves with the diameter, the areal resolution improves with the square of the diameter, and the image information content in some sense probably increases in similar proportion. Second, if perfect Strehl and 100% throughput could be achieved, then due to the combined benefits of increased light gathering power and suppression of background, the sensitivity of a measurement would gain with the square of the aperture diameter. Though this kind of gain has not been demonstrated yet, it stands as a goal for the future. Large telescopes will potentially gain more from adaptive optics than small telescopes.

At a good seeing site (large r_0) the wavefront sensor subapertures will be large, hence the photon rate per subaperture will be high. The large r_0 will also result in a longer atmospheric time constant, hence a longer integration time on the wavefront sensor detector. So a good seeing site will be able to employ adaptive optics with much fainter reference stars than a poor seeing site.

In laser beacon adaptive optics, it is still necessary to have a natural star reference for tilt and focus. The full telescope aperture can be used for these measurements. Thus a large telescope can use fainter tilt and focus reference stars than a small telescope.

In sum, the largest telescopes at the best sites will profit most from adaptive optics. The inevitable size-site pecking order in astronomy will be exaggerated as adaptive optics is implemented, especially if it is preferentially implemented on the large telescopes at good sites, as appears to be the case thus far.

13. How Much is Adaptive Optics Used

The two adaptive optics systems available to a wide user community on a competitive, peer reviewed basis, are ADONIS at the ESO 3.6m, and PUEO at the CFHT. Over the interval 1997-1999 PUEO has been used slightly less than 25% of the time, and ADONIS has been used slightly more than 25% of the time [5]. A 1/4th share is a huge fraction for a 4m class telescope serving a large community, and it highlights a strong user interest and a high level of satisfaction on the part of the Telescope Allocation Committees.

14. What Does Natural Guide Star Adaptive Optics Offer Astronomers?

The answer is improved angular resolution, of course, but actually quite a bit more. The additional benefits are less obvious, but in fact much more generally applicable.

---

[5] From the observatory WEB pages, counting scheduled AO nights as a fraction of the calendar period.

14.1. CORRECTION OF STATIC ABERRATIONS

Virtually no telescopes are fabricated to sufficient optical quality to take full advantage of an excellent site. Even those few which meet very high specifications in the lab rarely deliver the same performance for nightly use. Since the optical aberrations are usually dominated by low order errors, a low order correction can go a long way to remove those aberrations.

A natural guide star adaptive optics system can and must be designed to reduce fixed aberrations. Hopefully the aberrations will be relatively independent of, or systematically dependent on, zenith distance, in which case an offset matrix for the deformable mirror can be determined with high signal to noise on a bright star, and the fixed correction will then be applied even if no wavefront reference star is available. The aberrations may even vary non-reproducibly with telescope pointing, but if the aberrations vary slowly with time (the case for any credible telescope), then a relatively faint reference star can be used to provide a low bandwidth but continuously updated correction. Slow correction of this kind is conventionally called active correction rather than adaptive. However, use of a deformable mirror other than the primary is a non-standard means of achieving this active correction. Apparently it has only been implemented as a secondary feature of adaptive optics.

The active correction is potentially valid for a wide field of view, and adaptive optics has always been implemented for a narrow field of view appropriate to the small adaptively corrected field. It is currently unclear to what extent the use of active/adaptive optics can be a cure-all for telescopes with marginal optical quality, but experience shows that telescopes inevitably have residual optical aberrations that can be profitably removed.

The most modern telescopes include an active mirror support and active secondary alignment adjustment which is expected to reduce telescope aberrations to a low level. Whether or not an adaptive optics stage will provide further improvement is a technical question of time constants, gains, and systematic errors. Normally, a continuously updated closed loop correction can be expected to provide a better aberration reduction than a lookup table correction.

14.2. PAINLESS FOCUS

A very significant benefit of adaptive optics is providing good, unambiguous focus with zero overhead. Observers know that achieving focus without adaptive optics can be very frustrating. When focus involves real-time interaction, the very red power spectrum of atmospheric focus variation is a serious problem. In integrating systems, readout and analysis time can be significant. For programs involving observations through multiple filters, the focusing time can be increased by a corresponding factor. Most telescopes have focus variations, and often there is no unambiguous indicator of when refocus is required. Even in a telescope with focus stabilized by an athermal opto-mechanical structure, pooling of cool air in the primary mirror can cause unexpected focus variations.

In a closed-loop active or adaptive system, the focus is fixed automatically in a few time constants. This is a major factor in the report that at the CFHT PUEO system, imagery with adaptive optics involves less overhead than normal direct imaging.

14.3. TILT CORRECTION

Tilt correction without higher order correction is already interesting. In the infrared, it can provide much improved images, particularly for moderate aperture telescopes where D/r_o can be on the order of a few. Correction of atmospheric tilt is effective over a wider field (sometimes called the isokinetic angle) than is higher order correction.

In the visible, the gain due to correction of atmospheric tilt should be small, but correction of telescope drive errors and wind shake can be significant, and of course apply to a large field. A surprising number of the most modern telescopes have been plagued with image quality limiting vibrations due to mechanical resonances in the vicinity of 20 Hz. While proper mechanical solutions are preferred, adaptive tilt correction can also reduce the amplitude of such image motion. Nevertheless, the number of telescopes implementing tilt correction alone is relatively small. A high-bandwidth secondary mirror may be a expensive, though a low-bandwidth correction retrofit can be reasonablly priced. However, as soon as the tilt is removed, one sees the distorted PSF due to aberrations which were previously just below the seeing limit. This PSF may be unaesthetic, difficult to represent numerically, and even variable. Fixingthe PSF by brute force can involve replacing mirror supports, repolishing secondaries, and installation of active alignment capability, all at high cost. Hence achieving the potential benefits of tilt-correction can entail an open-ended effort.

The measurement of tilt is not entirely trivial. Tilt as commonly measured with a quadrant detector or a centroiding algorithm includes contributions from higher order terms (comatic aberrations). The tilt for which the field of view is optimized is the true or Zernike tilt, which can be more accurately measured with a wavefront sensor. Just distinguishing the tilt from the first order coma is the most valuable gain. Thus an adaptive optics system may give better tilt correction than a tilt only system, simply due to the wavefront sensor.

If the field of view is greater than the isokinetic angle, there is no guarantee that the tilt from a single star (on-axis or not) is the best estimate of tilt over the full field. Multiple reference stars could be used to give a better estimate for the mean tilt value, or to apply different tilt corrections to subfields of the full field. Neither of these uses of multiple reference stars has been reported. There is a close analog, however, in the use of correlation tracking in solar observations (November, 1986).

14.4. RESOLUTION IMPROVEMENT NEAR SUFFICIENTLY BRIGHT STARS

The first three points above are guaranteed benefits of adaptive optics, and are valuable in their right. From that perspective, then, high angular resolution is an added benefit which is available to a degree limited ultimately by the brightness of nearby reference stars.

15. Why Not Higher Order Correction?

A common definition of "diffraction limited" is an image Strehl ratio of about 0.8. This is slightly beyond the limit of what is currently achieved even in the near-infrared with a bright reference star during exceptional conditions. In the visible, typical Strehls are much smaller, a few percent to a few tens of percent. It is possible to build systems with more actuators and subapertures, but the reference star requirement will become rapidly brighter.

To understand this, consider the common modal control used in many adaptive optics systems, in which the wavefront error is decomposed into its Zernike components. A low order correction corresponds to correcting the tilt, focus, astigmatism, etc. These low order modes include most of the aberration. Correcting them gives the most aberration reduction per mode. For higher and higher modes, the improvement per mode decreases steadily. But this is only part of the problem. In order to determine higher order Zernike components of the the wavefront error, it is necessary to use smaller subapertures in the wavefront analyzer. The number of available photons per subaperture per time constant is on order d^{-3}, where d is the subaperture diameter. This includes a factor of d^{-2} for the collecting area, and d^{-1} for the time constant, which must be on the order of v/d, where v is the wind velocity in the turbulent layer). This combination of factors vigorously closes the window on very high order correction.

16. Laser Beacons

This entire ASI is dedicated to laser beacons, and the remaining lectures in this volume expand at length on many important issues concerning laser beacon capabilities and operations. Only the highest level, user viewpoint will be mentioned here.

If the laser beacon brightness is assumed to be large, the wavefront measurement and correction can be extended to higher orders. However, the lowest order aberrations, tilt and focus (and possibly higher), must still be determined from a stellar reference source. Since most of the aberration power is in tilt and focus, even small errors in correcting them can be extremely deleterious, especially in the visible. Also, single-beacon systems will leave some turbulence unmeasured (due to the cone effect), with the result that the best achieved Strehls with laser beacons will actually be lower than the best achieved Strehls with natural reference stars. Hence, the motivation usually cited for laser beacons is to extend the adaptive optics functionality to the vicinity of fainter reference stars, albeit at somewhat reduced performance.

Rayleigh beacons are formed by Rayleigh scattering of a bright laser beam. In order to obtain reasonable scattering at an altitude of 10–20 kilometers, a relatively high power at UV wavelength is chosen. Low altitude scattering is rejected by temporal range gating. These choices have many ancillary impacts. Among the benefits, the required lasers are relatively cheap. The hazard of the laser light at the telescope is significant, but once launched (at reduced brightness) the hazard is small, and minimal aircraft and satellite coordination is required.

The only Rayleigh beacon system that has been operated for science is at the SOR[6]. Utilized with a 1.5 meter telescope, this system maintained a small guest observer program for several years under an NSF sponsored program, though recently this activity has been terminated. The scientific utility has been limited due to the R&D priorities of the facility, and to the moderate aperture and mediocre natural site seeing. In spite of these limitations, scientific papers have been published based on data from the Starfire system. It is not entirely clear at this time, however, that these science programs could not have been accomplished with a natural reference star system.

The laser beacon adaptive optics community has, for the most part, rejected the Rayleigh beacon on the basis that the sodium beacon obviously suffers less from the cone effect. Only one program to implement a Rayleigh laser beacon system for astronomy is under way - the UnISIS[7] project at Mt. Wilson (Thompson, 1994).

Sodium beacons are formed by resonant scattering of a sodium wavelength beam at high altitude, approximately 90 kilometers. The higher altitude reduces the error due to unmeasured turbulence (smaller cone ef-

[6]U.S. Air Force Phillips Lab Starfire Optical Range
[7]University of Illinois Seeing Improvement System

fect). The sodium wavelength has some disadvantages. The lasers are not readily available and are still relatively expensive. The sodium wavelength penetrates aircraft and satellite windows and coordination with the FAA and the Air Force Space Command is required.

As the sodium beacon systems have been developed, it has become clear that there are problems which were initially underestimated. The continuously varying distance to the sodium layer is a significant complication. The non-circular, and changing, beacon image shape (due to the finite thickness of the sodium layer and related perspective effects) results in serious difficulties in estimation of wavefront error. Major problems remain which have not been fully resolved.

Several groups have formed sodium wavelength spots in the sodium layer. Three groups have achieved limited function of adaptive optics systems with sodium beacon wavefront reference. Each group has reported a gain in Strehl and in resolution, but none have yet approached the image quality and stability regularly achieved with natural guide stars. The few results published so far are more in the nature of technical demonstration than reports of scientific programs.

17. Partial Adaptive Correction

Although astronomers initially hoped that adaptive optics would completely compensate for atmospheric turbulence, the reality of bright limiting magnitudes for complete correction led quickly to an interest in partial correction. (Partial correction is not entirely synonymous with low order correction, since at longer wavelengths low order correction may be sufficient to achieve high image quality.)

There has been some uncertainty, and even misunderstanding, of what image quality would be achieved with partial compensation. An early, schematic theory predicted that a partially compensated image would consist of a diffraction limited image core, plus a halo at least as large as the seeing width and possibly larger. Such a core is certainly observed when the Strehl ratio is of order 0.1–0.2 or greater. In fact, such images may have a near-diffraction limited FWHM. This can lead to misunderstandings, since observers unfamiliar with adaptive optics may tend to associate the small FWHM with a small encircled energy criterion - but in this case the two image quality measures are decoupled.

At sufficiently low Strehl, the core is missing or not prominent. However, experience is showing that image quality improvement is often obtained at visible and even blue wavelengths. The nature of the improvement is reflected by significant FWHM reductions. Lacking a narrow image core, the FWHM can legitimately be compared to the natural seeing FWHM. Gains of order 2× have been demonstrated at Mauna Kea on the CFHT. Image improvement is achieved over a much larger field of view than expected for the classical isoplanatic angle. This is not surprising, since the low order aberration modes corrected have a larger correlation angle than the high order modes considered in the formal definition of isoplanatism.

18. Science results from Adaptive Optics

Adaptive optics has been in regular use at several facilities for at least five years, but scientific productivity has only taken off in the last two. The problem of the early years can be illustrated by a remark from an adaptive optics guru in the aerospace industry, who admitted that wavefront sensing was limited to V=5 stars at that time, but after all, "there are lots of fifth magnitude stars to study". (The problem, of course, is that a factor of 10–30× improvement in angular resolution contributes little to the study of V=5 stars – a topic which is in itself a small subfield of astronomy.)

There have also been natural difficulties with mastery of the hardware, the observing technique, and data analysis issues. There has been an initial mismatch between the actual technical capability which adaptive optics provided and the major research themes of the astronomy community.

The current limits (especially with respect to reference star magnitude, field of view and image quality achieved) still constrain greatly the options, but astronomers have begun to master the capabilities offered and to match them to interesting scientific opportunities. The evidence for the success of this effort is illustrated in Table 5, which shows the number of adaptive optics science papers published or in press at major refereed astronomy journals (science, here, means papers which present actual analysis and interpretation of data, rather than an illustration of achieved instrumental performance). Of course some of these observations are merely in the nature of proof of feasibility, and some are focused on one or a few particularly well suited sources, and may not be indicative of future programs, but this is far less true than it was just two years ago.

Considerably more insight into the science programs can be gained from the distribution of publications according to science topic. This is shown in Table 6, which includes additional refereed papers published in early 1999 or known to be in press. Studies of young stars, of solar system objects and of galaxies stand out as the areas where most adaptive optics observing activity is concentrated. For YSO's and solar system objects, the scientific objectives commonly require study of relatively bright sources, compact but with important image detail just finer than the seeing limit. In extragalactic studies, angular resolution is also critical especially in order to obtain photometry in crowded regions. Here, adaptive optics is hard up against the

TABLE 5. The number of science papers published through 1998, in refereed astronomy journals, based in substantial part on adaptive optics supported data.

Year	Number
1993	3
1994	2
1995	5
1996	8
1997	26
1998	20

TABLE 6. The distribution of adaptive optics science publications according to science topic, based on papers published or in press in refereed astronomy journals during the interval 1990–May, 1999.

Science Topic	Number
Young Stars and Vicinity	17
Solar System	13
Extragalactic and Galactic Center	18
Binary Stars	4
Star Clusters	6
Compact Nebulae	6

faint reference limit, with most programs limited to modest partial correction. The ESO conference *Astronomy with Adaptive Optics: Present Results and Future Programs*, held in fall of 1998, included 22 science papers. These were mostly works in progress, and may tend to indicate the direction of near-future AO activity. The most heavily represented topics were YSO's, extragalactic, and compact nebulae.

19. A Closer Look at Adaptive Optics Science

It is not possible to do justice to 60+ papers in a few pages of review, but there are some useful lessons to be gained from looking in closer detail at some of the science topics. In the following, reference will be made to papers counted in Table 6, and also to unrefereed papers and to images which have been made available on the WEB. Although the latter often lack descriptive or interpretive text, they are very valuable for forming an appreciation of the current state of the art opportunities which adaptive optics offers to astronomy.

19.1. YOUNG STARS

The stellar disks of young stars, for example T Tauri and Ae-Be stars, are much too small to resolve with the largest filled-aperture telescopes, but they frequently are accompanied by near circumstellar material on scales slightly smaller than 1 arcsec, and hence are well suited for study with adaptive optics.

Use of adaptive optics to study the multiplicity of young stars has been described by Monin and Geoffray (1997), and Brandner et al (1997a). Duchene et al (1999) detected 60 multiple systems among 350 young cluster stars observed, and found an inverse correlation between binary fraction and cluster density. Geoffray and Monin (1999) determine the spectral energy distributions of YSO pairs, and show by comparison with evolutionary tracks that wider pairs are not always coeval. Trouboul et al (1999) used AO to record resolution 4000 spectra of the components of Ae/Be binary systems.

Close et al (1997a,b,c) have studied multi-spectral infrared imagery of T Tauri stars GG Tau, HL Tau, UY Aur, and R Mon (also observed by Ageorges and Walsh, 1997). Quirrenbach and Zinnecker (1997) and C. Roddier et al (1999) studied T Tauri itself. McCullough et al (1995) imaged young stars with envelopes which are evaporating under the influence of stellar winds and radiation from nearby O stars. Observations have also been reported for possibly related source types, such as the hydrogen line source LkHα 198 (Koresko et al, 1997), Z CMa (Malbet et al, 1993), and NX Pup (Schoeller et al, 1996). Many new structural details of these systems have been determined, including identification of new stellar companions, discrimination of circumstellar and circumbinary shells, detection of disks, and investigation of scattering (from polarization maps) in the extended nebulosity. Ménard et al (1999) used AO (plus HST imagery) to study disks and jets around YSO's, detecting proper motions in the DG Tau jet just 2-3 arcsec from the star.

With respect to the study of young stars, adaptive optics seems particularly well suited. Many prototypical and representative sources are amply bright for natural guide star adaptive optics. There are numerous current research topics which gain immediately with improved angular resolution, including multiplicity of the stars, natal disks and related inward flows, jets and related outward flows, proto-planetary disks, planets and brown dwarfs, etc. The scientific issues are profound - the birth of stars, the origin of planets. The physics is complex. The possible investigations are almost unlimited. In this area, AO is not just an incremental gain, but a substantive advance in the research capability. The study of young stars and their surroundings will probably be dominated by adaptive optics equipped large telescopes within the near future.

Due to obscuring molecular clouds in regions of recent star formation, there is in these regions a paucity of stars sufficiently bright for visible wavefront sensors. Development of sensitive, infrared wavefront sensors (possible and perhaps imminent) will extend AO to much more obscured targets. Unfortunately, even the largest filled aperture telescopes will scarcely begin to resolve structure within a pre-planetary disk at typical distances of 200-500 parsecs. Distributed telescope arrays will be required.

19.2. SOLAR SYSTEM

The solar system is full of sources which have important image detail just beyond the seeing limit, and in many cases the information gain with increased resolution is quite dramatic. Many of these results were obtained by the U. Hawaii group, and were recently reviewed (F. Roddier et al, 1999).

Asteroids are an interesting example and have been the subject of extensive studies with several adaptive optics systems. The few largest asteroids have an apparent diameter of about 1 arcsec at closest approach to earth, so sub-arcsec resolution offers a unique opportunity to access a wealth of structural and compositional information. The shape of an asteroid or other small body reveals information about the formation history and structural integrity. Saint-Pé et al (1993a) studied the shape of Pallas. Drummond et al (1997) analyzed images of Ceres and Vesta for rotational poles and triaxial dimensions. It appears that similar measurements could be acquired for a large number of much smaller bodies. Saint-Pé et al (1993b) studied thermal infrared images of Ceres to determine the rotational axis and the thermal density of the regolith. Dumas & Hainault (1996) have carried out an extensive program of near-infrared, narrow band imagery of Vesta, deriving a map of the surface pyroxene distribution.

The atmospheres of the giant planets undergo extensive variations, analogous to weather on earth. Ground based adaptive optics offers an important combination of angular resolution with the possibility of monitoring temporal changes. Roddier et al (1997b) reported the infrared observation of stratospheric clouds on Neptune, including evidence for wind velocities from the rotational period. Rigaut and Arsenault (1996) observed a dramatic, transient storm on Saturn. An ESO observer obtained a very nice 2 μm image of polar haze on Jupiter (DESPA, 1997).

Planetary satellites are also well suited for adaptive optics observations. Titan has been observed in the infrared, at wavelengths chosen to optimize the penetration of the atmospheric haze and to distinguish surface features. Combes et al (1997) were able to rule out a global ocean from their Titan imagery. Infrared images of Io (Dumas et al, 1997) reveal volcanos, some previously known and some new, clearly showing the utility of adaptive optics for monitoring volcanic activity on this satellite of Jupiter.

Several groups studied the rings and satellites of Saturn at the recent ring crossing. Beuzit et al (1997b) and Roddier et al (1996a,b) and Roddier (1997) reported new discoveries of satellites and/or ring clumping. Roddier et al (1997a) reported observations of arc structure near the E ring, which they interpret as the result of an expected gravitational interaction of the ring with both Mimas and Enceladus. These observations profited from the unusual opportunity to view the rings edge-on, as well as choice of wavelengths at which the planetary disk is quite dark due to molecular absorption in a cold atmosphere.

There are also numerous adaptive optics images of comet Hale-Bopp (eg, Marco et al, 1997b), or on various WEB sites. The recent burst of results from infrared spectroscopy suggests that combining spectroscopy with high angular resolution would be a powerful observational technique for comets.

For solar system studies, adaptive optics equipped ground-based telescopes are surprisingly competitive with spacecraft. In spite of continuing NASA missions, it is likely that adaptive optics will continue to provide unique solar system observations. AO has a clear future in solar system observations, and the observing programs mentioned above probably provide an excellent guide to the opportunities. There can be extensive study of asteroids, considerable further detailed investigations of Titan, monitoring of the outer planets and Io, additional studies of planetary rings, and observations of comets as the opportunity arises.

19.3. EXTRAGALACTIC AND GALACTIC CENTER

The galactic center should be a challenging target for AO, at relatively low elevation from many sites, with a dearth of bright reference stars. However, infrared observations are critical, owing to the high intervening visible extinction, and the angular resolution of 4-11 meter telescopes turns out to be just about right to [artially relieve the confusion limit of the imagery, and to measure the transverse motions due to stellar orbits in the deep galactic center gravitational potential (Davidge et al, 1997a).

A number of studies have employed adaptive optics to resolve and study the bright stars in galactic and extragalactic populations. Davidge et al, 1997a,b) studied fields in the galactic bulge and near the nucleus of M31. Bedding et al, 1997) observed NGC5128, IC5152 and NGC200 in addition to a galactic bulge field, and concluded that it should be possible to obtain photometry of the brightest stars in the bulges of elliptical galaxies to a distance of about 3 Mpc.

Another extragalactic topic that attracts rapidly increasing attention for AO observations is the study of systems with AGN's[8]. Rouan et al (1997), Alloin et al (1997), and Marco et al (1997a) have reported on images of NGC 1068 and NGC 7469, studying details of the internal structure to refine constraints on physical models, and Alloin et al (1999) give an integrated overview of these results. Lai et al (1997) and Rouan et al (1999) observed the ultra-luminous infrared galaxy Mkr 231 and similar sources, showing that these galaxies are characterized by "super clusters" of starburst activity. Lai et al (1999) has extended AO observations of starburst galaxies to include mapping in molecular and atomic hydrogen emission lines.

The PUEO system has been used to study the region surrounding a selection of bright QSO's (that is, bright enough to use the QSO itself as wavefront reference). This is near the limit of current instruments, but has produced detection of possible companions, gravitational images and host galaxy (Theodore et al, 1999).

These studies are both resolution and reference star brightness limited (they usually employ the galaxy nucleus itself as a not-very-satisfactory wavefront reference). Successful operation of laser beacon systems could be the advance required to really open up extragalactic studies with adaptive optics.

19.4. BINARY STARS AND STAR CLUSTERS

The old standby of high resolution astronomy is the study of binary stars, which has the attractions of presenting a relatively easy observational problem. Binary studies also provide a unique service in astronomy – the observational determination of stellar masses – though interest is growing in such issues as binary statistics as evidence of star formation processes, and evolution of interacting systems. Speckle techniques have been quite successful in reaching the diffraction limited angular resolution, but have been most effective with relatively bright stars and modest magnitude difference between primary and secondary. Relative photometry has also been difficult to extract from speckle images. Adaptive optics reach considerably stars, larger dynamic range, and relative photometry should be more readily achievable.

Ten Brummelaar et al (1996, 1998) have employed adaptive optics to determine relative photometry for binary stars observed extensively with

[8]Active Galactic Nuclei

speckle measurements, and they have reported limitations in phototmetric precision, partially alleviated by data binning. Binary star deconvolution problems were further studied by Christou et al (1999). These programs gave relative photometry at about the 5% level. Bouvier et al (1997) carried out a survey of low-mass Pleiades stars to determine multiplicity for a study of the influence of companions on proto-planetary disks. Véran et al (1999) applied the use of wavefront sensor based PSF's to binary star studies, obtaining photometric errors of a few percent.

The star cluster R136 was first resolved by speckle, and subsequent HST and adaptive optics observations (Heydari-Malayeri et al, 1994; Brandl et al, 1996) have contributed to continuing studies of this remarkable, very rich group of early type stars. Adaptive optics has served as a diagnostic for observations of possibly related objects (Heydari-Malayeri et al, 1997a,b).

It is clear that studies of multiple stars and star clusters will be a staple of adaptive optics for many years. Here also, however, specialized optical arrays will be needed to pursue the important parameter space beyond the single aperture diffraction limit.

19.5. COMPACT NEBULAE AND CIRCUMSTELLAR SHELLS

Though young stars, which are commonly accompanied by circumstellar or nearby nebulae, have been discussed separately above, there are still many additional types of compact nebulae. As most of these are associated with stars, which may serve as wavefront reference, these should be in many cases well suited for adaptive optics studies.

Morossi et al (1995) detected rings and emission clumps in the shell of the Be star P Cyg. Mouillet et al (1997) employed a coronograph equipped adaptive optics system to image the disk of the main sequence star β Pic down to 2 arcsec from the star. Beuzit et al (1994) and Roddier et al (1995) observed a possible proto-planetary nebula, known informally as Frosty Leo. Brandner et al (1997b) used adaptive optics to compare a supernova remnant with the hourglass nebula, and Stecklum et al (1999) further explored the photometric and polarimetric properties. Rouan et al (1997) and Vannier et al (1999) exploited high angular resolution and narrow band imaging in molecular hydrogen to measure density contrasts and structures in the molecular cloud NGC2023. Feldt and Stecklum (1997) and Feldt et al (1999) obtained infrared images of ultra-compact HII regions for comparison with centimeter wave maps. Cruzalèbes et al (1999) studied circumstellar dust shells, believed associated with mass loss, around a number of luminous, evolved stars. These observations are difficult, due to the high dynamic range and the need for an excellent PSF compensation. An interesting combination of AO (for dynamic range) and aperture masking (for image detail) shows an apparent mass loss plume from the supergiant VY CMa (Monnier et al 1999).

The abundance and variety of compact nebulae, and the typical spatial scales, suggests that adaptive optics will support continuing advances in studies of their structure and evolution. Though little work has been done so far, this may be an increasingly active area in the future.

19.6. SUMMARY OF SCIENCE DIRECTIONS

In the areas of young star studies, solar system and extragalactic astronomy, select sources, often including the brightest prototypical examples, will be extensively studied with natural guide star adaptive optics.

Other areas, including the additional topics mentioned in Table 6, offer many suitable sources for adaptive optics, and the limited activity in these areas thus far probably is in part a legitimate indication of a somewhat lower level of research effort along these topical lines, and of limited research requirements for the kind of angular resolution improvement that adaptive optics offers. Furthermore, most observational studies of stars, clusters and nebulae are based on visible/UV observations, owing partly to tradition and largely to the tendency for ground state atomic transitions to occur in this wavelength region. HST still has a large performance edge in high resolution visible imaging. It may be that a major ramp-up of activity in these areas with ground-based adaptive optics would be facilitated by implementation of visible optimized adaptive optics systems on large ground based telescopes. This capability does not seem to be represented in the current active and proposed astronomy programs, and may first appear at the Starfire Optical Range, where access for astronomy programs is extremely limited.

Implementation of laser beacons offers advances particularly in extragalactic applications, and more generally in reaching beyond the most favorable few examples of a class of object to a more representative selection.

20. Future Directions in Adaptive Optics

The initial push to develop natural guide star adaptive optics technology appears to be tapering off, and the primary effort now is in copying known technical solutions to additional facilities. For scientific applications similar to the ones represented in published research, the performance can be predicted with reasonable confidence.

Laser beacon adaptive optics lag far behind expectations, and significant additional resources will be required if they are to be brought into operation.

In addition to these main lines of adaptive optics development, there are a number of less well explored paths that are also promising for the future – partial correction and adaptive secondaries.

20.1. NATURAL GUIDE STAR ADAPTIVE OPTICS

Natural guide star adaptive optics is well on its way to establishing a permanent niche in astronomy. The improved angular resolution in study of compact but resolvable sources is increasingly valuable with the advent of very large telescopes. For some science topics, the availability of adaptive optics is likely to be a prerequisite for carrying out competitive work. Thus far, this appears to be increasingly true for study of young stars and for some ground based solar system studies, and other areas are likely to move into this category in the future.

The performance of natural guide star adaptive optics will be augmented somewhat with improved wavefront sensor detectors, and especially with infrared wavefront sensors which will be well suited to study of young stars in heavily obscured regions. New tools for tracking the atmospheric conditions over an observatory, such as SCIDAR, can be utilized to improve the optimization of parameters and algorithms.

20.2. SODIUM BEACON ADAPTIVE OPTICS

Several major projects are underway, millions of dollars have been committed, and some of astronomy's most talented instrumental groups are involved. Whether any particular project succeeds or fails, a great deal will be learned from each that will impact the direction of future sodium beacon programs. There is little doubt that the laser beacon systems can be brought to a degree of functionality. Yet the problems to be solved are difficult and the most likely solutions complex. At this time, the cost effectiveness of sodium beacons remains rather uncertain.

20.3. RAYLEIGH BEACON ADAPTIVE OPTICS

Rayleigh beacons have been functional for military R&D for years, but thus far astronomers have very limited experience with the technique. Although clearly capable of producing science, the systems have not received the kind of systematic shakeout, documentation, and reporting of performance that is characteristic of, for example, ADONIS and PUEO. Consequently, experience does not strongly motivate additional development of this technique at this time. As a result, Rayleigh beacons may be currently somewhat undervalued by the astronomy community. The UnISIS project at the Mt. Wilson 2.5 meter could be an important guide to future work with this technology.

20.4. ADAPTIVE OPTICS FOR PARTIAL CORRECTION

The achievement of image improvement in the visible with partial correction, though not fully understood, is potentially quite interesting. Work at CFHT has shown reductions in visible image FWHM to 0.2–0.3 arcsec with partial correction [9]. While the corrected field of view is not understood rigorously, it seems likely that it will be on the same order as the corrected field of view in the infrared, hence of order 60 arcsec for 4m class telescopes, and larger for smaller apertures. Reduction in the FWHM of a factor of order 2× with a modest field of view can be very important and this opportunity is sure to be explored and developed further.

20.5. LOW ALTITUDE TURBULENCE

There are several techniques which enable a trade between Strehl and corrected field of view. These options arise from the observation that at typical, high quality astronomy sites, the turbulence is sharply layered. The principle contributors to seeing degradation are in the dome, in a low altitude boundary layer, and in one or a few discrete layers at much higher altitude, typically of order 10 kilometers.

The dome and boundary layer contributions are close enough to the telescope to have a high correlation over a large field of view. If it were possible to distinguish the wavefront aberration due to these components, and to correct for them alone, the resulting wavefront would have only the residual aberration from the high altitude turbulence. The fractional gain in image quality is of course site specific, and in fact is potentially larger at poorer sites. Though this is hardly a solution for poor site selection, it may be a counter-example to the thesis above that "the rich get richer". To the extent that this technique succeeds, a site will be limited primarily by high altitude turbulence.

There are several methods available to distinguish to some extent the low altitude turbulence from the high altitude turbulence. Using a low temporal bandwidth will tend to reject the high altitude contribution, which will commonly be associated with higher velocity winds than low altitude, or certainly than in-dome, turbulence. Some knowledge of the wind profile could be used to guide the selection of instrument parameters to optimize this discrimination. Currently, AO employs only a single wavefront reference star. Wavefront information from multiple reference stars could be combined to partially isolate the common component and correct for it alone.

[9]PUEO WEB pages.

Interestingly, a Rayleigh beacon might be well suited for discrimination of low altitude turbulence. Instead of forming the beacon at the highest possible altitude, as normally desired if the goal is the highest on-axis Strehl, it could be formed below the high altitude turbulence layers. This would still be high enough to well sample the low altitude turbulence. This concept has been explored theoretically (Ragazzoni & Marcheti, 1995) and experimentally (Chun, 1997; Chun et al, 1998), but is still undeveloped.

20.6. ADAPTIVE SECONDARIES

Two major impediments to achieving gains in sensitivity with adaptive optics are the losses in throughput and the increases in emissivity. The adaptive secondary can potentially reduce the number of extra optical surfaces to zero, and the number of adaptive optics specific light losses potentially to just one – a dielectric beamsplitter coating on the entrance window of the instrument. This is clearly a very powerful – and potentially expensive – technique. Much work has been done on control of large deformable mirrors (reviewed in numerous articles in Bonaccini and Tyson, 1998). Most work reported on adaptive secondary technology has emphasized relatively high order correction. Low order correction at the secondary could be an interesting option as well.

21. Summary

Adaptive optics offers to astronomy improved spatial resolution in the realm between the seeing limit and the diffraction limit. Natural reference star adaptive optics has largely fulfilled the hopes and expectations for improvements in angular resolution, and a rapid, recent increase in the rate of publications shows that astronomers are mastering the techniques and putting them to work for research. Experience is showing that studies of YSO's and some solar system bodies is advanced fundamentally by this capability. Significant applications are found in extragalactic studies and compact nebulae. Thus far, most AO work produces imagery, but a shift is beginning toward use of AO to obtain spatially resolved spectroscopy, and this will probably provide a greater scientific impact in the long run.

Some indirect benefits of improved angular resolution, especially increased sensitivity, have proven more elusive and may be less important than expected. Laser beacons, vital for the extension of adaptive optics image improvement to faint sources, are still in a difficult development phase and it is premature to draw conclusions about their eventual effectiveness or optimum utilization.

The overview of science results, most published or submitted for publication while this review was in preparation, would have been impossible without the assistance of many authors who provided preprints and often additional commentary. Discussions with J.E. Graves, O. Lai, C. Max, F. Rigaut, F. Roddier, and C. Shelton were particularly valuable in gaining an understanding of existing and future adaptive optics systems. The trend for AO groups to offer a richly illustrated WEB page is helpful to reviewers as well as publicizing results rapidly. Of course, the author takes full responsibility for any errors of fact or foolish opinions herein.

References

1. Ageorges, N., and Walsh, J.R. 1997, *Mssngr* **87**, 39.
2. Alloin, D., Marco, O. 1997, *ApSS* **248**, 237.
3. Alloin, D., Clenet, Y., Granato, J.L., Lagage, P.O., Marco, O., and Rouan, D. 1999, in *Astronomy with Adaptive Optics: Present Results and Future Programs*, ESO Conference and Workshop Proceedings No. 56, ed. D. Bonaccini, 21 (ESO Topical Meeting).
4. Angel, J.R.P. 1995, *ApSS*, **223**, 136.
5. Baudoz, P., Rabbia, Y., Gay, J., Rossi, E., Petro, L., Casey, S., Bely, P., Burg, R., MacKenty, J., and Fleury, B., Madec, P.-Y. 1998, *SPIE* **3353**, 455.
6. Bedding, T.R., Minniti, D., Courbin, F., and Sams, B. 1997, *AstronAstrophys* **326**, 936.
7. Beuzit, J.-L., Thiebault, P., Perrin, G., Rouan, D. 1994, *AstronAstrophys* **291**, L1.
8. Beuzit, J.-L., Mouillet, D., Lagrange, A.-M., Paufique, J. 1997a, *AstronAstrophysSup* **125**, 175.
9. Beuzit, J.-L., Mouillet, D., Lagrange, A.-M., and Le Mignant, D. 1999, ESO Topical Meeting (ibid), 471.
10. Beuzit, J.-L., Prado, P., Sicardy, B., and Poulet, F. 1997b, from the DESPA WEB pages at http://despa.obspm.fr
11. Bonaccini, D., and Tyson, R.K. 1998, *SPIE* **3353**.
12. Bouvier, J., Rigaut, F., Nadeau, D. 1997, *AstronAstrophys* **323**, 139.
13. Brandl, B., Sams, B.J., Bertoldi, F., and Eckart, A. 1996, *ApJ* **466**, 254.
14. Brandner, W., Alcalá, J.M., Frink, S., Kunkel, M. 1997a, *Mssngr* **89**, 37.
15. Brandner, W., Chu, Y.-H., Eisenhauer, F., Grebel, E., and Points, S.D. 1997b, *ApJ* **489**, L153.
16. Brandner, W., Bouvier, J., Grebel, E.K., Tessier, E., De Winter, D., and Beuzit, J.-L. 1995, *AstronAstrophys* **298**, 818.
17. Bruns, D.G., Barrett, T.K., Sandler, D.G., Brusa, G., Huber, M., and Angel, J.R.P. 1997, *SPIE*, **2871**, 890.
18. Chalabaev, A., Le Coarer, E., Rabou, P., Mignart, Y., Petmetsakis, P., amd Le Mignan, D. 1999a, ESO Topical Meeting (ibid), 61.
19. Chalabaev, A., Le Mignant, D., and Le Coarer, E. 1999b, ESO Topical Meeting (ibid),491.
20. Chapman, S., Walker, G., and Morris, S. 1999, ESO Topical Meeting (ibid), 73.
21. Christou, J.C., Ellerbroek, B., Fugate, R.Q., Bonaccini, D., and Stanga, R. 1995, *ApJ* **450**, 369.
22. Christou, J. C., Drummond, J. D. and Spillar, E. 1999, ESO Topical Meeting (ibid), 87.
23. Chun, M. 1997, in this volume.
24. Chun, M.R., Wild, W.J., Shi, F., Smutko, M.F., Kibblewhite, E.J., Fugate, R.Q., and Christou, J.C. 1998, *S.P.I.E.* **3353**, 384.
25. Close, L.M., Roddier, F., Northcott, M.J., Roddier, C., and Graves, J.E. 1997a, *ApJ* **478**, 766.
26. Close, L.M., Dutrey, A., Roddier, F., Guilloteau, S., Roddier, C., Duvert, G., North-

cott, M., Ménard, F., Graves, J.E., and Potter, D. 1997b, *ApJ* **499**, 883.
27. Close, L.M., Roddier, F., Hora, J.L., Graves, J.E., Northcott, M., Roddier, C., Hoffman, W.F., Dayal, A., Fazio, G.G., and Deutsch, L.K. 1997c, *ApJ* **489**, 210.
28. Close, L.M., Roddier, F., Potter, D., Roddier, C., Graves, J.E., and Northcott, M. 1999, ESO Topical Meeting (ibid), 109.
29. Combes, M., Vapillon, L., Gendron, E., Coustenis, A., Lai, O., Wittemberg, R., Sirdey, R. 1997, *Icarus* **129**, 482.
30. Cruzalèbes, P., Rabbia, Y., Monnier, J.D., and Lopez, B. 1999, ESO Topical Meeting (ibid),143.
31. Currie, D.G., Avizonis, P.V., Kissell, K.E., and Bonaccini, D. 1998, *SPIE* **3353**, 1049.
32. Davidge, T.J., Rigaut, F., Doyon, R., and Crampton, D. 1997a, *AJ* **113**, 2094.
33. Davidge, T.J., Simons, D.A., Rigaut, F., Doyon, R., Crampton, D. 1997b, *AJ* **113**, 2094.
34. Davies, R.I., Hackenberg, W., Ott, T., Eckart, A., Rabien, S., Anders, S., Hippler, S., and Kasper, M. 1999, ESO Topical Meeting (ibid), 153.
35. Dayton, D., Sandven, S., Gonglewski, J., Rogers, S., McDermott, S., and Browne, S. 1998, *SPIE* **3353**, 139.
36. DESPA, 1997 (An unattributed image from the DESPA WEB homepage, http://despa.obspm.fr/planeto/planeto.html)
37. Diolaiti, E., Bendinelli, O., Bonaccini, D., Parmeggiani, G., and Rigaut, F. 1999, ESO Topical Meeting (ibid), 175.
38. Drummond, J.D., Christou, J.C., and Fugate, R.Q. 1995, *ApJ* **450**, 380.
39. Drummond, J.D., Fugate, R.Q., Christou, J.C., and Hege, E.K. 1997, preprint.
40. Duchene, G., Bouvier, J., Simon, T., Close, L., and Eisloffel, J. 1999, ESO Topical Meeting (ibid), 185.
41. Dumas, C., and Hainault, O.R. 1996, *Mssngr* **84**, 13.
42. Dumas, C., Close, L., Connelley. M., Graves, R., Hainaut, O., Northcott, M., Roddier, F. 1997, private communication.
43. Esslinger, O., and Edmunds, M.G. 1998, *AstronAstrophysSupp*, **129**, 617.
44. Feldt, M., Stecklum, B., Henning, T., and Hayward, T.L. 1999, ESO Topical Meeting (ibid),513.
45. Feldt, M., Stecklum, B., Henning, T., Hayward, T.L., Lehmann, T., and Klein, R. 1998, *AstronAstrophys*, **287**, L17.
46. Geoffray, H. 1999, ESO Topical Meeting (ibid), 531.
47. Geoffray, H., and Monin, J.-L. 1999, ESO Topical Meeting (ibid), 203.
48. Glenar, D.A., Hillman, J.J., Lelouarn, M., Fugate, R., and Drummond, J.D. 1997, *PASP* **109**, 326.
49. Guyon, O., Roddier, C., Graves, J.E., Roddier, F., Cuevas, S., Espejo, C., Martinez, A., Gonzalez, S., Biseacchi, G., and Vuntesmeri, V. 1999, ESO Topical Meeting (ibid), 537.
50. Harder, S., and Chelli, A. 1999, ESO Topical Meeting (ibid), 217.
51. Heydari-Malayeri, M., and Beuzit, J.-L. 1994, *AstronAstrophys* **287**, L17.
52. Haydari-Malayeri, M., Rauw, G., Esslinger, O. and Beuzit, J.-L. 1997a, *AstronAstrophys* **322**, 554.
53. Heydari-Malayeri, M., Courbin, F., Rauw, G., Esslinger, O., Magain, P. 1997b, *AstronAstrophys* **326**, 143.
54. Kluckers, V.A., Wooder, N.J., Nicholls, T.W., Adcock, M.J., Munro, I., and Dainty, J.C. 1997, *AstronAstrophysSuppl*, **130**, 141.
55. Koresko, C.D., Harvey, P.M., Christou, J.C., Fugate, R.Q., and Li, W. 1997, *ApJ* **458**, 213.
56. Lagrange, A.-M., Beuzit, J.-L., and Mouillet, D. 1996, *JGeophysRes* **101**, 14831.
57. Lai, O., Rouan, D., and Alloin, A. 1999, ESO Topical Meeting (ibid), 555.
58. Lai, O., Rouan, D., Rigaut, F., Arsenault, R., Gendron, E. 1998, *AstronAstrophys* **334**, 783.

59. Langlois, M., Sandler, D., Ryan, P., and McCarthy, D. 1998, *SPIE* **3353**, 189.
60. Malbet, F., Rigaut, F., Bertou, C., and Léna, P. 1993, *AstronAstrophys* **271**, L9.
61. Marco, O., Alloin, A., and Beuzit, J.L. 1997a, *AstronAstrophys* **320**, 399.
62. Marco, O., Alloin, D. 1998, *AstronAstrophys* **336**, 823.
63. Marco, O., Encrenaz, T., Gendron, E. 1997b, *PlanetSpSc* **46**, 547.
64. Max, C.E., Macintosh, B., Olivier, S.S., Gavel, D.T., and Friedman, H.W. 1998, *S.P.I.E.*, **3353**, 277.
65. McCullough, P.R., Fugate, R.Q., Christou, J.C., Ellerbroek, B.L., Higgins, C.H., Spinhirne, J.M., Cleis, R.A., and Moroney, J.F. 1995, *ApJ* **438**, 394.
66. Ménard, F., Dougados, C., Lavalley, C., Cabrit, S., and Stapelfeldt, K. 1999, ESO Topical Meeting (ibid), 303.
67. Monnier, J., et al, 1999, *Ap.J* **512**, 351.
68. Morossi, C., Franchini, M., Ragazzoni, R., Sedmak, G., Suzuki, A., Restaino, S., Albetski, J., Africano, J., Nishimoto, D., and Sydney, P. 1995, *ESO ConfWorkshopProc* **54**, 329.
69. Monin, J.-L., and Geoffray, H. 1997, *Mssngr* **89**, 33.
70. Mouillet, D., Lagrange, A.-M., Beuzit, J.-L., and Renaud, N. 1997, *AstronAstrophys* **324**, 1083.
71. Mouillet, D., Larwood, J.D., Papaloizou, JCB., Lagrange, A.M. 1997, *MNRAS* **292**, 896.
72. November, L.J. 1986, *ApOpt*, **25**, 392.
73. NRC (National Research Council) 1991, *The Decade of Discovery in Astronomy and Astrophysics*, National Academy Press (Washington).
74. Quirrenach, A., and Zinnecker, H. 1997, *Mssngr* **87**, 36.
75. Quirrenback, A. 1999, ESO Topical Meeting (ibid), 361.
76. Ragazzoni, R. and Marchetti, E. 1995 *AtmosOceanOpt*, **8**, 174.
77. Roddier, C. 1996, *CRAcadSciParis* **325**, Ser. 11b, 109.
78. Roddier, C., Roddier, F., Graves, J.E., Northcott, M.J., Close, L., Surace, J., and Véran, J.P. 1999, ESO Topical Meeting (ibid), 389.
79. Roddier, C., Roddier, F., Northcott, M.J., Graves, J.E., and Jim, K. 1996, *ApJ* **463**, 326.
80. Roddier, C. et al. 1997, *IAU Circular* 6697.
81. Roddier, F., Roddier, C., Graves, J.E., and Northcott, M.J. 1995, *ApJ* **443**, 249.
82. Roddier, F., Roddier, C., Close, L., Dumas, C. Graves, J.E., Guyon, O., Han, B., Northcott, M. J., Owen, T., Tholen, D., Brahic, A. 1999, ESO Topical Meeting (ibid), 401.
83. Roddier, F. et al. 1996a, *IAU Circular* 6407.
84. Roddier, F. et al. 1996b, *IAU Circular* 6515.
85. Roddier, F. 1997, *CRAcadSciParis* **325**, Ser. 11b, 35.
86. Roddier, F., Roddier, C., Brahic, A., Dumas, C., Graves, J.E., Northcott, J.M., and Owen, T. 1997, *PlanSpSci* **45**, 1031.
87. Rouan, D., Clenet, Y., and Lai, O. 1999, ESO Topical Meeting (ibid), 411.
88. Rouan, D., Rigaut, F., Alloin, D., Doyon, R., Lai, O., Crampton, D., Gendron, E., Arsenault, R. 1998, **339**, 687.
89. Rouan, D., Field, D., Lemaire, J.-L., Lai, O., Pineau des Forets, G., Falgarone, E., and Deltorn, J.-M. 1997, *MNRAS* **284**, 385.
90. Ryan, P.T., Fugate, R.Q., Langlois, M., and Sandler, D.G. 1998, *SPIE* **3353**, 107.
91. Saint-Pe, O., Combes, M., Rigaut, F., Tomasko, M., and Fulchignoni, M. 1993a, *Icarus* **105**, 263.
92. Saint-Pe, O., Combes, M., and Rigaut, F. 1993b, *Icarus* **105**, 271.
93. Schoeller, M., Brandner, W., Lehmann, T., Weigelt, G., and Zinnecker, H. 1996, *AstronAstrophys* **315**, 445.
94. Schroeder, D. 1987, /it Astronomical Optics, Academic Press (San Diego).
95. Shelton, C. 1998 (private communication).
96. Stecklum, B., Hayward, T.L., and Hofner, P. 1999, ESO Topical Meeting (ibid), 421.

97. Ten Brummelaar, T.A., Hartkopf, W.I., McAlister, H.A., Mason, B.D., Roberts, L.C., and Turner, N.H. 1998, *S.P.I.E.* **3353**, 391.
98. Ten Brummelaar, T.A., Mason, B.D., Bagnuolo, W.G., Hartkopf, W.I., McAlister, H.A., and Turner, N.H. 1996, *AJ* **112**, 1180.
99. Tessier, E. 1997, *AstronAstrophysSuppl* **125**, 581-593.
100. Tessier, E., and Perrier, C. 1996, *OSA Topical Meeting on Adaptive Optics*, Optical Society of America (Washington DC).
101. Theodore, B., Petijean, P., Bremer, M., and Ledoux, C. 1999, ESO Topical Meeting (ibid), 429.
102. Thompson, L.A. 1994, *SPIE*, **2201**, 1074.
103. Trouboul, L., Bouvier, J., Chalabaev, A., Corporon, P., and Le Coarer, E. 1999, ESO Topical Meeting (ibid), 681.
104. Tyler, D.W., and Ellerbroek, B.L. 1998, *SPIE* **3353**, 201.
105. UHAO 1997, (Unattributed image from the University of Hawaii adaptive optics Web pages, http://queequeg.ifa.hawaii.edu)
106. Vannier, L., Lemaire, J.L., Field, D., Rouan, D., Pijpers, F.P., Pineau des Forets, G., Gerin, M., and Falgarone, E. 1999, ESO Topical Meeting (ibid), 687.
107. Véran, J.-P., Beuzit, J.-L., and Chaytor, D. 1999, ESO Topical Meeting (ibid), 691.
108. Véran, J.P., Rigaut, F.J., Rouan, D., and Maitre, H. 1997 J.Opt.Soc.Am. **A 14**, 3057.
109. Walker, G., Chapman, S., Manduchev, G., Racine, R., Nadeau, D., Doyon, R., and Véran, J.-P. 1999, ESO Topical Meeting (ibid), 449.
110. Woolf, N., Angel, J.R.P., and Black, J. (1995), *BAAS*, **185**, 42.07.

Evening session with (from left to right): A. Sheinis, B. Bigelow, E. Kibblewhite, T. Roberts, M. Chun & P. Mc Guire

Happy lecture with E. Gendron

CHAPTER 11
POLARIMETRIC MEASUREMENTS AND DECONVOLUTION TECHNIQUES

N. AGEORGES
European Southern Observatory
Casilla 19001
Santiago 19 - Chile

1. Introduction

Two different issues are treated in this Chapter. First, the problem of adaptive optics polarimetric data is considered. Although based on the experience acquired with one type of adaptive optics system, section 2 aims at being general enough to apply to all similar kind of data. The data reduction process is explained here in detail, stressing the points where careful attention is required and where difficulties may arise. Starting from raw data, this leads us to 'clean' undeconvolved data ready for scientific interpretation.

The second part of this Chapter deals with the deconvolution problem linked to adaptive optics data. Before pointing out why the deconvolution of these data is more demanding than 'classical' data deconvolution, a few existing methods are outlined. Section 3 then ends on the presentation of a new deconvolution technique based on blind deconvolution and especially developed for adaptive optics data. The aim is to inform astronomers of the deconvolution difficulties linked to the point spread function problem in adaptive optics data.

2. Polarimetric measurements

2.1. INTRODUCTION

High angular resolution (HAR) techniques have evolved rapidly and have begun to be applied, in the near-infrared (NIR), to two-dimensional polarimetric observations.

Since these wavelengths are less absorbed and scattered, observations of the detailed structures of the close environment of embedded sources become possible. In the near-infrared, scattering still dominates whilst at longer wavelengths emission dominates. In the UV, absorption dominates over scattering for small grains. In the optical and infrared, emission features from the dust, e.g. PAHs bands, which are unpolarized, are an important contributor.

The interest of combining adaptive optics with polarization measurements resides in the wish of astronomers to get the highest possible resolution in order to get information about the 'exact' distribution of scattering grains.

This unique combination of techniques permit to get information on the exact geometry of an extended source or to determine the individual polarization of multiple sources or, for large sample of sources to determine e.g. the magnetic field in the observed area.

2.2. OBSERVATIONAL MODE

Results of polarimetric observations with the ESO ADONIS adaptive optics system are presented and discussed here; however the general remarks can extend to any similar polarimetric observations.

Figure 1 represents the optical layout of the ADONIS adaptive optics (AO) system (Beuzit et al. 1997). The observational mode described here might be slightly different for other instruments, such as, e.g., for a system equipped with a Wolleston prism. In the present configuration, the data are acquired by rotating a wire grid, installed at the entrance of the camera (i.e. IR focus of f/45 in Fig. 1), used as polarizer, in an oversampling redundant way. The quality of the measurements directly relies on the photometric quality of the data.

In the present case, data have been acquired in chopping mode (between the source and a corresponding sky position) for each position of the polarizer and wavelength range defined by a filter. The same procedure has then been used for a polarimetric calibrator and its corresponding sky. This calibrator is either a source of known non-variable polarization (i.e. a polarized standard) or an unpolarized point source (i.e. an unpolarized standard), see e.g. Mathewson & Ford (1970), Tinberger (1979), Bailey & Hough (1982) and Turnshek et al. (1990) for a list of polarization standards. Even if the instrument is perfectly calibrated in terms of polarization, it is always better to regularly observe a polarization calibrator during the observations, owing to the variation of the sky contribution.

For each object, data cubes of N×N spatial pixels × M frames have been acquired, for each selected position of the polarizer. Observations have

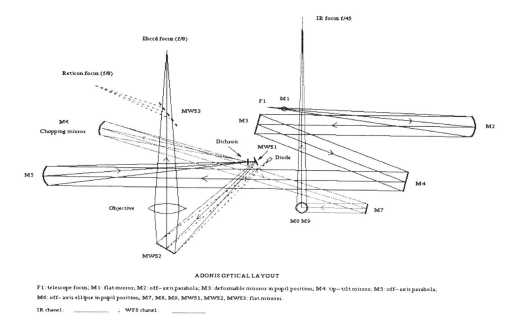

Figure 1. Optical layout of the ESO ADONIS adaptive optics system. The polarimeter, as a prefocal instrument, is installed at the entrance window of the dewar (IR focus f/45).

been done at 9 different positions of the polarizer, each 22.5° apart, to determine the three Stokes parameters (see e.g. the first chapter of Azzam and Bashara 1987): I, U and Q, which are used to determine the degree and angle of polarization of the object studied. Observations at 0, 45, 90 and 135° would have been enough to determine the degree and angle of polarization, but the polarization curve has been oversampled, on purpose, in order to reduce the error bars on the measurements. Observing with the polarizer at 0 or 180° position should give the same result; it is therefore a good way to monitor the photometry during the observations.

2.3. HOW TO CALIBRATE THE DATA

What is interesting is the degree of polarization of the source itself. However, the measurements are representative of the degree of polarization of the source plus a contribution from the interstellar medium and the instrumentation. To determine this last contribution, different possibilities exist:

- the instrumental polarization is known (through calibration of the instrument) and then one just needs to apply this correction factor to

the data or
- one has to determine this contribution in observing polarization calibrators (see 2.2) or
- the polarization of the target source is known in the literature and one can then autocalibrate the data and thereby derive the instrumental polarization or
- in the case of multiple sources, if one has no way to determine the instrumental contribution, at least relative polarimetric information between the sources can be derived.

The interstellar contribution to the measured polarization is harder to determine. It has either been determined by a large polarimetric survey (see e.g. Vrba et al. 1976) achieved in the close neighbourhood of the source of interest and is then known or is not determined, in which case the data cannot be fully calibrated.

Polarization of background starlight is no doubt still useful for studies of large-scale fields and of the properties of interstellar dust. However, subtle differences in the polarizing properties of grains tend to average out over large distances or large data sets. As a result, polarization maps of large regions (e.g whole dark cloud complexes) are likely to be simple projections of the magnetic field along the line of sight onto the plane of the sky (see Goodman 1996).

Once these two contributions (instrumental and interstellar) are known, the data are calibrated with the help of the Stokes parameters. Goodrich (1986) presents a clear and simple method to do this by using the Müller matrix.

2.4. ANALYSIS OF THE DATA

The complete basic data reduction procedure applied to the ADONIS adaptive optics data is represented in Fig. 2.

The first step of the data analysis consists in cleaning the frames, which means flat fielding, sky subtraction and bad pixel correction, in order to get the so-called 'raw cleaned data cubes'. These procedures are not presented in detail here but their effect on the polarization measurements is stressed hereafter. It is important that the photometry stays unmodified during these steps since the polarization is directly derived from the latter.

2.4.1. *How to determine the polarization*

There are two possibilities, linked to one another, to determine the polarization. Both methods imply that at least 4 measurements, 45° apart, have been made.

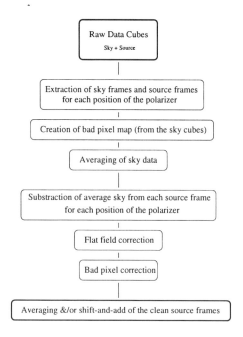

Figure 2. Schematic representation of the basic data reduction applied to the ADONIS adaptive optics data presented here

As mentioned in 2.2, one wants to derive the I, Q and U Stokes parameters. 'V' is characteristic of the circular polarization. It is not treated since it is not measurable here.

U and Q can be derived as follow:

$$U = I\sin(2\theta) = \frac{I_0 - I_{90}}{I_0 + I_{90}} \quad \text{and} \quad V - I\cos(2\theta) = \frac{I_{45} - I_{135}}{I_{45} + I_{135}}.$$

'I' can therefore be derived as : $\sqrt{U^2 + Q^2}$ and the angle of polarization as: $\frac{1}{2}\arctan\frac{U}{Q}$. Since, in this particular case (angles at 0, 45, 90 and 135°), I represents the total intensity, it can be found by adding all polarization maps independantly of the angle at which they have been measured.

The second method, applied here, consists of fitting a cosine curve through the 9 intensity values derived (one for each position of the polarizer). This intensity has been calculated in an aperture (few pixels) or over the complete frame. One way of presenting the results is then to plot the intensity derived, e.g. over the source, as a function of the polarizer angle and show the fitting cosine curve, as is done in the following figures.

The first legitimate question is: how much is the sky polarized in the near-infrared, since this has a direct impact on how the sky subtraction should be achieved. In the optical, during dark time (i.e. no moon), the sky polarization is typically of 3-4%. In the night we made the measurements, the degree of polarization of the sky has been found to be zero within the error bars, which are smaller than 0.01% (Fig. 3). In this case, the contribution of the sky to the global polarization measurements is low enough in order not to play a major role in the data reduction. It will however be shown now that the way the sky subtraction is done plays a major role in terms of precision of the measurements, in the case of noisy data.

2.4.2. *Different sky subtractions*

Different tests have been done to determine what is the best way to subtract the sky from the data and to apply the bad pixel correction.

The 'reference' method consists of, for each position of the polarizer, subtracting an average sky, acquired at the same polarizer position, from the source files and applying a bad pixel correction. In this case as well as the following one, the bad pixel map has been derived separately for each polarizer position.

For the second method, the difference lies in the sky used. Indeed this time, all sky data have been used to create a single 'averaged sky frame' that has been subtracted from the source for each position of the polarizer.

In the last possibility studied here, not only a single averaged sky has been used, but also a single bad pixel map derived from the latter.

Using sky data for each position of the polarizer or averaging these data permits to take the possible sky contribution into account without having to care about how much it might be. In the case of ADONIS, since observations are done in chopping mode, the necessary sky frames are easily acquired. None of these methods seem better than the others, mostly since they are very similar (all using sky data acquired at different position angles of the polarizer). The main difference is linked to the bad pixel correction. Indeed in the last method proposed, the huge error bars mostly come from poor bad pixel correction. This poor determination can be explained by a given polarization sensitivity of the pixels.

The results, presented in Fig. 4 show that not only the sky subtraction but also the bad pixel map determination can modify the photometry of the source. During the data 'cleaning' process, it is imperative not to modify the photometry in order not to affect the determination of the polarization.

In the data presented here, there is a discrepant point, at 157.5°, due to a technical problem with the polarizer, at the time of these observations. For the astronomical interpretation of the data this point will not be taken into account. However all degrees of polarization presented hereafter have been derived using this point, thus explaining the large error bars.

Figure 5 shows the polarization curves fitted to the above presented data. The polarization is 2.3 ± 1.5 % at $64.9 \pm 23°$, 4.25 ± 1.5 % at 67.25

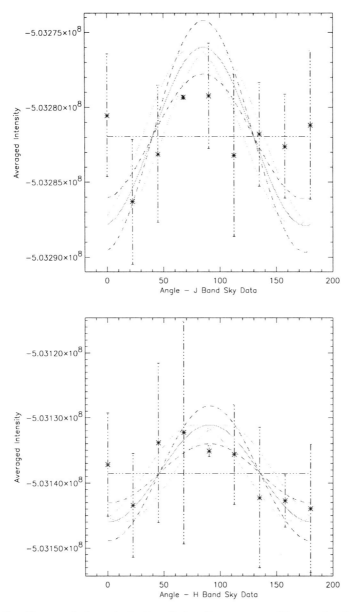

Figure 3. Curves of sky polarization (intensity versus position angle of the polarizer) in J (top) and H band (bottom); in both cases P = 0 ± 0.01 %. The negative values for the intensity (vertical axis) are due to an offset that has not been corrected here.

± 11.5° and 3.85 ± 1.7 % at 67.6 ± 15.1° for the different sky subtraction method and bad pixel correction methods presented.

On the previous plots, the error bars on the determination of the inten-

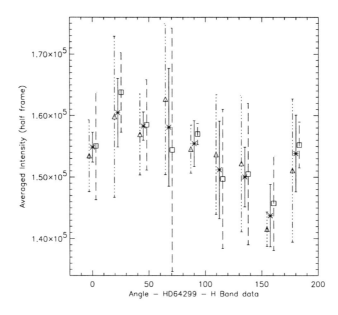

Figure 4. H band polarization curves obtained for HD 64299, depending on the mode of sky subtraction. The intensity derived while using the 'reference' method is represented by the crosses. The triangles are the results of the second method (using one single sky and individual dead pixel maps); and the squares correspond to the utilisation of a single sky and bad pixel map for all polarizer position angle.

sity are due to the photometric variation of the source during the observations. This is illustrated in Fig. 6. The overall distribution of points has a cosine form, characteristic of polarized data. Every ensemble of 200 points corresponds to frames acquired at the same position of the polarizer; those are the ones illustrating the photometric variation during observations. This variation is of the order of 5% (peak to peak) for the images considered, which have a saturated core and are thus not perfectly representative of the actual photometric conditions during the observations.

2.4.3. *Intensity variation and Strehl ratio*

The interest is now focused on a possible link between the intensity variation (independant of the polarization) and the Strehl ratio of the data. As a reminder, the Strehl is defined as the ratio of the central intensity of the PSF relative to the central intensity in the diffraction-limited PSF. Its theoretical limit is 1, ideal case where the complete flux is concentrated in the peak. The analysis proposed here is independant of the polarization measurements but typical for adaptive optics data, since the Strehl is a way of quantifying the quality of the correction applied to the data.

Figure 7 is a representation of the Strehl ratio versus intensity in the

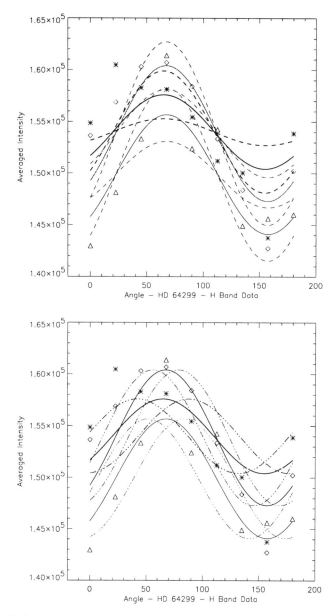

Figure 5. Polarization curves fitted to the above presented different methods of sky substraction. The mean polarization curves are represented with full lines (decreasing thickness from method 1 to 3 corresponding to crosses, triangles and squares respectively). Bottom plot: the dashed line curves represents the variations of the degree of polarization in terms of error bars. Same on the top but for the variations of the angle of polarization.

case of J, H and K band data of HD 64299, a non-polarized standard. No correlation can be seen. This is not a surprise since no theoretical correlation

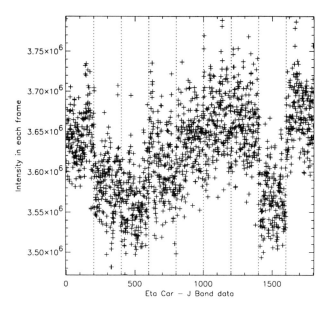

Figure 6. Illustration of the photometric variation of the data during acquisition. Every 200 frames, the polarizer has been rotated by 22.5° which explains the overall cosine shape of this curve. Vertical dashed lines indicate the change of polarizer position angle. The width of this curve is characteristic of the photometric variations. The discrepant point at 157.5° is attributable to an instrumental problem.

exists: the two quantities are independant of one another.

2.5. RESULTS AND CONCLUSION

Once the data have been reduced properly, the instrumental degree of polarization, which will be used to calibrate the data, can then be derived.

On average over the three nights where the observations, used here as demonstration material, have been acquired and over all polarization calibrators observed, the instrumental polarization, for the ESO ADONIS system, has been found to be 1.7% at 105°, 1.7% at 96° and 1.9% at 104° in J, H and K band respectively. Those values are accurate within 0.5% and 10°. These values have been derived without using the values at 157.5°, where there was a technical problem. For highly polarized sources (P$\geq$ 3%), the error bars quoted above are smaller, mostly for the position angle.

To summarise the important points of the data reduction:

- For proper derivation of the polarization, one should be very careful about the removal of the sky background, which can have a non-negligible degree of polarization (mostly at optical wavelengths).

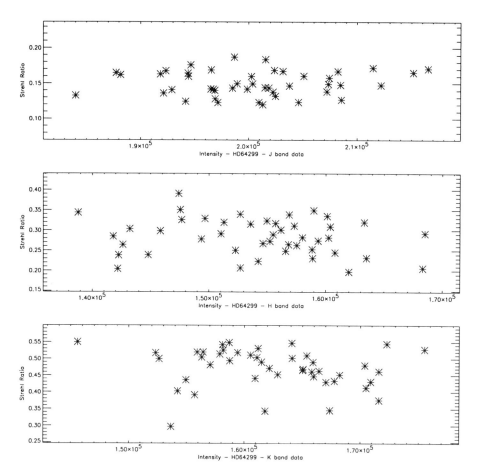

Figure 7. Strehl ratio in the J, H and K band (from top to bottom) versus intensity for the unpolarized standard star HD 64299

- One has to be very careful with the bad pixel correction especially when dead pixel clusters exist. One should never use polarization values derived close to these areas.
- For the calibration of the data, not only the determination of the instrumental degree of polarization is important, but also the angle of polarization, in order to recover absolute sky position angles.
- For ADONIS, a problem has been noticed (and illustrated here) for the polarizer position angle of 157.5°. The origin of this discrepancy remains unclear but other polarization measurements obtained, later on, with ADONIS, did not present this problem.

Results of adaptive optics polarimetric imaging have been recently obtained from two different systems:

- PUEO, the University of Hawaii AO system for the Canada-France-Hawaii telescope, working with a curvature sensor. Potter et al. (1999) obtained images of the circumbinary disk around UY Aurigae.
- ADONIS, the AO system installed on the ESO 3.6m telescope in La Silla, which uses a Shack-Hartmann sensor. Ageorges & Walsh (1999) explain extensively the detail of data acquisition and reduction; and Walsh & Ageorges (1999) present results obtained on the Homunculus nebula surrounding η Carinae.

3. Post-processing of adaptive optics data: Recovery of diffraction limited images

3.1. INTRODUCTION: WHY ARE DECONVOLUTION TECHNIQUES NEEDED?

In the isoplanatic approximation, the intensity (i) of the image can be expressed as the convolution of the object intensity distribution (o) with the point spread function (p) plus an additional noise term (n):

$$i(\vec{r}) = o(\vec{r}) * p(\vec{r}) + n(\vec{r}) \tag{1}$$

where p is the point spread function (PSF) of the telescope plus instrument plus atmosphere. The phase of the Fourier transform of the object intensity is lost due to the atmospheric perturbations. Most deconvolution techniques actually concentrate on recovering this phase information. In the case of adaptive optics however the system partially corrects on-line for these perturbations, so that in theory the PSF of interest is only the instrumental one.

The idea of deconvolving corresponds to the wish to improve the resolution in the data and to retrieve, when possible, the real object intensity distribution. In a first step, I present two existing methods and then demonstrate in Section 3.3 that things are more complicated than stated above and finally introduce a new deconvolution method developed especially for adaptive optics data.

3.2. WHAT DOES EXIST?

3.2.1. Deconvolution with a known PSF

There are two classes of deconvolution algorithms: linear and non-linear ones. The first ones conserve the photometry over the complete image and the latter tend not to preserve it at all.

- *Linear deconvolution algorithms*
  The method here simply consists in inverting Eq. 1; so that the object is then expressed as:

$$\hat{o}(\vec{r}) = FT^{-1}\left\{\frac{I(\vec{f})}{P(\vec{f})}\Phi(\vec{f})\right\} \quad (2)$$

In the case of inverse filtering, $\Phi(\vec{f})$ is a bandpass-limited attenuating filter. In the case of Wiener filtering (Wiener 1950; Bates, McKinnon & Bates 1982), this filter is noise dependent.
— *Non-linear Algorithms*
They all are iterative. Here are some examples of what exist:

- **Maximum entropy method**

  In Eq. 1, the data $i(\vec{r})$ can be satisfied by an infinity of maps that differ from each other at unmeasured spatial frequencies by arbitrary amounts and at measured frequencies by amounts consistent with the noise.

  The entropy of an image $i(\vec{r})$ is defined as:

  $$S = \int\int f[i(\vec{r})]d\vec{r} \quad (3)$$

  where f is a suitably selected function. This expression of the entropy is maximized subject to the constraints imposed to the data. The condition that the entropy be a maximum selects one among these many possible solutions and therefore regularizes the problem expressed by Eq. 1.

  Two forms have been proposed for the entropy function:

  $$f_1(i) = \ln i, \quad f_2(i) = -i\ln i \quad (4)$$

  Here are two important properties possessed by both forms:
  * they do not permit negative values of i, thus automatically imposing the positivity constraint on the image;
  * when the only available measurement is the total flux, both forms have maximum entropy for an uniform image with a constant intensity.

  The choice of the logarithm in these equations ensures the desirable property that the entropy behaves additively when one multiplies the propabilities for two independent systems.

  This is the basic idea of the maximum entropy methods, which have been developed extensively lately (see e.g. Narayan & Nityananda 1986 and Pantin & Starck 1996).

- **The Richardson - Lucy method**
  This is an estimation-maximization algorithm (Lucy, 1974) which converges to a maximum-likelihood solution.

  This method compares at each iteration a solution map $(o(x,y))$ convolved with the function of transfer of modulation $(s(x,y))$ with the measured data $(i(x,y))$. At the kth iteration:

  $$i_k(x,y) = o_k(x,y) * s(x,y) \qquad (5)$$

  The comparison between the solution and the observed image is done by calculating the quotient:

  $$t_k(x,y) = \frac{i(x,y)}{i_k(x,y)} * s(x,y) \qquad (6)$$

  The new solution map is obtained in multiplying the previous map with the above calculated correction factor $t_k(x,y)$:

  $$o_{k+1}(x,y) = o_k(x,y) \times t_k(x,y) \qquad (7)$$

  The intrinsic convergence criterion is: $|o_{k+1} - o_k| \leq \epsilon$, where ϵ is a fixed threshold.

  Recently, a generalized form of this method has been developed (Hook & Lucy 1994). It divides the images in the object plane into two channels. One of these contains point sources (δ funtions) and the other models a smooth background distribution. The latter image is regularised by the use of an entropy term and hence has enforced smoothness. This approach avoids problems encountered in the photometry of the results of non-linear restoration methods.

- **CLEAN** (Högbom 1974)
  This algorithm is extensively used in radio-astronomy. It exploits the fact that point sources dominate many radio images.

3.2.2. **The case of short exposure imaging** (*Speckle techniques*)
These methods are applied to compensate short-exposure data in the same manner as to speckle interferometric data.

– Shift-and-add
  This technique, working in the image plane is actually not a deconvolution method, but substantially improves the resolution of the image. It was introduced for the first time by Lynds, Worden & Harvey (1976).

A 'modern' version has been published by Bates & Cady (1980). The principle, with some slight variations, is to co-add short-exposure time speckle frames once recentered on the brightest speckle:

$$I(x,y) = \frac{1}{N}\sum_{i=1}^{N} Ni_i(x+x_i, y+y_i) \tag{8}$$

$$= \frac{1}{N}\sum_{i=1}^{N} o(x,y) * s_i(x,y) + \frac{1}{N}\sum_{i=1}^{N} n_i(x,y) \tag{9}$$

$$= o(x,y).PSF_{SSA}(x,y) + n_{SSA}(x,y) \tag{10}$$

where (x_i, y_i) is the position of the brightest speckle in the ith image and N the total number of short exposure time frames. The resulting long exposure image is composed by an image at the limit of diffraction sitting on a seeing background resulting from the co-add of all speckle except the brightest one.

Partially corrected data are typically dominated by a single bright speckle. Therefore, applying shift-and-add to short exposure adaptive optics data improves the Strehl ratio, and thus the correction of this data. However such data are still convolved with the modulation transfer function of the telescope.

- Knox-Thompson technique

  This technique, based on the computation of cross-spectra, which contain a phase information, was the first method dedicated to phase recovery in Fourier domain (Knox & Thompson, 1974).

$$\langle \tilde{I}(f)\tilde{I}^*(f+\Delta f)\rangle = \tilde{O}(f)\tilde{O}^*(f+\Delta f)\langle \tilde{S}(f)\tilde{S}^*(f+\Delta f)\rangle \tag{11}$$

Thus:

$$\tilde{O}(f)\tilde{O}^*(f+\Delta f) = \langle \tilde{I}(f)\tilde{I}^*(f+\Delta f)\rangle / \langle \tilde{S}(f)\tilde{S}^*(f+\Delta f)\rangle \tag{12}$$

But it can written:

$$\tilde{O}(f)\tilde{O}^*(f+\Delta f) = |\tilde{O}(f)|\exp(i\phi(f))|\tilde{O}(f+\Delta f)|\exp(-i\phi(f+\Delta f)) \tag{13}$$

which is known for all frequency f such that $|\Delta f| < r_o/\lambda$. In a similar way, it is known that:

$$\exp(i\phi(f))\exp(-i\phi(f+\Delta f)) = \exp(i(\phi(f)-\phi(f+\Delta f))) \tag{14}$$

So that:

$$\Delta\phi(f) = \phi(f+\Delta f) - \phi(f) \tag{15}$$

It shows how an average of the phase differences in the image spectrum is used to estimate phase differences in the object. The phase at one point in the Fourier space is the sum of the phase differences from the origin to this point.

- Triple correlation

  The terms of triple correlation and bispectrum appear in the so-called 'speckle masking' technique (Lohmann, Weigelt & Wirnitzer, 1983). The advantage of this technique is that it gives images with a resolution at the theorical limit of diffraction. The disadvantage is that it is very time consuming.

  The basic data are the speckle interferograms (hereafter the index n is not written to avoid unreadable writing), whose intensity distribution is:

  $$i(r) = o(r) * p(r) = \int o(r')p(r-r')dr' \qquad (16)$$

  where $o(r)$ is the intensity of the observed object, $p(r)$ the PSF and r a two-dimensional vector. The Fourier transform of the intensity, I, is:

  $$I(u) = O(u)P(u) \qquad (17)$$

  The averaged triple correlation is defined as:

  $$\langle T(x,y) \rangle = \langle \int i(x')i(x'+x)i(x'+y)dx' \rangle \qquad (18)$$

  where $\langle ... \rangle$ is the average over an ensemble of N speckle interferograms and $T(x,y)$ a four-dimensional function.

  The bispectrum $B(u,v)$ is the four-dimensional Fourier transform of $T(x,y)$, i.e.:

  $$B(u,v) = \int\int T(x,y) \exp[-2\pi i(u.x + v.y)]dxdy \qquad (19)$$

  The mean bispectrum can be written as:

  $$\langle B(u,v) \rangle = \langle I(u)I(v)I^*(u+v) \rangle \qquad (20)$$

  where X^* is the complex conjugate of X and

  $$I(u) = \int i(x) \exp(-2\pi i u.x)dx \qquad (21)$$

  $$I(v) = \int i(x) \exp(-2\pi i v.x)dx \qquad (22)$$

$$I^*(u+v) = \int i(x)\exp[2\pi i(u+v).x]dx \qquad (23)$$

Since I=OP (Eq. 17), the average of the bispectrum can be written:

$$\begin{aligned}\langle B(u,v)\rangle &= \langle O(u)P(u)O(v)P(v)O^*(u+v)P^*(u+v)\rangle & (24)\\ &= O(u)O(v)O^*(u+v)\langle P(u)P(v)P^*(u+v)\rangle & (25)\end{aligned}$$

$\langle P(u)P(v)P^*(u+v)\rangle$ is the transfer function peculiar to speckle masking (SMTF). Dividing by the SMTF implies for the bispectrum $B_o(u,v)$ of the object $o(x)$:

$$\begin{aligned}B_o(u,v) &= O(u)O(v)O^*(u+v) & (26)\\ &= \langle I(u)I(v)I^*(u+v)\rangle/\langle P(u)P(v)P^*(u+v)\rangle & (27)\end{aligned}$$

The modulus and phase of the object Fourier transform $O(U)$ will now be calculated from the object bispectrum $B_o(u,v)$. Let $\phi(u)$ and $\beta(u,v)$ be the phase of the object spectrum and bispectrum respectively. So,

$$O(u) = |O(u)|\exp[i\phi(u)] \qquad (28)$$

and

$$B_o(u,v) = |B_o(u,v)|\exp[i\beta(u,v)] \qquad (29)$$

In substituting Eqs. 28 and 29 in Eq. 26 one gets:

$$\begin{aligned}B_o(u,v) &= |B_o(u,v)|\exp[i\beta(u,v)] & (30)\\ &= |O(u)|\exp[i\phi(u)]|O(v)|\exp[i\phi(v)]|O(u+v)|\exp[i\phi(u+v)] & (31)\end{aligned}$$

where

$$\exp[i\beta(u,v)] = \exp[i\phi(u)]\exp[i\phi(v)]\exp[-i\phi(u+v)] \qquad (32)$$

i.e.

$$\beta(u,v) = \phi(u) + \phi(v) - \phi(u+v) \qquad (33)$$

So that:

$$\phi(u+v) = \phi(u) + \phi(v) - \beta(u,v) := \phi(w = u+v) \qquad (34)$$

which is called a phase closure relation.

For the recursive calculation of the object phase $\phi(w) = \phi(u+v)$, not only the object bispectrum phase but also starting points are needed: $\phi(0,0)$, $\phi(0,1)$ et $\phi(1,0)$. Since $o(x)$ is real, $O(u) = O^*(-u)$, $O(0,0) =$

O*(0,0) and thus $\phi(0,0) = 0$. $\phi(0,1)$ and $\phi(1,0)$ are set egal to zero since the exact position of the object in the image is not of interest. This method has the advantage of offering several different recursive ways for each object phase element. It is thus possible to improve the signal to noise ratio in averaging over all $\phi(w)$. Since all recursive ways for a same $\phi(w)$ do not give the same signal to noise ratio, a weighting function is used to make the average.

3.3. WHY IS IT DIFFERENT FOR ADAPTIVE OPTICS?

Some of the additional problems linked to AO PSF are linked to the fact that:

– the AO correction is not perfect because of the finite number of actuators (undermodelling or fitting error), the finite size of the subapertures (spatial aliasing), the finite time lag between the time of the measurement and the action on the deformable mirror (servolag) and the noise on the measurements.
– there is a spatial variation of the PSF (linked to the anisoplanatism problem; see e.g. Christou et al. 1998, for a study of the PSF variation in a crowded field).
– the shape of the PSF is variable in time (since the correction varies as the seeing changes with time).

In summary, the measurement of the reference star's PSF can approximate the system's performance on the target object but not give an exact measurement.

3.4. NEW METHODS WELL ADAPTED TO ADAPTIVE OPTICS DATA

3.4.1. Deconvolution with poorly determined PSF
It is here that the expression of 'blind deconvolution' first shows up. These methods permit to recover both the object distribution and the PSF simultaneously. They make use of the physical nature of both, e.g. the fact that they are positive definite. I give here only some references and an 'Ariane thread' through the history, since '*idac*', presented more extensively in section 3.4.3, is one of these methods.

Blind deconvolution has been introduced by Ayers & Dainty in 1988. Since then various improvement have been attached: the error matrix minimization (e.g. Lane (1992), Jefferies & Christou (1993) or Thiébaut & Conan (1994)) or the maximum likelihood (Holmes 1992, Schulz 1993), which makes use of the physical nature of the imaging process, i.e. uses a Poisson or Gaussian model.

There exists also what is called 'myopic deconvolution', which is a derivative of blind deconvolution that uses poorly determined estimates of the object and the PSF (Fusco et al., 1999).

Parametric 'blind' deconvolution assumes the nature of the object it tries to recover: it extracts parametric information about the target and the PSF based on models.

3.4.2. Deconvolution using the wavefront sensor informations

Astronomical images obtained with adaptive optics systems can be enhanced using image restoration techniques. However, this usually requires an accurate knowledge of the system point spread function which is variable in time. The method here estimates the PSF related to each image using data from the adaptive optics control computer, namely the wavefront sensor measurements and the commands to the deformable mirror, accumulated in synchronization with the acquisition. This method requires no extra observing time and has been successfully tested on PUEO, the Canada-France-Hawaii telescope adaptive optics system, data (Véran et al., 1997). One should mention that it works with bright sources and has proven to be efficient with an adaptive optics system equipped with a variable curvature mirror. For an adaptive optics system based on a Shack-Hartman sensor, good results are harder to produce because of the noise on the CCDs and therefore the poor PSF estimation that follows; see Harder & Chelli (1998) for an application of this method to the ESO ADONIS adaptive optics system. Another point to be mentioned is that this method does not recover details of the PSF at high spatial frequency. It relies on statistical properties, i.e. long exposures ($\tau \geq 20$ sec). Following paragraph states the basic theoretical idea behind this method.

Assuming a long exposure AO acquisition has been performed at wavelength λ, in the isoplanatic approximation, the residual optical aberrations in the integrated image are perfectly characterized by the integrated image of a point source (PSF). The PSF depends on how well the AO correction worked during the acquisition; that is on the phase fluctuations of the residual wavefronts. The phase of the residual wavefront at instant t, $\phi(\vec{x}, t)$, is defined on the pupil of the telescope. The only available information on it is the measurement that was performed by the WFS at instant t. In most AO systems, the WFS measurements can be saved for the full length of the exposure time and thus are available for post-acquisition analysis. Generally speaking, a WFS provides a measurement of the fluctuations of $\phi(\vec{x}, t)$, at different locations (measurement points) across the pupil. Because there can be only a finite number of measurement points, only the low frequency fluctuations can be accurately measured. A basic requirement for an AO system is to have a WFS that is matched to the DM, that is the WFS able to measure accurately at least the subset of all the phase functions that the DM may achieve.

3.4.3. IDAC (Iterative Deconvolution Algorithm in C)

This blind deconvolution code has been developed for use with post-processing astronomical images and tested widely on adaptive optics data.

A basic description of what is presented here has been given in Jefferies & Christou (1993). The algorithm is based on a conjugate gradient minimisation technique. *idac* uses conjugate gradient driven least-squares relaxation technique, i.e. an error metric minimisation. The error matrix to be minimized is defined as :

$$\epsilon = E_{conv} + E_{BL} \qquad (35)$$

The individual components of the error metric are the convolution constraint, E_{conv}, and the band-limit constraint, E_{BL}. The former can be expressed as:

$$E_{conv} = \sum_{\vec{r} \in R} \sum_k [i_k(\vec{r}) - \hat{o}(\vec{r}) \star \hat{p_k}(\vec{r})]^2 \qquad (36)$$

The convolution constraint takes into account multiple observations of the same object as for the speckle imaging case although, one single observation can be used as well. It ensures that the target and PSF estimates at each iteration convolve to the measured data. R represents the extent of the convolution image and f_c (limit in the Fourier domain) represents the highest spatial frequency passed by the pupil. This constraint is very powerful when applied with multiple observations. The reason is that the solution assures that there is a common object to the set of observations, each with a different PSF. If the PSF differences are large, then the object solution is easier and more quickly reached.

The band-limit constraint, defined as:

$$E_{BL} = \sum_k \sum_{\vec{f}} \mid \hat{P}_k(\vec{f}) \mid^2 B(\vec{f}) \qquad (37)$$

is important in preventing the trivial solution of a δ-function and the convolution image. Strict positivity of both the object and the PSF's is determined by defining both of them as the squares of another variable, i.e. $o(\vec{r}) = \alpha(\vec{r})^2$ and $p_k(\vec{r}) = \beta_k(\vec{r})^2$.

The object estimate, given as input, can be a co-add of the data or a shift-and-add image, if available. The other input, a data cube of the source, a same dimension data cube for the PSF, the support of these images (which answer the question: is the information to be found everywhere in the image?) and the number of iterations. Last, but not least, a number corresponding to a binary code is given; this is the bandpass constraint flag. This indicates which options of the program have been selected. The possibilities are to give a band limit or aperture constraint, to fix the PSF or the object and to give a sky frame to be subtracted. As an output, an image of the source is given, as well as a data cube of the PSF (each frame being the PSF in the corresponding input data frame), a cube of residuals, which are the difference between the input object frame and the convolution of the output source image with the corresponding PSF, and the error matrix that gives all informations about the way the program converged.

An illustration of this method is now presented on the eclipsing symbiotic binary R Aqr (see Fig. 8). This source is surrounded by a morphologically complex nebulosity which extends to ∼ 1 arc-minute from the central source. At smaller scales, there is an elongated jetlike emission feature (Hollis et al. 1985).

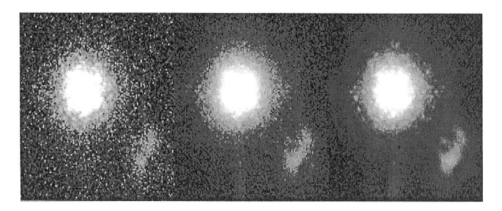

Figure 8. Brγ images of R Aqr. From the left to the right: a single frame, the average of 8 frames and a 80 frame average image, showing the improvement of the signal to noise ratio and the clear appearance of the nebulosity.

It was observed in December 1996 with the ESO ADONIS system equipped with a Fabry-Perot in both the 2μm continuum, the hydrogen emission line, and Brγ (2.1655μm) in order to detect and resolve the structures of the nebulosity. The image scale was $0.05''$/pixel and $t_{exp} = 0.6$s for a total integration of 48s for the Brγ observations presented here. The individual frame SNR was computed to be ∼ 2000 as measured by the ratio of the peak to the rms background, but for the nebulosity, ∼ 3 $''$ from the central source, it was only ∼ 5. In order to improve the signal to noise the data was binned down from 80 to 10 frames, increasing the signal to noise ratio to ∼ 15 for the nebulosity. These binned data were reduced using the PSF constraint where the estimated PSF was obtained from the co-added central source for all 80 initial frames. After convergence the deconvolution was continued with the known PSF constraint switched-off. Figure 9 compares the nebulosity for a raw data frame, an 8 frame average showing the improved SNR and the 80 frame sum showing the initial object estimate for the deconvolution algorithm. Also shown are the results obtained using the PSF constraint and after relaxation. They show the nebulosity to have structure on the scale of $\sim 0.01''$, i.e. diffraction limited.

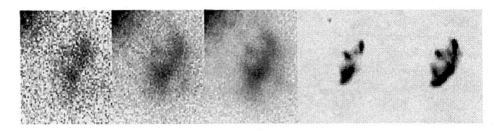

Figure 9. Brγ images of R Aqr. From the left to the right: a raw frame, the input object estimate, the results while using the PSF constraint and then with the PSF constraint relaxed after 200 and 400 iterations respectively.

3.4.4. *Conclusion of the deconvolution problem with AO systems*

Adaptive optics data do not have a stable PSF. The reference PSF are good first estimate but not necessarily accurate (and observing them - if they are not within the frame of the object of interest is time consuming at the telescope). **Deconvolution is necessary, if the full diffraction limit is to be recovered.** In some cases, like for fields of point sources, deconvolution are not necessary, mostly if the information searched are only the position and exact photometry of the sources (see e.g. Véran & Rigaut (1998) for a deconvolution method for accurate astrometry and photometry of adaptive optics images of stellar fields). Most of the deconvolution methods presented above, with the exception of the linear ones, do not preserve the photometry.

For deconvolution, one wants to use as much a priori information as possible, i.e. the non-negativity and the restricted support of the image, use the average PSF as a first estimate and put constraint on the bandpass. The blind deconvolution presented in 3.4.3 has demonstrated that the object recovery with a poor PSF estimate is possible; further illustrations of this method can be found in Christou et al. (1998).

For a good comparison of few reconstruction algorithms using AO instrumentation, see Tyler et al. (1998) and references therein.

Acknowledgements

I would like to thank D. Bonaccini, J. Christou, F. Rigaut and J. Walsh for their help and collaboration.

References

1. Ageorges N. & Walsh J.R., 1999, Acquisition and analysis of adaptive optics polarimetric data, accepted for publication in A&A Supp. Series
2. Ayers G.R. & Dainty J.C., 1988, Opt. Lett. 13, 547
3. Azzam R.M.A. & Bashara N.M., 1987, Ellipsometry and polarized light, Elsevier Science, Amsterdam
4. Bailey J.A. & Hough J.H., 1982, PASP 94, 618
5. Bates R.H.T. & Cady F.M., 1980, Opt. Comm. 32, 365
6. Bates J.H.T., McKinnon A.E. & Bates R.H.T., 1982, Optik 62,1
7. Beuzit J.-L. et al., 1997, Experimental Astronomy 7, 285
8. Christou J.C., Marchis F., Ageorges N., Bonaccini D. & Rigaut F., 1998, Proc. SPIE Vol 3353, 984-993
9. Fusco T., Véran J.-P., Conan J.-M., Mugnier L.M., 1999, A&AS 134, 193
10. Goodmann A.A., 1996, The interpretation of polarization position angle measurements. In W.G. Roberge & D.C.B. Whittet (eds) ASP Conf. Series 97, Polarimetry of the interstellar medium, 325
11. Goodrich R.W., 1986, ApJ 311, 882
12. Harder S. & Chelli A., 1998, Proc. SPIE Vol 3353, 1022-1029
13. Högbom J.A., 1974, AASS 15, 417
14. Hollis J.M., Kafatos M., Michalitsianos A.G. & McAlister H.A., 1985, AJ 289, 765
15. Holmes T.J., 1992, JOSA A 9,1052
16. Hook R.N. & Lucy L.B., 1994, Image restorations of high photometric quality. II. Examples. In Hanisch R.J. & White R.L. (eds) Space Telescope Science Institute, The restoration of HST images and spectra II, 86
17. Jefferies S.M. & Christou J.C., 1993, ApJ 415, 862
18. Knox K.T. & Thompson B.J., 1974, ApJL 193, L45
19. Lane R.G., 1992, JOSA A 9, 1508
20. Lohmann A.W., Weigelt G. & Wirnitzer B., 1983, Appl. Opt. 22, 4028
21. Lucy L., 1974, AJ 79, 745
22. Lynds C.R., Worden S.P. & Harvey J.W., 1976, ApJ 207, 174
23. Mathewson D.S. & Ford V.L., 1970, Mem. R. Astr. Soc. 74, 139
24. Narayan R. & Nityananda R., 1986, ARAA 24, 127
25. Pantin E. & Starck J.L., 1996, AASS 118, 575
26. Potter D.E. et al., 1999, A high resolution polarimetry map of the circumbinary disk around UY Aur, accepted for publication in Ap.J.
27. Schulz T.J., 1993, JOSA A 10, 1064
28. Thiébaut E. & Conan J.M., 1994, JOSA A 12, 485
29. Tinbergen J., 1979, AASS 35, 325
30. Turnshek D.A., Bohlin R.C., Williamson I.I. & Lupie O.L., 1990, AJ 99, 1243
31. Tyler D.W. et al., 1998, Proc. SPIE Vol 3353, 160-171
32. Véran J.P., Rigaut F., Maître H. & Rouan D., 1997, JOSA A 14, 3057
33. Véran J.P. & Rigaut F., 1998, Proc. SPIE Vol. 3353, 426-437
34. Vrba F.J., Strom S.E. & Strom K.M., 1976, AJ 81, 958
35. Walsh J.R. & Ageorges N., 1999, High resolution near-infrared polarimetry oh η Carinae and the Homunculus nebula, submitted to A&A
36. Wiener N., 1950, *Extrapolation, interpolation, and smoothing of stationary time series*, Wiley, New York

Lunch break: R. Ragazzoni (left), A. Touchant (foreground), N. Ageorges & T. Roberts (right)

CHAPTER 12
MEASURING ASTEROIDS WITH ADAPTIVE OPTICS

JACK D. DRUMMOND
Air Force Research Laboratory
Starfire Optical Range
3550 Aberdeen Ave SE
Kirtland AFB, NM 87117-5776 - USA

Parametric Blind Deconvolution (PBD) is described to extract the semimajor axis, semiminor axes, and orientation of the ellipse projected by a triaxial ellipsoid asteroid, as well as the point spread function, from the convolution of the two. In turn, the formulation relating these observables as a function of rotation back to the underlying ellipsoid is recounted, which allows the determination of the triaxial dimensions and orientation of the spin vector to be made from a series of images. Such analysis is well-suited for Adaptive Optics (AO) studies of asteroids because these objects are always smaller than the seeing disk, yet many are larger than the diffraction limit of even modest telescopes. As opposed to photometric techniques which require many lightcurves from many apparitions, PBD of AO images can lead to absolute dimensions (rather than axial ratios) and a rotational pole in only one or two nights.

1. Introduction

Towards the end of graduate school at New Mexico State University, around 1979, I tried to express the relation between an ellipsoid asteroid and its projected ellipse. Somewhere I had learned that any cut through an ellipsoid produced an elliptical outline, but I had not been able to uncover the relations between the ellipsoid and the apparent ellipse. Although I still feel that it has been done before, it was not until a post-doctoral position at the University of Arizona in 1981 led me to a collaboration with John Cocke that I was able to achieve my mathematical goal. Keith Hege had hired me to work on asteroid data taken with the emerging technique of speckle interferometry, and at that time there were no methods for determining pole orientation and axes dimensions of asteroids, and no space platforms to make above-atmosphere images of these minor planets. Asteroids were still only unresolved points of light.

Speckle provided hope for obtaining resolved images, and I realized that by following these changing, resolved images, it might be possible to find the direction of an asteroid's spin vector and its three dimensions. Because of the extreme difficulty in dealing with the speckle transfer function, instead of images we had to restrict our analysis to the power spectra or autocorrelation of images. Nevertheless, these were sufficient to determine an asteroid's dimensions and rotational pole, even without references to features in an image. Following features as they move across the face of a resolved object is the normal procedure in analyses of spacecraft images of planetary satellites. Thomas et al. (1995), for example, analyzed Voyager images of Saturn's moon Hyperion to determine its rotational pole and shape from visible features, and Davies et al. (1994) and Thomas et al. (1997) have recently applied the technique to asteroids imaged from space – Galileo images of the 951 Gaspra, and Hubble Space Telescope (HST) images of Vesta, respectively. By following features as they move across the disk of an object, it is possible to obtain the same information, and with fewer assumptions, as obtained by following the changing size, shape, and orientation of the projected ellipses.

However, in many cases features may not be visible because the asteroid may have been painted uniformly gray over its collision dominated history, or it may be too small or faint for feature analysis. In these cases, the method that was first described by Drummond et al. (1985a) in application to the very elongated Earth approaching asteroid 433 Eros, is available. Since that first paper, poles and dimensions were obtained with this method from the power spectra of asteroids 532 Herculina (Drummond et al., 1985b), 511 Davida (Drummond and Hege, 1986), 4 Vesta (Drummond et al., 1988a), and 2 Pallas and 29 Amphitrite (Drummond and Hege, 1989), and represented the first analyses that produced both the triaxial dimensions and pole direction from direct measurements from one or two night's data. Similar analysis has been performed on images of Vesta reconstructed with speckle data (McCarthy et al., 1994).

Concurrent with the advent of adaptive optics, which eliminates the speckle transfer function, the analysis, with the same mathematical foundation, has been applied to asteroids 1 Ceres and Vesta (Drummond et al., 1998). The speckle transfer function has been replaced by a much simpler point spread function in AO, and so the measurements of the Fourier transform of each image of an asteroid is simultaneously fit for the PSF and the asteroid. Julian Christou and I have collaborated on this new procedure called PBD, Parametric Blind Deconvolution, a technique to extract information about both without having to make reference to a PSF observation made at a different time and under perhaps different conditions. Although the procedures outlined herein apply to asteroids, the more general lesson to be learned is that if astronomical objects can be parameterized, then the best use of AO images is in the least squares extraction of these parameters.

At about the time of the initial studies of asteroids with speckle interferometry, photometric lightcurve techniques began to mature that took advantage of the movement of the sub-Earth point across lines of longitude on the asteroid to yield poles from timings of lightcurve features (Taylor and Tedesco, 1983), and across lines of latitude to give both poles and ellipsoid axial ratios from lightcurve amplitudes and brightness estimates (Magnusson 1986; Drummond *et al.*, 1988b). Magnusson *et al.* (1989) gives a good summary and comparison of the various techniques to obtain poles and axial ratios. Notice, however, that lightcurve techniques do not give true axes dimensions, only their ratios, and that it is necessary to accumulate lightcurves with good absolute photometry over decades, whereas true dimensions and the rotational pole can be obtained on essentially one night with resolved image analysis.

Although the formulation of obtaining ellipsoid information from its projected ellipses has been available since 1985, other than back-calculating the appearance of the nucleus of comet Iras-Araki-Alcock by Sekanina (1988) from a postulated pole and dimensions, the procedure seems to have gone largely unrecognized. Asteroids, because they are always sub-seeing in size make ideal targets for the new technology of AO, since there are so many asteroids and so few HST's. In general, images of asteroids can be well-modelled as ellipses, especially main belt objects which are observed at small solar phase angles. Therefore, application of PBD to asteroids is a project well-suited for emerging adaptive optics programs, and perhaps now others will find useful the ellipses to ellipsoid mathematics that I struggled to develop.

So that the procedures and techniques do not disappear with me, Section 2 describes, without derivations (but given by Drummond *et al.*, 1985a), the mathematical relation between an ellipsoid and its projected ellipse, Section 3 outlines a method for measuring the image of an ellipse in the Fourier domain, and Section 4 discusses non-linear least squares fitting of images to obtain solutions for the six unknowns. The Summary section contains a brief description of the entire procedure to obtain ellipsoid dimensions and spin vector direction from images of an asteroid made throughout its rotation.

2. Ellipses from a Triaxial Ellipsoid

2.1. BASIC FORMULAE

A triaxial ellipsoid asteroid with semiaxes $a \geq b \geq c$ viewed from any direction looks like an ellipse with apparent semiaxes $\alpha \geq \beta$. Define ω to be the solar phase angle (Sun-object-Earth) and ρ to be the angle, as seen from the Earth, from the position angle of the Sun (NTS) to the position angle of the node – the intersection of the plane of the sky with the asteroid's equatorial plane. Using Euler angles θ and ψ as defined by Goldstein (1950), with the third angle fixed at $\pi/2$, area vectors to the asteroid from the Earth and Sun are

$$\mathbf{E} = [\, b\,c\,i_E \,;\, a\,c\,j_E \,;\, a\,b\,k_E \,] \qquad (1)$$

$$\mathbf{S} = [\, b\,c\,i_S \,;\, a\,c\,j_S \,;\, a\,b\,k_S \,] \qquad (2)$$

where the unit vectors for the Earth are

$$i_E = -\sin\psi_E \cos\theta_E = \cos\lambda_E \cos\theta_E$$

$$j_E = -\cos\psi_E \cos\theta_E = \sin\lambda_E \cos\theta_E$$

$$k_E = \sin\theta_E \qquad (3)$$

and for the Sun they are

$$i_S = -\sin\psi_S \cos\theta_S = \cos\lambda_S \cos\theta_S$$
$$= -\cos\omega \cos\theta_E \sin\psi_E - \sin\omega\,(\sin\rho\sin\theta_E\sin\psi_E - \cos\rho\cos\psi_E)$$

$$j_S = -\cos\psi_S \cos\theta_S = \sin\lambda_S \cos\theta_S$$
$$= -\cos\omega \cos\theta_E \cos\psi_E - \sin\omega\,(\sin\rho\sin\theta_E\cos\psi_E + \cos\rho\sin\psi_E)$$

$$k_S = \sin\theta_S = \cos\omega \sin\theta_E - \sin\omega \sin\rho \cos\theta_E \qquad (4)$$

where the relations between the Earth and Sun vectors are also given.

For an ellipsoid spinning about axis c (the usual case for an asteroid), with its north pole defined by the right hand rule, $\theta_{E(S)}$ corresponds to the sub-Earth (-Sun) asterocentric latitude, ψ corresponds to the rotational phase angle as seen by the Earth (or Sun), which always increases with time, and λ corresponds to the sub-Earth (or Sun) longitude as measured from the tip of the long axis and decreases with time. The rotational phase angle and the longitude are related as $\lambda = 270° - \psi$. ψ is defined to be zero when the maximum cross sectional area is seen. At this instant, the semimajor axis of the ellipse, α, is equal to a of the ellipsoid, which lies unprojected along the node in the plane of the sky. λ is defined to be zero a quarter of a rotation earlier, when the minimum cross sectional area is seen, b lies unprojected in the plane of the sky, and $\beta = b$.

The transformation from the triaxial ellipsoid to the apparent ellipse (Fig. 1) as seen by the Earth is given by setting ω, the solar phase angle, equal to zero in the following formulae:

$$\alpha^2 = \frac{B + \sqrt{B^2 - 4A^2/C}}{2} \quad ; \quad \beta^2 = \frac{B - \sqrt{B^2 - 4A^2/C}}{2} \tag{5}$$

where

$$A = \mathbf{E} \cdot \mathbf{S} = b^2 \, c^2 \, i_E \, i_S + a^2 \, c^2 \, j_E \, j_S + a^2 \, b^2 \, k_E \, k_S$$

$$C = \mathbf{S} \cdot \mathbf{S} = (b \, c \, i_S)^2 + (a \, c \, j_S)^2 + (a \, b \, k_S)^2$$

$$D = a^2(1 - i_E^2) + b^2(1 - j_E^2) + c^2(1 - k_E^2)$$

$$B = D + (a \, b \, c \, \sin\omega)^2 / C. \tag{6}$$

The position angle of α, measured from the node (the intersection of the plane of the sky and the ellipsoid's equatorial plane), is found by setting $\omega = 0$ in the following:

$$\tan 2\gamma = \tag{7}$$

$$\frac{\sin\theta_E \sin 2\psi_E (a^2 - b^2) + \sin 2\rho \, (abc \sin\omega)^2/C}{\cos 2\psi_E (a^2 - b^2) + (ai_E)^2 + (bj_E)^2 + (ck_E)^2 - c^2 - \cos 2\rho \, (abc \sin\omega)^2/C}.$$

When the solar phase angle is not zero, the *apparent* unilluminated area is bounded by the edge of the apparent ellipse (Eq. 5 with $\omega = 0$) and the terminator. The terminator ellipse has semiaxes α_t and β_t as given by Eq. 5 using the appropriate ω and ρ, with α_t oriented at a position angle from the node of γ_t – Eq. 7 with the appropriate ω and ρ.

2.2. USEFUL RELATIONSHIPS

The solar phase angle must satisfy

$$\cos\omega = i_E i_S + j_E j_S + k_E k_S.$$

Note that as $\omega \to 0$,

$\mathbf{S} \to \mathbf{E}$; A and $C \to [\mathbf{E} \cdot \mathbf{E}]$; $B \to D$; and therefore $[\alpha_t; \beta_t] \to [\alpha; \beta]$.

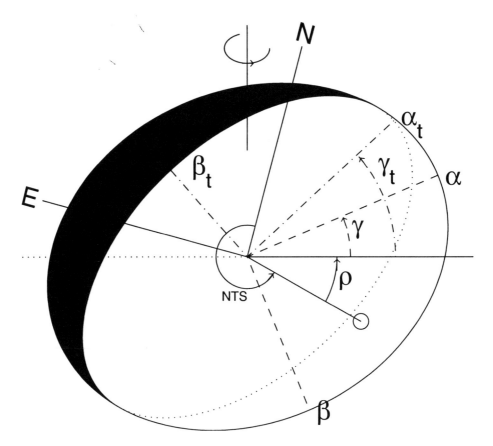

Figure 1. The projection of a 5x4x3 ellipsoid. All quantities are defined in the text. The terminator ellipse is the dotted line and the edge of the area in shadow. The projected ellipse values for $NTS = 264.9°$, $\omega = 46.4°$, $\rho = 29.1°$, and a sub-Earth point at $[\lambda; \theta] = [40; +40]$, are $\alpha = 4.57$, $\beta = 3.59$, $\gamma = 22.6°$, and $\nu = 26.0°$. Therefore, the sub-Sun point, indicated by the small circle, is at $[80; +10]$, and the terminator ellipse parameters are $\alpha_t = 4.38$, $\beta_t = 2.52$, $\gamma_t = 42.2°$, and $\nu_t = 6.4°$.

The obliquity ρ can be found from

$$\sin\rho = \frac{k_E \cos\omega - k_S}{\cos\theta_E \sin\omega}; \quad \cos\rho = \frac{i_S \cos\psi_E - j_S \sin\psi_E}{\sin\omega}.$$

The position angle of the Sun, NTS, in the direction of the asteroid, is a known quantity for each observation. Given the position of both the Sun and asteroid, $[RA_S; \delta_S]$ and $[RA_a; \delta_a]$, then for

$$d = \arccos[\cos(RA_S - RA_a) \cos\delta_S \cos\delta_a + \sin\delta_S \sin\delta_a],$$

$$NTS = \arccos(\frac{\sin\delta_S - \sin\delta_a \cos d}{\cos\delta_a \sin d}).$$

If $\sin(RA_S - RA_a) > 0$, then change the sign of NTS.

It is often more tractable to treat situations numerically rather than analytically. In scattering, for example, each point on the asteroid's surface needs to be examined, and the surface normal has to be calculated. The equation for any point on the surface of an ellipsoid is

$$x^2/a^2 + y^2/b^2 + z^2/c^2 = 1 \qquad (8)$$

where

$$x = R_p i_p \; ; \; y = R_p j_p \; ; \; z = R_p k_p. \qquad (9)$$

Substituting Eq. 9 into Eq. 8, and rearranging, shows that the radius vector from the center to a point on the surface having asterocentric latitude θ_p and longitude λ_p, has length

$$R_p = 1/\sqrt{(\frac{i_p}{a})^2 + (\frac{j_p}{b})^2 + (\frac{k_p}{c})^2}. \qquad (10)$$

From the gradient of Eq. 8, the unit surface normal at this point has components

$$N_p = [\frac{i_p}{a^2}; \frac{j_p}{b^2}; \frac{k_p}{c^2}] / \sqrt{(\frac{i_p}{a^2})^2 + (\frac{j_p}{b^2})^2 + (\frac{k_p}{c^2})^2}. \qquad (11)$$

If R_p is the sub-Sun point R_S, then the total area illuminated by the Sun is

$$\pi \sqrt{C} = \pi \, a \, b \, c / R_S = \pi \, \|\mathbf{S}\|. \qquad (12)$$

By multiplying α_t and β_t from Eq. 5, the area of the terminator ellipse seen by the Earth is

$$\pi \, \alpha_t \, \beta_t = \pi \, A/\sqrt{C} = \pi \, (\mathbf{E} \cdot \mathbf{S})/\|\mathbf{S}\|, \qquad (13)$$

and by addition and subtraction

$$\alpha_t^2 + \beta_t^2 = B \; ; \; \alpha_t^2 - \beta_t^2 = \sqrt{B^2 - 4A^2/C} = \sqrt{B^2 - 4(\mathbf{E} \cdot \mathbf{S})^2/(\mathbf{S} \cdot \mathbf{S})}. \qquad (14)$$

If R_p is the sub-Earth point R_E, then the total area visible to the Earth is

$$\pi \, \alpha \, \beta = \pi \, a \, b \, c / R_E = \pi \, \|\mathbf{E}\|, \qquad (15)$$

and furthermore,

$$\alpha^2 + \beta^2 = D \; ; \; \alpha^2 - \beta^2 = \sqrt{D^2 - 4(\mathbf{E} \cdot \mathbf{E})}. \qquad (16)$$

A line connecting the cusps goes through the center and divides both the apparent and terminator ellipses in half. Therefore, the total illuminated area seen by the Earth is the sum of the terminator and apparent ellipses both divided by two, while the unilluminated area is the difference:

$$\text{ill} = (\pi\alpha\beta + \pi\alpha_t\beta_t)/2 \; ; \; \text{un} = (\pi\alpha\beta - \pi\alpha_t\beta_t)/2. \tag{17}$$

If $A < 0$, then β_t should be considered negative, the asteroid will appear crescent rather than gibbous, and the unilluminated area will be greater than the illuminated area.

The cusps are two points that lie on both the projected and the terminator ellipses, marking the points where the terminator ellipse is internally tangent to the projected ellipse. The angle, ν, from the tip of the long axis of the projected ellipse to the first cusp is given by

$$\cos^2\nu = \frac{\alpha^4(\beta^2 - \beta_t^2)(\alpha_t^2 - \beta^2)}{(\alpha^2 - \beta^2)[\alpha^2\beta^2(\alpha_t^2 + \beta_t^2) - \alpha_t^2\beta_t^2(\alpha^2 + \beta^2)]}. \tag{18}$$

If $\tan(\gamma_t - \gamma) < 0$, then change the sign of ν. The angle from the tip of the terminator ellipse to this cusp is found from

$$\gamma + \nu = \gamma_t + \nu_t. \tag{19}$$

If we want to draw the two half ellipses that demarcate the visible, illuminated portion of the projected ellipse, then if $\beta_t \sin(\rho + \gamma_t + \nu_t) < 0$, 180° must be added to both ν_t and ν, allowing that β_t can be negative when the terminator appears concave. Plotting only the first 180° of the terminator ellipse, from $\nu_t + \gamma_t$ to $\nu_t + \gamma_t + \pi$, draws the terminator. Continuing with the projected ellipse from this cusp, $\nu + \gamma + \pi$, to $\nu + \gamma + 2\pi$, in a positive direction (NESW) if $\beta_t > 0$, or in the negative direction if $\beta_t < 0$, completes the boundary of the visible, illuminated part of the projected ellipse.

3. Fitting the Ellipse

3.1. A MEAN ELLIPSE

In practice, a single observation would be measured over the illuminated part and fit against an average theoretical ellipse. This could be obtained from the average of α and α_t, β and β_t, and γ and γ_t, with some kind of weighting. Or a mean ellipse could be made with half of the apparent ellipse outline and half of the terminator ellipse, joined at the cusps. This has been done in Fig. 2, where the mean ellipse is drawn over an ellipsoid with Lambertian scattering – the surface brightness is proportional to the cosine of the angle of incidence between the surface normal and sunlight, but not to the angle of emergence to the Earth. The mean ellipse has been obtained from a non-linear least squares fit of 180 points of the apparent

Figure 2. Mean ellipse. On the ellipsoid of Fig. 1, with Lambertian scattering, a mean ellipse is drawn having $\alpha = 4.45$, $\beta = 3.09$, $\gamma = 35.7°$, with a center at $x = 0.50$ and $y = -0.60$ units. The sub-Earth and -Sun points are indicated.

ellipse from one cusp to the other, plus 180 points of the terminator ellipse from there back to the original cusp. This works well, especially at the low solar phase angles for main belt asteroids, illustrated in Fig. 3, but at more moderate phase angles, such as illustrated in Figs. 1 and 2, the surface brightness gradient violates our assumption of a smooth uniform ellipse.

3.2. THE FOURIER DOMAIN

For the usual assumptions that an asteroid is a smooth, featureless, triaxial ellipsoid rotating about its shortest axis, its image can be modelled as a step function ellipse, with a height of A_E corresponding to its albedo or mean surface brightness. Since this function is not amenable to least squares fitting, and since the image of the asteroid is a convolution of the ellipse and the point spread function (PSF), least squares fitting for the parameters describing the asteroid is best done in the Fourier domain where the convolution becomes a multiplication of two functions.

The general equation for a normalized variable, $r' = r/h$, of an elliptical

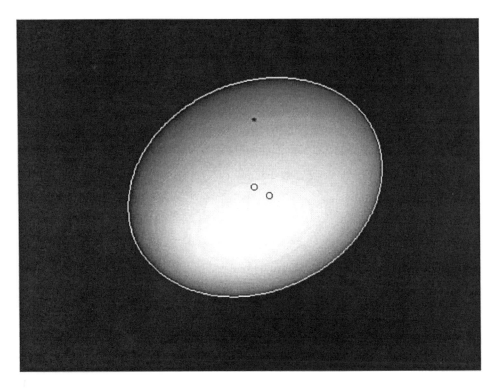

Figure 3. Mean ellipse, small solar phase angle. The geometry is the same as in Figs. 1 and 2, except that the Sun is located such that $\omega = 9.4°$, while ρ remains 29.1°. The mean ellipse has $\alpha = 4.56$, $\beta = 3.57$, $\gamma = 23.3°$, with a center at $x = 0.02$ and $y = -0.01$ units. The north pole, marked near the top of the ellipse, is also barely visible in Fig. 1.

distribution function, with semi-axes of α and β, and oriented at an angle of ϕ between α and the X axis, is

$$\left(\frac{r}{h}\right)^2 = \left[\left(\frac{\cos\phi}{\alpha}\right)^2 + \left(\frac{\sin\phi}{\beta}\right)^2\right](x - x_0)^2$$

$$+ \left[\left(\frac{\sin\phi}{\alpha}\right)^2 + \left(\frac{\cos\phi}{\beta}\right)^2\right](y - y_0)^2$$

$$+ \sin(2\phi)\left(\frac{1}{\alpha^2} - \frac{1}{\beta^2}\right)(x - x_0)(y - y_0). \tag{20}$$

The locus of points where the left hand side is equal to unity, $r'^2_E = (r/h)^2 = 1$, describes an elliptical contour marking the boundary of the asteroid ellipse, E, in the image domain.

We have found that a Lorentzian is a suitable image domain analytic function that describes the AO PSF at the Starfire Optical Range (SOR).

A Lorentzian is described as

$$L = \frac{A_L}{1+(r/h)^2} \quad (21)$$

where A_L is the amplitude, r is the pixel radius and h is the elliptical contour of half width at half maximum (HWHM). For our purposes the Lorentzian PSF is parameterized by α_L, β_L and ϕ_L.

The AO image of the asteroid is then the convolution of an ellipse and a Lorentzian. The Fourier transform ($\mathcal{F}$) of this image is the product of the Fourier transform of the ellipse, E, and the transform of the Lorentzian, L. The former involves a Bessel function of the first kind of order one, J_1, and the latter a Hankel function, specifically a modified Bessel function of the second kind of order zero, K_0. Since the derivatives of the transforms are needed to perform least squares fitting, the following are listed for convenience:

$$\mathcal{F}\{A_E E\} = \frac{A_E \pi \alpha_E \beta_E 2 J_1(\pi r'_E)}{\pi r'_E}$$

$$d\mathcal{F}\{A_E E\} = \frac{-A_E \pi \alpha_E \beta_E 2 J_2(\pi r'_E)\pi}{\pi r'_E} dr'_E$$

$$\mathcal{F}\{L\} = A_L 2\pi \alpha_L \beta_L K_0(\pi r'_L)$$

$$d\mathcal{F}\{L\} = -A_L 2\pi \alpha_L \beta_L K_1(\pi r'_L)\pi \, dr'_L \quad (22)$$

where r' is the normalized variable (Eq. 20) in the frequency domain. In the discrete case of an FFT, the relation between the image and frequency domain semi-axes variables are $\alpha_f = (S/2)/\alpha_i$; $\beta_f = (S/2)/\beta_i$, S being the length of the side of the array. Since the Fourier transform of the observation is a product of two analytic functions in the frequency domain, it can be fit with a non-linear least squares routine that solves for the three parameters defining the Lorentzian, the four defining the flat-topped ellipse, and the two locating the image center.

Although the integral of a one dimensional Lorentzian is convergent, the integral of a two dimensional Lorentzian (Eq. 21) is not, and thus the center of its transform is infinity, $K_0(0,0) = \infty$. In the case of the discrete Fourier transform, however, the center pixel in the frequency domain is an integral over the infinity, which yields a finite number:

$$\int_0^{2\pi} \int_0^{1/\sqrt{\pi}} A_L 2\pi \alpha_L \beta_L K_0(\pi r'_L) r \, dr \, d\theta = A_L S^2 [1 - \frac{\sqrt{\pi \alpha_L \beta_L}}{S/2} K_1(\frac{\sqrt{\pi \alpha_L \beta_L}}{S/2})]. \quad (23)$$

K_0 is normalized by this number, giving unit volume for the Lorentzian. The total flux in the image, corresponding to the center pixel in the frequency domain, is thus attributed to the asteroid. In practice, after the normalization, this center pixel is excluded from the fit, which also has the advantage that any flat background residual in the image has no effect in the frequency domain fitting. The asteroid's mean albedo, A_E, can then be found by dividing the flux (determined from either the frequency domain fit or from standard aperture photometry in the image domain) by the area of the ellipse. The derivation of the integral, along with further details, are given by Drummond (1998).

While we have used a Lorentzian PSF, the same procedure can be followed for any analytic point spread function to derive the ellipse and PSF parameters in the frequency domain. If the Fourier transform of the PSF is analytic, then non-linear least squares can be performed to find the relevant parameters. Compared to a technique such as Iterative Blind Deconvolution (IBD), as implemented in the appendix of Drummond et al., 1998, for example, considerable computer time is saved by searching for only nine relevant unknown parameters as opposed to finding $2S^2$ unknowns in IBD, where each pixel in an SxS array is an unknown for both the PSF and object arrays.

4. A Triaxial Ellipsoid from Ellipses

4.1. RESIDUALS FOR LEAST SQUARES FITTING

Once a series of ellipses have been determined from observations distributed over a rotation, the triaxial ellipsoid that gives rise to them can be determined from least squares. Each observation provides a measured α, β, and position angle, pa, as a function of the independent variable time, expressed as the true rotational phase, $\psi_E = \psi_{rel} - \psi_0$, where the relative rotational phase is constructed with the usually well known rotational period. To describe the triaxial ellipsoid and the direction of its rotational pole requires six parameters: three constants a, b, c, and three angles that can be considered constant over several nights, θ_E, ψ_0, the zero point for the relative rotational phase, and ρ, which can also be considered a relative zero point in the measured position angles. Referring to Fig. 1, each measured position angle of the long axis is related to the other angles by

$$pa - NTS = \rho + \gamma \tag{24}$$

The left hand side of Eq. 24 involves known (NTS, the position angle of the Sun) or measured (pa) constants, and the right hand side involves two unknowns.

Minimizing the sum of the residuals squared from several observations yields the least squares solution to $[a, b, c, \theta_E, \psi_0, \rho]$, all as a function of ψ_E. The residuals can be the differences between each measured α, β, and $pa - NTS$ (as determined in the frequency domain, Sec 3.2) and the calculated mean α, β, and $\rho + \gamma$ (Sec 3.1). Since the three observed quantities are two linear measures and an angle, the angle has to be placed on a linear scale by weighting it with $\sqrt{\alpha^2 - \beta^2}$, the distance from the center to the focus of the ellipse. When the eccentricity of the ellipse is zero, this distance becomes zero, the position angle loses its meaning, and the weight is zero. Thus, a mean ellipse is formed by using Eqs 5 and 7 with $\omega = 0$ (the projected ellipse) and $\omega \neq 0$ (the terminator ellipse), and initial guesses for the six unknowns. The residuals to minimize would then be between the calculated quantities

$$\alpha_c = f(a, b, c, \theta_E, \psi_0, \rho)$$

$$\beta_c = f(a, b, c, \theta_E, \psi_0, \rho)$$

$$\gamma_c + \rho = f(a, b, c, \theta_E, \psi_0, \rho) + \rho \tag{25}$$

and the measured parameters α_m, β_m, and $pa_m - NTS$, with the angular residuals weighted by $\sqrt{\alpha_m^2 - \beta_m^2}$. Care should taken to ensure that the angular residuals are between $-\pi/2$ and $\pi/2$. This can be accomplished by adding or subtracting π to γ when necessary, since $\gamma \pm \pi$ and γ are equivalent.

Alternatively, the residuals can be the differences between the measured and calculated moments, the coefficients F, G, and H of $(x - x_0)^2$, $(y - y_0)^2$, and $(x - x_0)(y - y_0)$, respectively, in Eq. 20. Any convenient coordinate system can be used. The ϕ in Eq. 20 could be, for example, the difference between the position angle of the ellipse and the position angle of the Sun, $\phi = pa - NTS = \rho + \gamma$, (McCarthy et al., 1994). Then the residuals to minimize would be between the calculated coefficients

$$F_c = (\frac{\cos(\gamma_c + \rho)}{\alpha_c})^2 + (\frac{\sin(\gamma_c + \rho)}{\beta_c})^2$$

$$G_c = (\frac{\sin(\gamma_c + \rho)}{\alpha_c})^2 + (\frac{\cos(\gamma_c + \rho)}{\beta_c})^2$$

$$H_c = \sin(2(\gamma_c + \rho))(\frac{1}{\alpha_c^2} - \frac{1}{\beta_c^2}) \tag{26}$$

and the measured coefficients

$$F_m = (\frac{\cos(pa - NTS)}{\alpha_m})^2 + (\frac{\sin(pa - NTS)}{\beta_m})^2$$

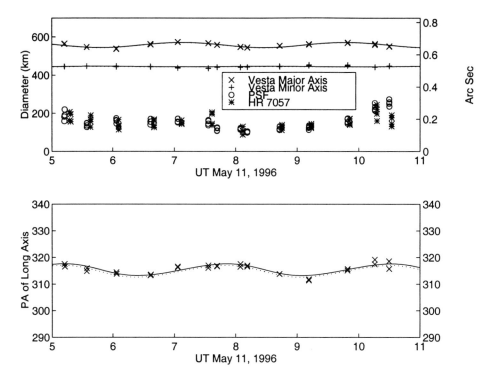

Figure 4. A fit of α and β (top), and pa (bottom) to Vesta images. Also illustrated in the top panel are the sizes of the PSF as obtained simultaneously with sizes of the asteroid as explained in Sec. 3.2, and of the star HR7057 obtained independently between Vesta observations. Although barely visible because of the low solar phase angle of 5.2°, the terminal ellipse parameters are indicated by dotted lines.

$$G_m = \left(\frac{\sin(pa - NTS)}{\alpha_m}\right)^2 + \left(\frac{\cos(pa - NTS)}{\beta_m}\right)^2$$

$$H_m = \sin(2(pa - NTS))\left(\frac{1}{\alpha_m^2} - \frac{1}{\beta_m^2}\right). \tag{27}$$

Notice that Eq. 27 involves measured (α_m, β_m, and pa) or known (NTS) quantities, while Eq. 26 contains quantities that are calculated from the six unknowns. Figure 4, taken from Drummond *et al.*, 1998, illustrates the fit to data obtained from AO images of Vesta obtained at the SOR.

4.2. DERIVATIVES FOR LEAST SQUARES FITTING

It is necessary to take the partial derivatives of each observable with respect to each of the unknowns, a sometimes complicated, but not impossible, project. For example,

$$\frac{\partial \mathcal{F}\{A_E E\}}{\partial a} = \frac{-A_E \pi \alpha \beta 2 J_2(\pi r'_E) \pi}{\pi r'_E} \frac{\partial r'_E}{\partial \alpha} \frac{\partial \alpha}{\partial a} \quad (28)$$

where

$$\frac{\partial r'_E}{\partial \alpha} = \frac{1}{2 r'_E}[\cos^2 \phi (x-x_0)^2 + \sin^2 \phi (y-y_0)^2 + \sin 2\phi (x-x_0)(y-y_0)]\frac{2\alpha}{S/2}$$

and

$$\frac{\partial \alpha}{\partial a} = \frac{1}{4\alpha}\{\frac{\partial B}{\partial a} + \frac{1}{2\sqrt{B^2 - 4A^2/C}}[2B\frac{\partial B}{\partial a} - 4(\frac{2A}{C}\frac{\partial A}{\partial a} - \frac{A^2}{C^2}\frac{\partial C}{\partial a})]\}$$

and continuing with the partials of A, B, and C with respect to a.

These partials must be performed for both the projected ellipse and the terminator ellipse, and combined in a way that reflects the relative contribution of α and α_t to the mean ellipse. If, as in Sec 3.1, we fit for a mean ellipse that results in $\overline{\alpha}$, we might construct the weights for the derivatives as in the following:

$$\frac{\partial \overline{\alpha}}{\partial a} = \frac{(\overline{\alpha} - \alpha_t)}{(\alpha - \alpha_t)}\frac{\partial \alpha}{\partial a} + \frac{(\alpha - \overline{\alpha})}{(\alpha - \alpha_t)}\frac{\partial \alpha_t}{\partial a}. \quad (29)$$

The partials of β and $\gamma + \rho$ with respect to a must be similarly calculated, and likewise the partials of α, β, and $\gamma + \rho$ with respect to the other five unknowns. For each observation, then, 36 sets of derivatives must be calculated, two sets (projected and terminator ellipses) of derivatives of the three observables with respect to six parameters each. Taking advantage of the many symmetries in the formulation leads to some savings in programming and computer time. For example, Eq. 5 shows that once A, B, C, and D (Eq. 6) are calculated, α and β only differ by a sign, and the code for $\partial r'_E/\partial \beta$ can be extracted from $\partial r'_E/\partial \alpha$ merely by interchanging $\cos^2 \phi$ and $\sin^2 \phi$ and changing the sign of $\sin 2\phi$ in Eq. 28. Changing one sign in $\partial \alpha/\partial a$ leads to $\partial \beta/\partial a$.

4.3. FINAL CALCULATIONS

Once the six unknowns are found, one of the angles is used to find the time when the maximum cross sectional area occurs: $\psi_E = \psi_{rel} - \psi_0 = 0$. The other two angles, θ_E and ρ, together with the position of the asteroid, $[RA_a; \delta_a]$, are converted to the rotational pole coordinates $[RA_p; \delta_p]$ by

$$\sin \delta_p = -\sin(NTS + \rho) \cos \theta_E \cos \delta_a - \sin \delta_a \sin \theta$$

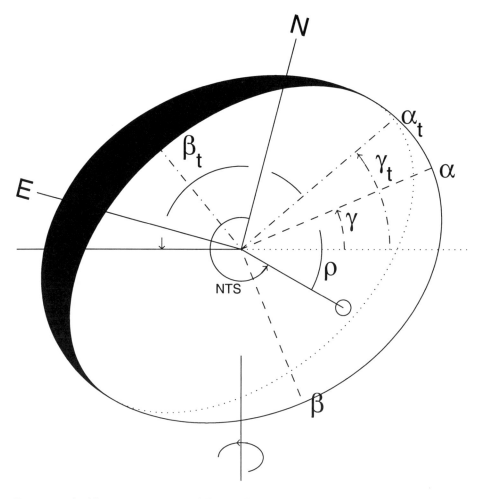

Figure 5. Ambiguous orientation of Fig. 1 ellipsoid. The sub-Earth point, the projected ellipse values, ω, and NTS are the same as Fig. 1, but the obliquity, ρ, differs from Fig. 1 by 180°. The sub-Sun point is at [335.5; +45.5] and the terminator ellipse parameters are $\alpha_t = 4.40$, $\beta_t = 2.78$, $\gamma_t = 39.8°$, and $\nu_t = 8.8°$.

$$RA_p = RA_a + \arccos(\frac{-\sin\theta_E - \sin\delta_a \sin\delta_p}{\cos\delta_a \cos\delta_a}). \qquad (30)$$

If $\cos(NTS + \rho) < 0$, then subtract rather than add the arccos to RA_A.

There is normally a two-fold ambiguity in the determination of the direction of the pole that arises from the fact that the ellipsoid's north pole shown in Fig. 1 could have been pointing down and the node to the left, as shown in Fig. 5, and yet the projected ellipses would appear exactly the same throughout the rotation. It should be emphasized that the difference between Figs. 1 and 5 is *not* that the south pole is seen in one and the north pole in the other, but in fact, the north pole, as defined by the right hand rule, is visible in both figures and is tipped toward the viewer by the same amount. The discrepancy between the two figures is caused by the difference of 180° in ρ. Although the sub-Earth quantities and projected ellipses would be the same in the two orientations, the sub-Sun quantities and the terminator ellipses would differ. However, at low solar phase angles, when the terminator is near the edge of the disk, this difference is difficult to detect. Therefore, two similar least squares solutions will be found with the two ρ's differing by $\approx 180°$. If $\omega = 0$, then the two solutions will be identical. At low solar phase angles, even if there is a slight difference in the final standard error of fit to the observations for the two cases of ρ, the solutions should be considered equally likely. Only at moderate and high phase angles can the difference in terminators lead to a meaningful distinction between solutions, especially if one orientation results in the Earth and Sun straddling the asteroid's equator and the other does not. If features can be followed across the disk, then the ambiguity can also be broken.

Other pole finding techniques also lead to ambiguities. Poles that result from following the changing brightnesses and amplitudes of lightcurves usually result in a four-fold ambiguity. Poles that result from timing lightcurve features usually have either a two- or four-fold ambiguity. However, the two-fold ambiguity that results from our work here, where we measure the changing size, shape, and orientation of a resolved but featureless ellipsoid, is a completely different ambiguity. Our ambiguity is with respect to the current position of the asteroid, while the others are with respect to the asteroid's orbital plane. If we were to make another set of observations when the asteroid is in a different part of the sky, then we should be able to break our ambiguity. Alternatively, comparing our solutions from one set of observations to the solutions from lightcurve photometry or from lightcurve timings is usually sufficent to break the ambiguity. See Magnusson *et al..* (1989) for a further discussion of rotational pole ambiguities.

5. Summary

To the extent that an asteroid can be modelled as a smooth, featureless, triaxial ellipsoid, rotating about its axis of maximum moment of inertia, throughout its rotation it will project an image of an ellipse. Except for the changing sub-longitude, the geometry for main belt asteroids does not change over a few rotations. Therefore, the three axes dimensions and the three angles that describe the direction of the ellipsoid's rotational pole can be obtained by following the changing ellipse parameters throughout the rotation. By measuring the major and minor semiaxes, and the position angle of each ellipse at each rotational phase, and comparing them to the calculated projected ellipse parameters, a non-least squares solution can be made for the six unknowns. The three equations of observation for each rotational phase are given by Eqs. 5 and 7. Each equation has to be calculated twice, one for $\omega = 0$ and one for $\omega \neq 0$, and combined using a weighting scheme such as Eq. 29, for constructing residuals with the observed ellipse parameters.

If the asteroid can be observed from space, and its outline measured directly, then least squares comparison of measured to calculated parameters can be performed immediately. However, even in this case, finding the ellipse parameters from the step function image ellipse is not straight forward since step functions are not amenable to least squares fitting. Instead, ellipse parameters can be obtained by fitting the Fourier transform of the ellipse, Eq. 22, and then compared to calculated parameters. From the ground, with adaptive optics, for example, it will also be necessary to account for the point spread function, either through deconvolution – a terrible thing to do because of small number divides at high spatial frequencies – or by simultaneously fitting for both the ellipse and the PSF in the Fourier domain, provided that the PSF can be described by an analytic function. For the SOR in its present configuration, the PSF appears to be Lorentzian in shape, but other possible functions which might describe PSF's are Gaussians or Airy patterns, both of which can be expressed analytically in the image and Fourier domains.

The three observed ellipse parameters, then, should be obtained by fitting elliptical distributions (Eq. 20) in the Fourier domain. At every rotational phase, the calculated ellipse parameters are found by making initial guesses for $[a, b, c, \theta_E, \psi_0, \rho]$, and the resulting $[\alpha, \beta, \gamma]$ compared to the observed parameters. The guesses are iterated until convergence is reached – the minimum in the sum of the square of the residuals. Although there are six unknowns, and each observation provides three measurements, certainly many more than the mathematically minimum of two observations should be made for meaningful results to be obtained. Ideally, half a dozen or more observations should be obtained equally spaced over the rotation, even if spread over three or four nights. Once the least squares solution for the six unknowns is found, Eq. 30 converts the angles to rotational pole coordinates.

Acknowlegements

I have been fortunate to have been associated with many individuals who have helped me understand asteroids and the mathematics of it all, or have allowed me to work on their high resolution projects. I would like to acknowlege Ed Tedesco, John Cocke, Keith Hege, Julian Christou, Bob Fugate, and *in memoriam*, Charles Higgins (1947-1997), who have all played key roles at various stages in my career. Laurie Wells and Julian Christou made conscientious reviews of this manuscript before submission.

References

1. Davies, M.E., T.R. Colvin, M.J.S. Belton, J. Veverka, and P.C. Thomas 1997. The direction of the North Pole and the control network of asteroid 951 Gaspra. *Icarus*, **107**, 18-22.
2. Drummond, J.D. 1998. The adaptive optics Lorentzian point spread function *SPIE* **3353** 1030-1037.
3. Drummond, J.D., R.Q. Fugate, and J.C. Christou 1998. Full Adaptive Optics Images of Asteroids Ceres and Vesta; Rotational Poles and Triaxial Ellipsoid Dimensions. *Icarus* **132**, 80-99.
4. Drummond, J.D. and E.K. Hege 1986. Speckle interferometry of asteroids. III. 511 Davida and its photometry. *Icarus* **67**, 251-263.
5. Drummond, J.D. and E.K. Hege 1989. Speckle interferometry of asteroids. In *Asteroids II* (R.P Binzel, T. Gehrels, and M.S Matthews, Eds.), pp 171-191. Univ of Arizona Press, Tucson, AZ.
6. Drummond, J.D., W.J. Cocke, E.K. Hege, P.A. Strittmatter, and J.V. Lambert 1985a. Speckle interferometry of asteroids. I. 433 Eros. *Icarus* **61**, 132-151.
7. Drummond, J.D., E.K. Hege, W.J. Cocke, J.D. Freeman, J.C. Christou, and R.P. Binzel 1985b. Speckle interferometry of asteroids. II. 532 Herculina. *Icarus* **61**, 232-240.
8. Drummond, J.D., A. Eckart, and E.K. Hege 1988a. Speckle Interferometry of Asteroids. IV. Reconstructed images of 4 Vesta. *Icarus* **73**, 1-14.
9. Drummond, J.D., S.J. Weidenschilling, C.R. Chapman, and D.R. Davis 1988b. Photometric geodesy of main-belt asteroids. II. Analysis of lightcurves for poles, periods, and shapes. *Icarus* **76**, 19-77.
10. Goldstein, H. 1950. *Classical Mechanics*. Addison-Wesley, Reading, Mass.
11. Magnusson, P. 1986. Distribution of spin axes and senses of rotation for 20 large asteroids. *Icarus* **68**, 1-39.
12. Magnusson, P., M.A. Barucci, J.D. Drummond, K. Lumme, S.J. Ostro, J. Surdej, R.C. Taylor, and V. Zappala 1989. Determinatons of pole orientations and shapes of asteroids. In *Asteroids II* (R.P Binzel, T. Gehrels, and M.S Matthews, Eds.), pp 66-97. Univ of Arizona Press, Tucson, AZ.
13. McCarthy, D.W., J.D. Freeman, and J.D. Drummond 1994. High resolution images of Vesta at 1.65μm. *Icarus* **108**, 285-297.
14. Sekanina, Z. 1988. Nucleus of comet Iras-Araki-Alcock 1983VII. *Astron. J.* **95**, 1876-1894.
15. Taylor, R.C. and E.F. Tedesco 1983. Pole orientation of asteroid 44 Nysa via photometric astrometry, including a discussion of the method's application and its limitations. *Icarus* **54**, 13-22.
16. Thomas, P.C., G.J. Black, and P.D. Nicholson 1995. Hyperion: Rotation, shape, and geology from Voyager images. *Icarus*, **117**, 128-148.
17. Thomas, P.C., R.P. Binzel, M.J. Gaffey, B.J. Zellner, and A.D. Storrs 1997. Vesta: Spin pole, size, and shape from HST images. *Icarus*, **128**, 88-94.

V. Parfenov (left), T. Roberts & Claudia Rola

Week-end relaxation

CHAPTER 13
THE ENVIRONMENTS OF YOUNG STARS

T.P. RAY

Dublin Institute for Advanced Studies
5 Merrion Square, Dublin 2, Ireland

1. Introduction

In writing this set of lectures I am conscious of the fact that many participants to this NATO school will not be aware of the detailed processes underlying star formation. I therefore intend to spend at least half of my time giving background information which I feel is vital if one is to appreciate the important role that adaptive optics (AO) can play in understanding not only how stars form but also the way in which young stars interact with their immediate environment. Moreover as the theme of the school is AO, and because AO techniques are only really useful in the optical and near-infrared, I will take a somewhat jaundiced view of my subject by concentrating on what the optical/near-infrared bands have to tell us.

2. The Early Years

As you know, stars begin their lives in the dark dusty environments of molecular clouds. There are two main types of molecular clouds that need concern us: giant molecular clouds (or GMCs), such as the Orion Complex and small molecular clouds (or SMCs) such as Taurus-Auriga or Ophiuchus (see, for example, Blitz 1993 and Lada, Strom & Myers 1993). Typical masses of GMCs are around 10^5 to 10^6 $M_\odot$ and their average diameters are 50–100 pc. SMCs, on the other hand, are less massive by 1–3 orders of magnitude, and are about 10–20 pc in size. The division of molecular clouds into SMCs and GMCs is somewhat arbitrary, as there appears to be a continuous spectrum of cloud sizes and masses (Blitz 1993). At the same time the split is useful since both groups tend to have different properties. For example, GMCs are largely confined to the arms of our galaxy whereas SMCs are interspersed throughout the disk of the Milky Way. Moreover, groups of massive young stars (OB associations) are produced only in GMCs.

It is not known how molecular clouds form from the interstellar medium (ISM) although various theories exist: suggestions include spiral density waves, instabilities, and compression of the ISM by multiple supernovae (see, for example, Shu, Adams & Lizano 1987 and McCray & Kafatos 1987). While it is likely that spiral density waves play a major role in triggering their formation, other processes must also be as work. This is evident from the fact that even in galaxies without spiral arms, stars can be made very efficiently! To give a feel for the numbers involved in our own galaxy, the average rate at which gas is consumed in star formation is about $3\,M_\odot$ per year. A large fraction of this ends up as low mass stars (with $M_\star \lesssim 1\,M_\odot$) and most of the star formation seems to be going on in GMCs.

Observations at millimeter wavelengths, for example using CO or NH_3 transitions, show that molecular clouds are clumpy. In particular they contain high density cores, ranging in mass from 1 to $10^3\,M_\odot$, which appear to be the actual sites where stars form (Blitz 1993). Soon after their discovery, it was realized that these cores cannot be supported by thermal pressure alone. This is illustrated, for example, by the fact that their CO line widths are always broader than expected on the basis of their temperature. There must be additional forces present, such as those due to magnetic fields and/or turbulence to broaden the lines. One, perhaps major, source of turbulence may well be the outflows from young stellar objects (or YSOs) which we will discuss shortly (see also Margulis, Lada & Snell 1988).

While magnetic fields can help support cores against gravitational collapse they can only do so up to a critical mass for a given amount of magnetic flux (see, for example Mestel 1985). If insufficient magnetic flux is present, the core is super-critical and gravity will ensure that it collapses rapidly. Observations would suggest that, in the solar neighbourhood at least, massive cores on scales of a few parsecs should be super-critical and thus obvious potential sites for star formation (Bally, Morse & Reipurth 1996). Moreover because of fragmentation, such cores are expected to form dense clusters of stars of varying mass and, in fact, such a "cluster mode" of star formation is probably the dominant one in our galaxy (Lada, Strom & Myers 1993). An example of a young cluster that no doubt formed from a massive core is the one surrounding the Trapezium stars in Orion (Fig. 1). The density of YSOs at its core approaches an amazing 50,000 stars per cubic parsec! These clusters, however, are gravitationally unbound and survive only for a typical stellar crossing time which, in the case of the Trapezium Cluster, is about a million years. Their "evaporation" is due to the fact that, while their stellar velocities are close to what was the initial escape velocity of the cloud, a significant portion of the cloud's binding

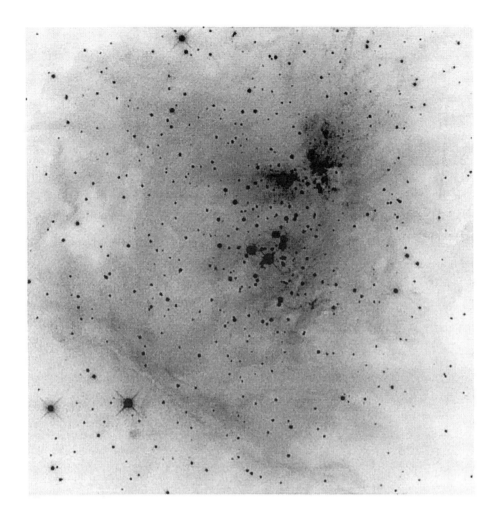

Figure 1. The Trapezium Cluster in K as imaged using MAGIC on the Calar Alto 3.5m Telescope from McCaughrean et al. (1996). North is up in this 5'x5' image and the bright Trapezium stars can easily be seen towards the centre.

mass has been removed through the processes that accompany the birth of stars, e.g. outflows and the formation of HII regions. That said, the cluster mode of star formation is still thought to be very efficient in the sense that a large fraction of the original core's mass would appear to have ended up in the form of stars.

Even if a core is magnetically sub-critical, it may still eventually collapse, i.e. the collapse may only be delayed. This is because the magnetic field acts directly only on the ions in the core and the ion fraction is quite small ($\approx 10^{-6}$). While collisional coupling between the neutrals and the ions ensures that magnetic forces are felt indirectly by the neutrals, in time the

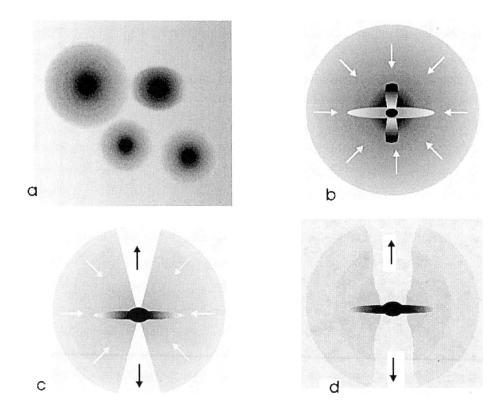

Figure 2. The various stages of star formation following the model of Shu, Adams & Lizano (1987). Sub-condensations (i.e. starless cores) begin to form inside a molecular cloud (a) which then subsequently collapse from the inside-out to produce a protostar and accretion disk. It is at this stage that a bipolar outflow emerges from the vicinity of the star/disk interface for the first time (b). The envelope is gradually depleted as the flow advances into the surrounding medium (c). Finally most of the envelope is eventually either accreted or blown away (d) to reveal a classical T Tauri or Herbig Ae/Be star.

neutrals drift with respect to the ions through the process known as ambipolar diffusion (see Mestel 1985). Sub-critical collapse may be important in low density star forming regions such as the Taurus-Auriga cloud where the star formation efficiency is quite low (Shu, Adams & Lizano 1987).

The basic stages leading to the birth of a star are thought to be well understood even though the detailed processes are certainly not! Fragments of the cloud cores are expected to gravitationally collapse from inside out (Shu, Adams & Lizano 1987; Shu et al. 1993) giving rise to an inner region containing a proto-star plus an accretion disk (Fig. 2). The outer envelope of gas mixed with dust continues to shower down material both on the star and its disk. In this way the accretion phase for a YSO can last a relatively long time, approximately a million years for a solar mass star, even though the

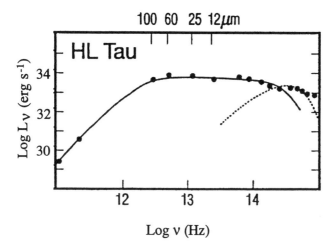

Figure 3. The spectral energy distribution of the classical T Tauri star HL Tau, a known jet source (Mundt et al. 1990). While the optical emission can be well fitted by a blackbody-like spectrum, the excess at infrared/mm and uv wavelengths cannot. These excesses are most readily explained as arising from accretion through a disk. The processes of disk accretion and the generation of jets/outflows seem to be intimately linked (see Edwards, Ray & Mundt 1993). Adapted from Sargent (1996).

initial collapse to form a protostar occurs quickly. During this early stage the embedded object is not optically visible and can only be studied using infra-red, sub-millimeter, and millimeter techniques (see Fig. 2). What came as a total surprise to theorists and observationalists alike was however the discovery that the accretion phase is simultaneously accompanied by the outflow of matter as well. Such outflows manifest themselves in a number of ways: they are seen not only at millimeter wavelengths (Padman, Bence & Richer 1997) but also in the infra-red (Eislöffel 1997) and optical regimes (see Edwards, Ray & Mundt 1993 or Ray 1996). While there has been continuous debate in the literature regarding the source of these outflows, the current consensus is that they arise from the accretion disc and not from the star itself (see Ray 1996).

It is ironic that the detection of outflows from young stars, while surprising, has proved to be a much easier task than finding evidence for the infall that must accompany star formation. Only in recent years has the picture begun to change with the discovery of inverse P-Cygni profiles for a number of molecules, e.g. HCO^+ (Ward-Thompson et al. 1996) in YSO environments. The observed velocities are, however, very low as one might expect given the location of the emission regions.

In any event, at some stage the supply of material from the surrounding envelope will cease, either because the envelope is wholly depleted or because it is dispersed due to the effects of an outflow. It is around this time that the star becomes optically visible for the first time. Depending on its mass we see either a T Tauri or a Herbig Ae/Be star. All of these various stages are illustrated in Fig. 2. As one might expect the spectral energy distributions (SED) of these different phases are dramatically different. In the early stages only the warm dust that surrounds the proto-star is detected. The spectrum of such a YSO peaks in the sub-millimeter range; and the YSO is known as a Class 0 source (Bontemps, André & Tereby 1996). Gradually, over a period of several hundred thousand years for a solar mass star, the veil of dust begins to lift, and ones sees an SED that is a combination of the heavily extincted stellar photospheric spectrum plus that of an accretion disk. An example, that of HL Tau, is illustrated in Fig. 3. We shall return to the subject of disks later but it is interesting that until recently there was essentially no direct evidence for their existence. Instead indirect methods were used to infer their presence such as the form of the SED of a young star. The Hubble Space Telescope has changed all this with the discovery of the proplyds in Orion and, for example, the imaging of disks associated with outflow sources such as HH 30 (see §5).

It is perhaps worth reminding the reader at this point that all YSOs, derive virtually all of their energy from gravitational contraction. Normal fusion does not commence until the star has joined the main sequence, although YSOs do burn a certain amount of deuterium especially in the very early phase (Stahler & Walter 1993). The contraction time to the main sequence is given essentially by the Kelvin-Helmholtz time-scale:

$$t_{KH} = GM_\star^2/R_\star L_\star$$

and if we substitute solar values into the above equation, we obtain a contraction time of about 3×10^7 years. Given that the main sequence lifetime of the Sun is about 10 Gigayears, we would thus expect, and of course find, that main sequence stars are correspondingly rare.

3. The T Tauri and Herbig Ae/Be stars

The T Tauri stars were the first pre-main sequence stars to be identified. Joy (1945) noted their strong Hα emission (equivalent width of their Hα line in the range 10-200 Å) while carrying out an objective prism survey of the Taurus region. Later Ambartsumian (1947) realized that they are often found in association with groups of OB stars and correctly surmised that, like the OB stars, T Tauri stars are young. Their masses, deduced from a comparison of their actual position on the HR diagram with their theoretical tracks, are typically about a solar mass or less. In a number of cases these estimates were confirmed by surface gravity measurements (e.g. McNamara 1976).

We have already alluded to some of the characteristics of T Tauri stars, in particular their infra-red excesses. This excess is attributed to the presence of a disk and, more usually, a combination of a disk plus a dusty halo. T Tauri stars, however, also have excess continuum emission in the ultraviolet and their spectra are crossed not only by a number of strong permitted emission lines, e.g. Balmer and Ca II lines, but in many cases strong forbidden lines, e.g. [SII]$\lambda\lambda$6716,6731, [OI]λ6300. Moreover where absorption lines are present they are often partially filled-in, a phenomenon known as veiling. All of these features, at least in a qualitative way, can be understood in terms of the disk paradigm (e.g. Bertout 1989) although it is not possible to go into details here.

So far we have only been describing what are normally referred to as classical T Tauri stars. There is another group of T Tauri stars with rather different properties; these are the so-called weak-line T Tauri stars which, as their name suggests, show only feeble emission lines. They were first identified through their X-ray emission and found to lie in approximately the same locations in the sky as the classical T Tauri stars (Feigelson et al. 1993). Unlike the classical T Tauri stars, however, their spectral energy distributions show little or no evidence for the presence of a disk. They occupy similar positions in the HR diagram as, and are probably coeval with, the T Tauri stars, but they seem to rotate faster (Edwards et al. 1993), a fact that has been attributed to the lack of strong MHD breaking which a disk can provide. Unlike the case of classical T Tauri stars, weak-line T Tauri stars have emission line fluxes which are consistent with active stellar chromospheres. Moreover their X-ray emission is probably best explained by enhanced magnetic activity (see Feigelson et al. 1993 and references therein). In any event, because weak-line T Tauri stars are effectively devoid of any circumstellar matter, they are of very little interest from the perspective of AO-type studies. The only possible exception to this is in the field of binary/multiple system statistics. We will therefore say no more about them!

The Herbig Ae/Be stars are thought to be the intermediate mass counterparts of the T Tauri stars (Corcoran & Ray 1997a). Like the T Tauri stars, they are found in regions of active star formation and show Hα in emission. There are many other resemblances, for example the presence of infrared/mm excesses and outflows in a number of cases. Despite such similarities, there has nevertheless been considerable debate in the literature as to what fraction of these stars are surrounded by disks. The consensus seems to be emerging that disks are common, at least amongst the Herbig Ae/Be stars with strong forbidden line emission (Corcoran & Ray 1998).

When the forbidden line emission of both classical T Tauri and Herbig Ae/Be stars is examined using intermediate dispersion spectroscopy, it is

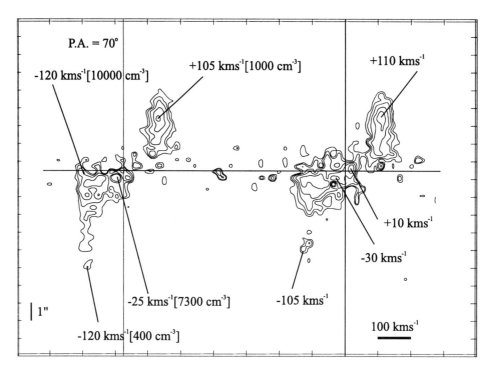

Figure 4. Position-velocity diagram, based on [SII]$\lambda\lambda$6716,6731 emission, for the Herbig Ae/Be star LkHα233. Both a high and a low velocity blueshifted component can be traced right back to the source whereas the corresponding redshifted flow is seen only at distances more than 1″ from the YSO. This "gap" in the redshifted emission is attributed to the occulting effects of the LkHα233 disk (see text). Note that the high velocity component is more extended than the low velocity component and that it has a lower electron density far away from the source. In this figure the continuum emission from the star has been subtracted, although its location is marked by the central horizontal line. From Corcoran & Ray (1997b).

often found that there are two or more velocity components present (Hirth, Mundt & Solf 1997; Corcoran & Ray 1997a). The high velocity component typically has a velocity of a few hundred kms^{-1} and is normally seen to be extended using long-slit spectroscopy (see Fig. 4). In contrast the low velocity component, with velocities closer to that of the systemic velocity of the star, is usually associated with a denser region and is quite compact. Kwan & Tademaru (1995) have proposed that the low velocity component is a disk wind perhaps centrifugally flung from it at distances of a few AU from the star. The high velocity component, in contrast, they suggest originates much closer to the star and in fact represents the core of a jet that in some cases (see below) extends outwards for many parsecs. Before going on to discuss outflows, it is worth noting one other remarkable observational property of the forbidden lines and that is the clear blue asymmetry in velocity that is normally seen for both the high and low velocity emission (see Fig. 4). The obvious explanation for this asymmetry, as realized many years ago by Appenzeller, Jankovics & Östreicher (1984), is that a disk is present that obscures our view of the redshifted flow but permits us to see those portions of the flow that point towards us. Of course once we look beyond the disk, we can see the redshifted flow once again (see Fig. 4). The spatial extend of the asymmetry, seen for example in Fig. 4, suggest disks with diameters of tens of AU. It is worth pointing out that although the low velocity forbidden emission line regions can just about be spatially resolved from the ground under good seeing conditions, the scales are such ($\lesssim 1''$) that their study is an ideal area for exploiting AO techniques.

Recently it has been demonstrated that the strength of the forbidden line emission, in both T Tauri and Herbig Ae/Be stars, scales with the size of the infra-red excess (see, for example, Edwards, Ray & Mundt 1993 and Corcoran & Ray 1998). This suggests a link between the outflow component, represented by the forbidden line emission and accretion, assuming the excesses in large part derives from the latter. Typical estimates of accretion rates ($M_{acc} \approx 10^{-7}$–$10^{-8} M_\odot$/yr) are around 10 times larger than those for mass loss (Hartigan, Edwards & Ghandour 1995).

4. HH Flows from Young Stars

The realization that YSOs have strong outflows can be traced back to the discovery by George Herbig and Guillermo Haro (Herbig 1951; Haro 1952) of faint emission line nebula, or Herbig-Haro (HH) objects as they became to be known, in a number of star forming regions. Many years were to pass, however, before Schwartz (1975) noticed the similarity of their spectra to those of supernova remnants (SNRs) leading him to propose that, as in the case of SNRs, their emission arose from radiative shocks. At that time there was already very strong evidence for mass loss from YSOs (e.g., Kuhi, 1964) prompting Schwartz (1978) to suggest that HH objects might arise as a result of shocks caused by "blockages" in a stellar flow. The "blockages" he proposed were caused by cloudlets that got in the way of the supersonic wind from the young star.

The finding that many HH objects either mark the region where a jet "terminates" (the reason from the quotation marks will be made clear later), or are actual parts of a jet, came with the discovery of YSO jets in the early 1980s. Prior to that a number of observations were made of YSO jets although they were not identified as such (e.g. Dopita, Schwartz & Evans 1982). The first HH flows to be recognised as jets were from HL Tau, HH 30 and DG Tau B (Mundt & Fried 1983). Many more were then found (Mundt et al. 1984; Strom et al. 1985; Reipurth et al. 1986; Ray 1987). Today a large number of HH flows from young stars are known

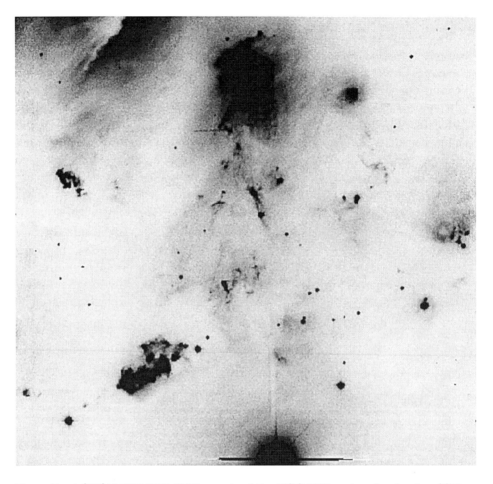

Figure 5. A [SII]λλ6716,6731 CCD mosaic of the NGC 1333 region showing in addition to the well known outflow HH 7–11 (to the centre and pointing towards the bottom left) a large number of other flows in its vicinity. This tendency for flows to occur in groups obviously follows from the fact that the clustering is common in star formation. Image courtesy of Jochen Eislöffel.

(Reipurth 1994) and although some just consist of a few disparate knots or bow shocks, others are made up of long linear chains of knots/jets.

As pointed out by Bally, Morse & Reipurth (1996), because star formation is clustered, HH flows tend to occur in groups as well. An example, the NGC 1333 region, is illustrated in Fig. 5 (see also Bally, Devine & Reipurth 1996). While in some cases there appears to be a tendency for groups of flows to be aligned (e.g. the HL Tau region, Mundt et al. 1990) this is not always the case.

To illustrate the properties of Herbig-Haro outflows, we will use as an example HH 34. While HH 34 was known for many years, and is listed in the early catalogue of Herbig (1974), it only attracted attention with the

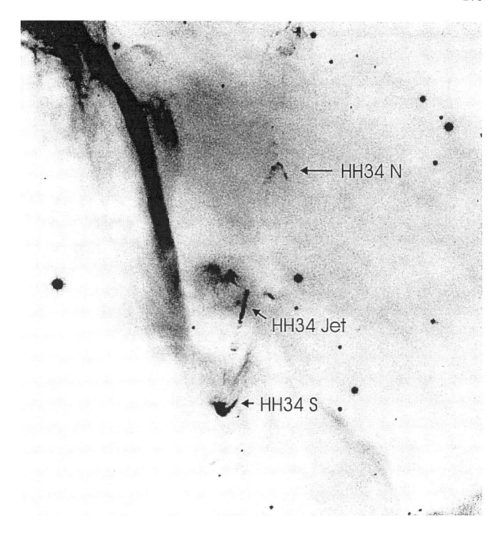

Figure 6. The centre of the HH 34 flow as seen through a [SII]$\lambda\lambda$6716,6731 filter. The HH 34 jet emanates from an embedded infrared source, HH 34IRS, at its northern apex and points towards HH 34 S. Both bow shocks are approximately 100″ away from the infrared source. The HH 34 flow is itself much more extensive (Bally & Devine 1994; Eislöffel & Mundt 1997) as already hinted at by this frame since additional parts of the flow can be seen beyond HH 34 N and S. This image, courtesy of J. Eislöffel and R. Mundt, was taken with the 3.5-m telescope on Calar Alto, Spain. North is to the top and east is to the left.

discovery of a jet pointing towards it (Mundt 1986; Reipurth et al. 1986; Bührke, Mundt & Ray 1988). It was then realized that HH 34 is in fact the bow shock of the jet as it ploughs its way though its surroundings (see Fig. 6). A counter bow shock was discovered to the northeast of the HH 34 jet (Bührke, Mundt & Ray 1988) at roughly the same distance as the HH 34 bow shock is from the jet source. Rather curiously, however, no counter-jet has ever been seen. In the past few years it has been shown that the HH 34 flow is much more extensive than previously thought, stretching almost 0.5° in projection on the sky (see Fig. 6, Bally & Devine 1994 and Eislöffel & Mundt 1997). Spectroscopic measurements of the different parts of this flow have shown that its southern part, including the jet, are blueshifted while the northern section is redshifted. The typical radial velocity of the jet is around -100 kms^{-1} (Bührke, Mundt & Ray 1988) and similar radial velocities are found for its associated bow shock, (sometimes referred to as HH 34 S) suggesting that the jet is meeting little resistance from the material ahead of it. Of course these are only radial velocities; to get the full kinematic picture so-to-speak one has to combine these velocities with the tangential velocities of the various outflow features deduced from proper motion studies. The typical shifts are small (of order a tenth of an arcsecond per year) but readily measurable, especially using AO techniques. In the case of the HH 34 flow, studies have shown that both the HH 34 N and S bow shocks (see Fig. 6 for the nomenclature) are moving at about 330 kms^{-1} along an axis which makes an angle of approximately 60–70° to our line of sight (Heathcote & Reipurth 1992). If this was the velocity of the shocks at the apexes of these bows, we would expect much higher excitation lines than is observed. The only way of avoiding this problem, and reducing the shock velocity, is to assume that the bows are not in fact moving into a stationary ambient medium but instead into material that is already moving outwards away from the source. That is to say that the bows are *internal shocks* in the flow. Such a finding is consistent with the discovery, mentioned above, that outflows like HH 34 are much extended than previously thought (Bally & Devine 1994; Eislöffel & Mundt 1997) and that they can persist for 10^5 years or so in the case of low mass stars.

The HH 34 jet illustrates a common feature of HH jets in that they often consist of a series of quasi-periodically spaced knots (see, for example Ray et al. 1996). A number of ideas have been put forward to explain the origin of such knots including the non-linear growth of hydrodynamic instabilities (Ferrari et al. 1996) and that they are internal "working surfaces" (Raga & Kofman 1992) caused by faster jet material catching up with previously ejected matter. As will be shown in §6, when we consider high resolution observations of HH jets, the data, particularly from the HST, favors the internal working surface hypothesis. While on the subject of jet morphology, it is worth mentioning that a number of HH jets/flows are seen to oscillate in the transverse direction to the flow, i.e. "wiggle", like HH 30 (e.g., Lopéz et al. 1995; Mundt et al. 1990) and RNO 15 (Davis et al. 1997a) although the source of these wiggles is much less certain than that of the knots. Instabilities once again might be an explanation, although precession of the flow, perhaps caused by the effects of a companion star on the source's disk,

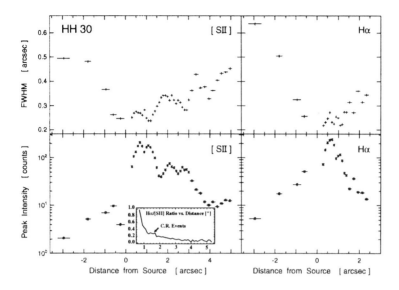

Figure 7. This figure (from Ray et al. 1996) illustrates how the (Full Width Half Maximum) diameter (top) and intensity (bottom) of the HH 30 bipolar jet changes with distance from the source. The data are derived from Wide Field Planetary Camera 2 HST images taken through red [SII] doublet and Hα line filters. The intensity variations (in the positive direction) correspond to the first few knots in the jet.

is perhaps more likely. In principle both ideas are testable by measuring how the wiggles evolve with time; again AO may be able to help here to distinguish between these possibilities.

A rather serendipitous feature of YSO jets, that makes them ideal for studying using AO techniques, is that their diameters are typically around $0''.5-1''.0$ in the case of the nearest flows. It follows that if we want to investigate sub-structure within jets, for example to observe the detailed morphology of their constituent knots, we have to resort to using either the Hubble Space Telescope (HST) or high resolution techniques from the ground. Early observations (e.g. Mundt, Ray & Raga 1991) carried out in conditions of very good seeing, showed a somewhat surprising variation in the diameter of some jets with distance from their source. What was found is that the jet diameter may decrease more slowly than r^{-1} with distance from the source. It follows that the effective opening angle, defined by the diameter (Full Width Half Maximum) of the jet divided by the distance to the source, *increases*. This behaviour was subsequently confirmed by the HST (Ray et al. 1996). and is illustrated, for the case of HH 30 in Fig. 7. Here one sees that the initial jet opening angle in the first few arcseconds is much larger than observed further out (see Mundt, Ray & Raga 1991 for the opening angle of HH 30 further out).

Examining what happens to the jet very close to the YSO is, however, normally very difficult because of scattered light. Moreover there is often a reflection nebula present to complicate matters. The case of HH 30 is better than the norm: the flow is virtually in the plane of the sky and we see the disk of HH 30 almost edge-on (Fig. 8). The net effect is that the disk acts rather like an occulting bar.

One of the beautiful characteristics of HH jets, from an observationalist's point of view although not necessarily from that of a theorist's (!), is that, because they are emission line objects, most of their physically important parameters can be derived spectroscopically. Typical temperatures and electron densities are found to be 5×10^3–10^4 K and 10^2–10^4 cm^{-3} respectively. Moreover it would appear that only a small fraction of a typical HH jet is ionized: neutrals may constitute as much as 90% (e.g., Bacciotti, Chiuderi & Oliva 1995) or more of the flow. We mentioned already, in reference to HH 34, that the velocity of a jet can be derived by combining its spectroscopically determined radial velocity with its tangential velocity deduced from the proper motion of its knots. The observed velocities, as one might expect, depend to some degree on the luminosity of the source but are normally in the range 200-1000 kms^{-1}. Given that the temperature, and hence the sound speed, for these flows are known, we can therefore determine their Mach numbers. Typical values ($M_{jet} \approx 20$–100) imply these flows are hypersonic. Finally the mass fluxes are estimated to be 10^{-8}–$10^{-6} M_\odot$yr^{-1} again, to some degree, depending on the luminosity of the source Edwards, Ray & Mundt 1993).

Before considering the rôle AO might play in investigating HH flows, it is worth pointing out that such flows may be observed in the near-infrared given that, for some AO systems, this may be the better option. In addition to optical lines, HH objects can, if the shock velocity is not too high, excite vibrational transitions of molecular hydrogen, e.g. the 2.12μm line (e.g., Eislöffel 1997). The cooling times for the molecule are quite short (of order a few years) so metaphorically the H$_2$ emission traces, what one author has described as, the "sparks" of the flow as it ploughs it way through the ambient medium. The obvious benefit, from a star formation perspective, in working in the infrared is of course the ability to observe much more highly embedded, and presumably younger, flows. An example of a HH flow seen at 2.12μm, Lynds 1634, is shown in Fig. 8. In recent years a number of HH flows, including jets, have been discovered in the infrared with no optically visible counterparts (see, for example, Zinnecker, McCaughrean & Rayner 1996). The study of such flows, for example to determine proper motion, is an obvious area where AO can make an important contribution.

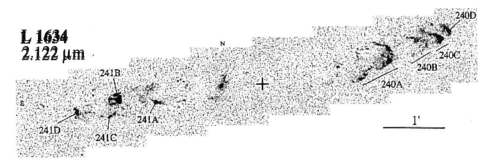

Figure 8. A H$_2$ 2.12μm image of the Lynds 1634 (HH 240 -241) outflow from Davis et al. (1997b) based on United Kingdom Infrared Telescope data. The highly embedded source of this flow (IRAS 05173-0555), which can only be seen in the far-infrared, is in the centre of the frame and its position is marked with a cross. Note the extensive bow shocks that are observed in both outflow directions. These features disappear in nearby infrared continuum images (see Davis et al. 1997b) implying that they are emission line objects. North is up and east is to the left.

5. Disks Around YSOs

We have already mentioned some of the indirect evidence for disks around young stars including infrared/mm excesses and obscuration of the redshifted forbidden line emission. Direct imaging of disks, however, has only been achieved very recently. In the past a number of flattened structures, observed using optically thick low rotational transitions of the CO molecule, were thought to be disks (e.g., Sargent & Beckwith 1991; Koerner, Sargent & Beckwith 1993) but their sizes ($\approx$ 1000 AU), and thus low rotational velocity, made it difficult to check whether they were in Keplerian rotation around a young star. A much more promising approach is to observe the optically thin millimeter dust emission around YSOs. Using millimeter wave interferometers (e.g. Lay et al. 1994) it has been shown directly that disks surround a number of T Tauri stars with diameters of under 100 AU. As I wish to concentrate on what can be seen optically and/or in the near-infrared, the reader is referred to the excellent review by Chandler (1998) for more details of such millimeter wave studies.

An area where I think AO can play a very important role is in studies of the so-called proplyds. We have already mentioned the cluster surrounding the Trapezium in Orion. This cluster consists of around 700 stars, a large number of which are contained within the HII region ionized, primarily, by the most powerful of the Trapezium stars, θ^1C Orionis. Recent HST observations (O'Dell & Wong 1996) have shown that at least 150 of the stars in the Trapezium Cluster are surrounded by extended structures with typical sizes of about 1''. Often they have cometary-like shapes with a tail pointing away from θ^1C Orionis and, in a number of cases, an associated ionization front on the side opposite the tail (see Fig. 9). As shown by McCaughrean & Stauffer (1994), YSOs are detected in the near-infrared at the centre of virtually all of these objects. Given the masses of these nebulae of around $0.1 M_\odot$, deduced on the basis of their present mass loss rates and the age of the Trapezium HII region, the material cannot be distributed spherically but must instead be in a flattened structure. If this were not the case it is readily shown that we would not see the YSO at their centre (see Koerner 1997 and references therein). The standard explanation for these cometary-like nebulae is that they are ionized proto-planetary disks or, as they have been nicknamed, "proplyds". Their title, however, may be something of a misnomer since these disks appear to be ablating so rapidly that there is probably insufficient time for large planetary bodies to form (e.g. Johnstone, Hollenbach & Bally 1998). In any event the shape of a proplyd seems to depend on its relative position with respect to the primary ionizing source. For example close to θ^1C Orionis, where there is a strong flux of Lyman continuum EUV photons, photo-evaporated gas from the disk appears to interact with the wind from this star and produce the bow shocks seen in Fig. 9. Far from the Trapezium, where the UV flux is not so strong, the disks themselves can sometimes been observed in silhouette against the backdrop of the Orion Nebula (see inset in Fig. 9).

The proplyds can be seen not only in the optical but in the near-infrared as well. In particular recent HST NICMOS observations (Chen et al. 1998) have shown that the surfaces of these disks emit copiously in the H_2 $2.12\mu m$ line. The $2.12\mu m$ line, and others, in turn are excited by the FUV photons from the Trapezium. Obviously given the advantages of AO systems in the infrared, and the very limited lifetime of NICMOS (!), further high spatial resolution studies could, and should, be conducted from the ground.

Since disks around young stars are usually optically thick, it follows that we can only detect them in the optical or infrared bands if something forces their surface to emit, such a nearby group of OB stars as in the case of the proplyds, or their surface scatters light. A very nice example of the latter is the disk around HH 30. We have already referred to this source in connection with its rather beautiful bipolar jet. Imaging with the HST (Ray et al. 1996; Burrows et al. 1996) shows that the jet emerges perpendicular to two oppositely curved cusp-like reflection nebula (see Fig. 10). Since the flow, as we have already mentioned, is virtually in the plane of the sky, its disk is almost edge-on. Note that in Fig. 10, the disk is the dark area *in-between* the two cusps. Again the scale of this disk, i.e. approximately $1''$, is ideal for studying using AO. Additional examples should exist, of comparable angular dimensions, since a number of other YSO outflow sources are known at similar distances. It should be emphasized, however, that objects like

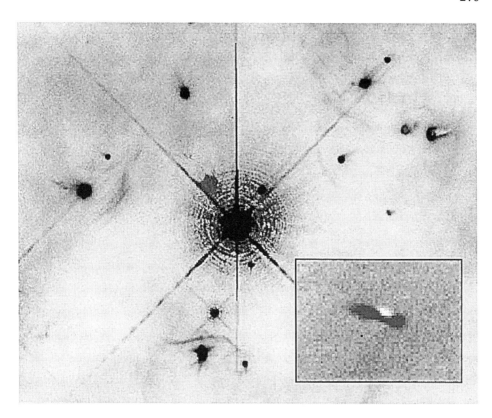

Figure 9. HST images of proplyds in the Orion Nebula. The main figure shows a collection of proplyds centred on θ^1C Orionis and is courtesy of John Bally, David Devine and Ralph Sutherland. Note the bow shocks where photo-evaporated gas from the disks are interacting with the wind from this O-type star. The tails are probably dust that is driven away from the photo-evaporated gas because of its strong interaction with soft UV photons. Inset is an image of a disk seen in silhouette against the backdrop of the Orion Nebula. This object is outside the HII region and scattered light can be seen from the central YSO.

HH 30 are amongst the nearest outflow sources to us. Obviously this is important if we wish to resolve their detailed structure.

6. A Few Examples of How AO Could Help Our Understanding of YSOs

Adaptive optics, I believe, could make important strides in our understanding of YSO jets and disks. As I have already mentioned both the angular widths of jets and disks are approximately $0''.5-1''$, for the nearest sources; thus AO is well suited to the study of their detailed morphology. To demonstrate this point, Fig. 11 shows a HST image of the HH 34 jet (see Fig. 6 for an overview of this flow) taken through a red [SII]$\lambda\lambda 6716,6731$ filter. The image was processed from the HST data archive (see Ray et al. 1996 for

Figure 10. A 3-colour composite image of HH 30 based on HST data from Ray et al. (1996). Note that the image has been rotated so that the blueshifted jet is oriented upwards. Here blue light represents continuum emission, which in this case is scattered starlight from the "top" and "bottom" of the disk. The source itself is highly embedded and the disk can be seen in silhouette as the dark lane bisecting the nebular cusps. Red light is [SII]$\lambda\lambda$6716,6731 emission and green represents Hα. Note the asymmetry in excitation conditions in the flow and the counterflow. Image reconstruction done by C.R. O'Dell and S.V.W. Beckwith.

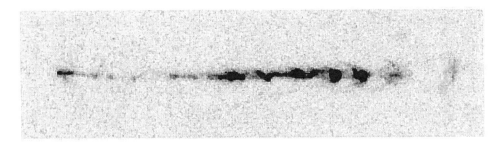

Figure 11. A HST [SII]$\lambda\lambda$6716,6731 image of the HH 34 jet. Here it can be seen that several individual knots, previously unresolved from the ground, are discerned as bow shocks. Faint shocks are also seen to propagate into the surrounding medium. Such images suggest that temporal variability of the flow, and in particular faster material catching up with previously ejected material, is responsible for most of the observed knots. The image has been rotated so that the source is on the left and the flow is horizontal.

details) and was obtained by the Wide Field Planetary Camera 2 Team. It is immediately seen that several of the jet knots are resolved into "mini-bow shocks" (or "working surfaces") supporting the idea that these knots arise as a result of temporal variations in the outflow and that they are shocks caused by faster jet material catching up with previously ejected slower gas. Moreover it is seen that the bows are curved away from the source, suggesting that there is a shear layer across the jet with the jet velocity peaking towards the central axis. Recent images, again from the HST, point to a similar origin for the knots in the HH 111 jet (Reipurth et al. 1997). To test this idea further, however, AO observations, both imaging and spectroscopy, are needed since although working surfaces may be an explanation for many of the observed knots in YSO jets, this may not be the case for all. Consider for example the HH 30 flow (Fig. 11). Here none of the well resolved knots close to the source appear to be bow-shaped although admittedly limited spatial resolution may play a rôle here. More importantly a steady decrease in the [SII]/Hα ratio is observed for the first few knots, whereas this ratio might be expected to show variations at knots if material is being re-shocked (Ray et al. 1996). The knots in this case could be mild condensations in the flow.

Apart from helping us understand how YSO jets propagate, AO might potentially tell us about the origins of outflows. To do this we must observe the flows close to their actual source. Of course many jet sources are highly embedded and so it is often impossible to study their immediate environments in the optical or near-infrared. Nevertheless there are a number of jet sources that are classical T Tauri stars or Herbig Ae/Be stars which, by definition, are optically visible. Moreover these stars are often very bright and and could thus act as natural guide stars. Sometimes the associated jets may only extend for a few arcseconds (hence their nickname "microjets", see Solf 1997) but they can be traced very close to their origin. Note also that, as we have previously mentioned, these flows can be studied in the near-infrared, for example using the [FeII]1.64μm line, and this may be the better option for many AO systems. At the same time it has to be emphasized that, even in those cases where a jet can be traced right back optically to the source, there are a number of problems in studying it. In particular the contaminating effects of light from the YSO has to dealt with. In addition scattered light from any associated reflection nebula in the vicinity of the source can cause difficulties even with narrow-band imaging. As demonstrated by Böhm & Solf (1994) however these problems can be overcome to some degree by the use of intermediate dispersion long-slit spectroscopy. The stellar continuum, along with any contribution from a surrounding reflection nebula, can be subtracted very effectively. The resultant spectra are then effectively 1-D position velocity diagrams and the various line ratios can give us information on physical conditions in the emitting gas.

One area where intermediate dispersion long-slit spectroscopy with an AO system could be very useful is in the study of the different outflow components from a YSO. We have already mentioned the low and high velocity components of the forbidden line emission and the suggestion of Kwan & Tademaru (1995) that these components are respectively due to a jet and a disk wind. It is certainly true that the gas in in the two components usually have very different excitation conditions and also different spatial properties (see, for example, Hirth, Mundt & Solf 1997). Moreover the low velocity component is just about spatially resolved under normal seeing conditions. Now giving the velocity of the low velocity component, i.e. a few kms^{-1}, we would naturally expect the acceleration region of this flow to occur on scales of tens of AU or so which, in the case of the nearest sources, corresponds to a few tenths of an arcsecond. Thus there is the real and exciting (!) possibility of using AO techniques to investigate directly the detailed behaviour of a YSO disk wind.

Acknowledgements

The author wishes to thank the organizers of the Cargèse school on adaptive optics, and in particular Chris Dainty and Renaud Foy, for their kind invitation to lecture to the school and for all the hospitality he was shown while there.

References

1. Ambartsumian, J.A., 1947, in *Stellar Evolution and Astrophysics*, Erevan: Acad. Sci. Armen. SSR
2. Appenzeller, I., Jankovics, I., Östreicher, R., 1984, A&A, 141, 108
3. Bacciotti, F., Chiuderi, C., Oliva, E., 1995, A&A, 296, 185
4. Bally, J., Morse, J., Reipurth, B., 1996, in *Science with the Hubble Space Telescope II*, eds. P. Benvenuti, F.D. Macchetto, E.J. Schreier, (Space Telescope Science Institute), p491
5. Bally, J., Devine, D., 1994, ApJ, 428, L65
6. Bally, J., Devine, D., Reipurth, B., 1996, ApJ, 473, L49
7. Bertout, C., 1989, ARA&A, 27, 351
8. Blitz, L., 1993, in *Protostars and Planets III*, eds. E. Levy and J. Lunine, (University of Arizona Press), p125
9. Bührke, T., Mundt, R., Ray, T.P., 1988, A&A, 200, 99
10. Burrows, C.J., et al., 1996, ApJ, 473, 437
11. Chandler, C.J., 1998, in *Proceedings of the International Origins Conference*, eds. C.E. Woodward, J.M. Shull and H.A. Thronson, (ASP), p237
12. Chen, H., Bally, J., O'Dell, C.R., McCaughrean, M.J., Thompson, R.L., Rieke, M., Schneider, G., Young, E.T., 1998, ApJ, 492, L173
13. Corcoran, M., Ray, T.P., 1997a, A&A, 321, 189
14. Corcoran, M., Ray, T.P., 1997b, in *Low Mass Star Formation - from Infall to Outflow*, Poster Proceedings of IAU Symp. 182, eds. F. Malbet and A. Castets, p82
15. Corcoran, M., Ray, T.P., 1998, A&A, 331, 147
16. Dopita, M.A., Schwartz, R.D., Evans, I., 1982, ApJ, 263, L73
17. Edwards, S., Ray, T.P., Mundt, R., 1993, in *Protostars and Planets III*, eds. E. Levy and J. Lunine, (University of Arizona Press), p567
18. Edwards, S., Strom, S.E., Hartigan, P., Strom, K.M., Hillenbrand, L.A., Herbst, W., Attridge, J., Merril, K.M., Probst, R., Gatley, I., 1993, AJ, 106, 372
19. Eislöffel, J. 1997, in *Herbig-Haro Outflows and the Birth of Low Mass Stars*, IAU Symposium No. 182, eds. B. Reipurth and C. Bertout, (Dordrecht: Kluwer Academic Publishers), p93
20. Eislöffel, J., Mundt, R., 1997, AJ, 114, 280
21. Feigelson, E.D., Casanova, S., Montmerle, T., Guibert, J., 1993, ApJ, 416, 623
22. Ferrari, A., Massaglia, S., Bodo, G., & Rossi, P., 1996, in *Solar and Astrophysical Magnetohydrodynamic Flows*, Proceedings of the NATO ASI, Heraklion, Crete, ed. K. Tsinganos, (Dordrecht: Kluwer Academic Publishers), p607
23. Haro, G. 1952, ApJ, 115, 572
24. Hartigan, P., Edwards, S., Ghandour, L., 1995, ApJ, 452, 736
25. Heathcote, S., Reipurth, B., 1992, AJ, 104, 2193
26. Herbig, G.H., 1951, ApJ, 113, 697
27. Hirth, G.A., Mundt, R., Solf, J., 1997, A&AS, 126, 437
28. Johnstone, D., Hollenbach, D., Bally, J., 1998, ApJ, 499, 758
29. Joy, A.H., 1945, ApJ, 102, 168
30. Koerner, D.W., 1997, Origins of Life and Evolution of the Biosphere, 27, 157
31. Koerner, D.W., Sargent, A.I., Beckwith, S.V.W., 1993, Icarus, 106, 2
32. Kuhi, L.V., 1964, ApJ, 140, 1409
33. Kwan, J., Tademaru, E., 1995, ApJ, 454, 382
34. Lada, E.A., Strom, K.M., Myers, P.C., 1993, in *Protostars and Planets III*, eds. E. Levy and J. Lunine, (University of Arizona Press), p245
35. Lay, O.P., Carlstrom, J.E., Hills, R.E., Philips, T.G., 1994, ApJ, 434, L75
36. López, R., Raga, A., Riera, A., Anglada, G., Estalella, R., 1995, MNRAS, 274, L19
37. Margulis, M., Lada, C.J., Snell, R., 1988, ApJ, 333, 316
38. McCaughrean, M.J., Stauffer, J.R., 1994, AJ, 108, 1382
39. McCaughrean, M.J., Rayner, J.T., Zinnecker, H., Stauffer, J.R., 1996, in *Disks and*

Outflows around Young Stars, eds. S.V.W. Beckwith, J. Staude, A. Quetz and A. Natta, Lecture Notes in Physics, Vol. 465, (Heidelberg: Springer), p33
40. McCray, R., Kafatos, M., 1987, ApJ, 317, 190
41. McNamara, B.J., 1976, AJ, 81, 845
42. Mestel, L., 1985, in *Protostars and Planets II*, eds. D.C. Black and M.S. Matthews, (University of Arizona Press), p320
43. Mundt, R., 1986, Can. J. Phys., 64, 407
44. Mundt, R., Bührke, T., Fried, J.W., Neckel, T., Sarcander, M., Stocke, J., 1984, A&A, 140, 17
45. Mundt, R., Fried, J.W., 1983, ApJ, 274, L83
46. Mundt, R., Ray, T.P., Bührke, T., Raga, A.C., Solf, J., 1990, A&A, 232, 37
47. Mundt, R., Ray, T.P., Raga, A.C., 1991, A&A, 252, 740
48. O'Dell, C.R., Wong, S.K., 1996, ApJ, 111, 846
49. Padman, R., Bence, S., & Richer, J. 1997, in *Herbig-Haro Outflows and the Birth of Low Mass Stars*, IAU Symposium No. 182, eds. B. Reipurth and C. Bertout, (Dordrecht: Kluwer Academic Publishers), p123
50. Raga, A.C., Kofman, L., 1992, ApJ, 386, 222
51. Ray, T.P., 1996, in *Solar and Astrophysical MHD Flows*, NATO ASI, Heraklion Crete, ed. K. Tsinganos, (Kluwer Academic Publishers), p539
52. Ray, T.P., 1987, A&A, 171, 145
53. Ray, T.P., Mundt, R., Dyson, J., Falle, S.A.E.G., Raga, A., 1996, ApJ, 468, L103
54. Reipurth, B., 1994, A General Catalogue of Herbig-Haro Objects, electronically published using anonymous ftp at ftp.hq.eso.org, directory /pub/Catalogs/Herbig-Haro
55. Reipurth, B., Bally, J., Graham, J.A., Lane, A.P., Zealey, W.J., 1986, A&A, 164, 51
56. Reipurth, B., Hartigan, P., Heathcote, S., Morse, J., Bally, J., 1997, AJ, 114, 757
57. Sargent, A.I., 1996 in *Disks and Outflows Around Young Stars*, eds. S.V.W. Beckwith, J. Staude, A. Quetz and A. Natta, Lecture Notes in Physics, Vol. 465, (Heidelberg: Springer), p1
58. Sargent, A. I., Beckwith, S.V.W., 1991, ApJ, 328, L31
59. Schwartz, R.D., 1975, ApJ, 195, 631
60. Schwartz, R.D., 1978, ApJ, 223, 884
61. Schwartz, R.D., Cohen, M., Jones, B.F., Bohm, K., Raymond, J.C., Hartmann, L.W., Mundt, R., Dopita, M.A., Schultz, A.S.B., 1993, AJ, 106, 740
62. Shu, F.H., Adams, F.C., Lizano, S., 1987, ARA&A, 25, 23
63. Shu, F.H., Najita, J., Galli, D., Ostriker, E., Lizano, S., 1993, in *Protostars and Planets III*, eds. E. Levy and J. Lunine, (University of Arizona Press), p3
64. Solf, J., 1997, in *Herbig-Haro Flows and the Birth of Low Mass Stars*, IAU Symposium No. 182, eds. B. Reipurth and C. Bertout, (Dordrecht: Kluwer Academic Publishers), p63
65. Stahler, S.W., Walter, F.M., 1993, in *Protostars and Planets III*, eds. E. Levy and J. Lunine, (University of Arizona Press), p405
66. Strom, S.E., Strom, K.M., Grasdalen, G.L., Sellgren, K., Wolff, S., 1985, AJ, 90, 2281
67. Ward-Thompson, D., Buckley, H.D., Greaves, J.S., Holland, W.S., André, P., 1996, MNRAS, 281, L53
68. Zinnecker, H., McCaughrean, M.J., Rayner, J., 1996, in *Disks and Outflows Around Young Stars*, eds. S.V.W. Beckwith, J. Staude, A. Quetz and A. Natta, Lecture Notes in Physics, Vol. 465, (Heidelberg: Springer), p236

CHAPTER 14
DISTANT (RADIO) GALAXIES: PROBES OF GALAXY FORMATION

HUUB RÖTTGERING
Leiden Observatory
P.O. Box 9513
2300 RA Leiden
The Netherlands

With the advances in the capabilities of both ground and space based telescopes it is now possible to study the universe at an epoch when its age was only 10 % of its present value. In this contribution, I will discuss two classes of objects that are known to be present at these extreme distances: the UV-dropout galaxies, which are likely to evolve into normal present day galaxies, and the powerful distant radio galaxies, which seem to be the precursors of nearby brightest-cluster-galaxies (BCGs). The observational characteristics of these two populations are reviewed and I comment on how some of these characteristics constrain one of the most important problems in current astrophysics: How do galaxies form? Finally, I make a number of comments on the importance of high resolution imaging for understanding distant galaxies.

1. Introduction

One of the most important topics in astrophysics for the coming decade undoubtedly will be galaxy formation. How and when did the galaxies we know from our local universe form? For two reasons, great advances in our understanding of the formation of galaxies are expected. First, the new observational facilities, especially the many 10-metre class telescopes that will come into operation within the next few years, will allow observational studies of large samples of galaxies when the age of universe was only 5–15 % of its present value. Second, the techniques for modelling galaxy formation that are currently being developed, combined with the most powerful computers, seem very promising. It is expected that the next generation of models will be sophisticated enough to allow for detailed comparisons with many observations.

It is clear that sensitive high resolution imaging has an important role to play in this area. Distant galaxies have angular sizes of order a few arcseconds. To properly study their structural parameters, and how these evolve during the history of the universe, observations have to be carried out with resolutions of order 0.1 arcsec and, of course, preferably even better. In this contribution, I will try to briefly review some issues in the field of galaxy formation and how these relate to the importance of high resolution imaging.

The outline of this contribution is as follows. In Section 2, I would like to briefly discuss the expansion of the universe and how the redshift of a galaxy is related to far back in time it is located. Subsequently, I would like to discuss a number of outstanding questions in the field concerning the formation of galaxies and AGN (Section 3). It is interesting to note that most of these questions are obvious, but that none of them have well established answers. The population of very distant galaxies that have recently been found will be discussed in Section 4. I will review their basic observational properties and discuss how some of these properties can be successfully modelled. In Section 5 basic properties of distant radio galaxies are discussed. I will concentrate on these objects rather than on the general AGN population, since these objects are relatively well studied, even at very high redshifts. As an introduction to the subject of distant radio galaxies, I will briefly review some properties of Cygnus A, since this is the only nearby radio galaxy with a radio power comparable to that of the very distant radio galaxies; it is the local analogue of the distant radio galaxies and understanding this object is therefore important. Finally, in Section 6, I will briefly discuss the important role that high resolution imaging is expected to play in the field of the formation and evolution of galaxies and AGN.

2. Lookback time and redshift

The main reason that observing distant galaxies is so crucial for our understanding of how galaxies form is that we are observing back into the history of the universe. How far we are looking back into time for an individual galaxy is defined as its "lookback time".

Our current best model of the global dynamics of the universe is that of an expanding universe that started with a "Big Bang". For a Friedmann universe, the expansion can be characterised by two parameters, the density parameter Ω and the present rate of expansion, the Hubble constant H_0 (for reviews and formulae: Condon 1988; Peebles 1993; Weinberg 1972; Zombeck 1990). Current estimates are that $0 < \Omega \lesssim 1$ and $H_0 = 50 - 100$ km s^{-1} Mpc^{-1}. Due to this global expansion, the observed wavelength of

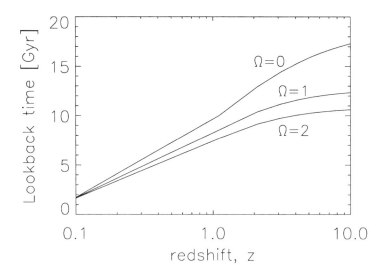

Figure 1. The lookback time as function of redshifts for $H_0 = 50$ km s^{-1} Mpc^{-1}, and for $\Omega = 0, 1, 2$.

TABLE 1. The ratio of lookback time to the age of the universe for $H_0 = 50$ km s^{-1} Mpc^{-1}, for various values for the redshift and Ω.

	$\Omega = 0$	$\Omega = 1$	$\Omega = 2$
z=1	65%	50%	68%
z=2	81%	66%	83%
z=3	87%	75%	89%
z=5	93%	83%	94%

spectral features of distant galaxies are shifted towards the red with respect to their rest-wavelengths. The magnitude of the relative shift, the redshift z, is defined as $z = (\lambda_{\rm obs} - \lambda_{\rm rest})/\lambda_{\rm rest}$ and is a measure of the distance of an object and hence its lookback time. The exact values for lookback times are dependent on the assumed values for the Hubble constant and the density parameter. In Fig. 1 the lookback time is plotted as a function of redshift for $H_0 = 50$ km s^{-1} Mpc^{-1}, and for $\Omega = 0, 1, 2$. In Table 1. the ratio of lookback time to the age of the universe is given for the same set of parameters.

During the last decade significant numbers of galaxies, quasars and radio galaxies have been found with redshifts larger than 3. This is at an epoch when the age of the universe was only 10 − 20 % of its present age and therefore the objects we are observing must be close to their epoch of formation.

3. Main questions

3.1. GALAXY EVOLUTION AND FORMATION

There are many questions on the issue of galaxy formation that might be solved by thoroughly studying very distant galaxies.

We would like to know when the first stars formed and how they subsequently settled into galaxies. Nearby galaxies have typical profiles that come in two classes, elliptical galaxies that have de Vaucouleurs profiles and spiral galaxies that have an exponential law. When did these profiles form? Is that at the same epoch during which the first generation of stars formed? Another interesting question is why are there ellipticals and spirals? These are such fundamentally different classes that there must be a universal explanation for the existence of these two classes. An important clue is the fact that ellipticals are preferentially located in clusters of galaxies. This indicates that merging of galaxies might be an important mechanism for creating elliptical galaxies. It is, however, not clear whether this is the only way in which ellipticals are formed and that spirals are objects that never have accreted another (dwarf)galaxy.

3.2. ACTIVE GALACTIC NUCLEI: QUASARS, RADIO GALAXIES

Another fascinating class of object in the universe is the class of the Active Galactic Nuclei. In these objects we are witnessing explosive events in the centres of galaxies. These events are presumably powered by massive black holes.

For a number of reasons it seems that galaxy formation and AGN activity are closely related (Efsthatiou and Rees 1988; Loeb 1993; Haehnelt and Rees 1993; Eisenstein 1995; Haehnelt et al. 1998; Silk and Rees 1998). A nice illustration of this relation has recently been published by Dunlop (1998). From several surveys, including the Canada-French Redshift Survey and the Hubble Deep Field, the number of stars that are being formed per unit volume, $\dot{\rho}_{star}$, can be estimated as a function of redshift (Madau et al. 1996). The results of this analysis are plotted in Fig. 2. Since the number of powerful AGN as a function of redshift is well known, we can estimate the rate per unit volume in the universe, $\dot{\rho}_{BH}$, at which mass is consumed by black holes at the centres of giant elliptical galaxies. Two estimates for this rate are indicated by the solid and dashed lines in Fig. 2. The func-

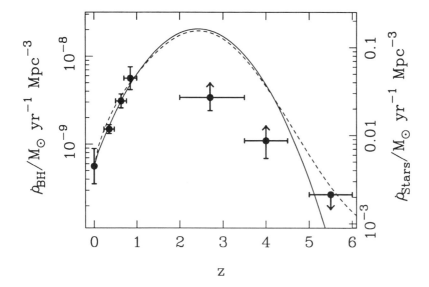

Figure 2. The comparison by Dunlop (1998) of the redshift dependence of the rate per unit volume, $\dot{\rho}_{BH}$, at which mass is consumed by black holes at the centres of giant elliptical galaxies (the solid and dashed line) with the rate per unit volume, $\dot{\rho}_{stars}$, at which mass is converted into stars (data points).

tional form of the rate at which black holes are consuming and the rate at which stars are forming are remarkably similar, indeed indicating the close relation between galaxy formation and AGN activity. However, there are a number of assumptions in estimating both rates. Especially at high redshifts ($z > 3$) the uncertainties are too large to say that the two rates have indeed a very similar functional form.

An intriguing question is what are the first sources of light in the universe. Did the first light come from activity associated with the formation of the first massive black holes or was it from the first generation of stars at extreme redshifts? There are a number of other interesting issues concerning the relation between galaxies and massive black holes. Why is it that all the powerful radio sources are located in ellipticals and not in spiral galaxies? There are now clear indications that the mass of the black hole roughly scales with the mass of the hosting galaxy. Does this relation hold up with new observations? Does this also imply that every galaxy has a black hole, most of them being quiescent? Why is that, and what does it elucidate on how black holes and galaxies are formed? Many interesting phenomenon are associated with AGN, including radio jets, accretion disks and photon beams. However, many important questions on the origin of these phenomenon have not been answered satisfactorily.

4. Normal galaxies at $z \sim 3$

Until recently, it was not clear whether a significant population of galaxies at $z > 2$ existed. Due to a combination of two techniques it is now possible to obtain large samples of galaxies at these extreme redshifts. The first step is to select candidate distant galaxies from deep multicolour surveys. The basic idea is that starforming galaxies at high redshift have a very characteristic UV spectrum with the UV flux being heavily absorbed below the Lyman continuum break at $\lambda = 912$ Å. Since the observed wavelength for this break shifts into the optical window for galaxies at $z \sim 3$, faint galaxies with extremely red colours in the optical spectral region around 4000 Å are good candidates for such distant galaxies. Since the UV spectra of young galaxies are expected to be fairly flat redward of Lyα (1216 Å), this selection can be further improved by selecting only galaxies with such a flat spectrum. The second step is to measure the redshift of the good candidates with a 10-m class telescope.

This technique was pioneered by Steidel and co-workers using three bands, U_n (3570 Å), G (4830 Å) and $\mathcal{R}$ (6930 Å), and produced significant numbers of very good candidates with a surface density of order 0.5 galaxies arcmin^{-2} down to a magnitude $\mathcal{R} = 25$ (Steidel and Hamilton 1992; 1993, Steidel et al. 1995). Spectroscopic confirmation of these candidates had to await the availability of a 10-m class telescope. The Keck telescope has now confirmed that indeed the vast majority of these galaxies are at these high redshifts (Steidel et al. 1996). At the time of writing almost a thousand of these galaxies are known and their numbers are still growing.

The UV-spectra of these "drop-out" galaxies are remarkably similar to those of present-day starbursts, the dominant features being low ionization interstellar absorption lines and high ionization stellar lines. The galaxies are generally compact in HST images, with typical half-light radii of only ~ 0.3 arcsec (~ 2 kpc for $H_0 = 50$ km s^{-1} Mpc^{-1} and $\Omega = 1$; Giavalesco et al. 1996). Typical star formation rates (SFRs) can be inferred from the redshifted UV continuum properties of the sample and are of order 10 M$_\odot$ yr^{-1}. Pettini et al. (1998) estimate typical dust extinction $\simeq 30\%$, $E(B-V) \simeq 0.1$; however, extinctions estimated from the UV continuum of star-forming galaxies are notoriously unreliable. Estimating ages for the stellar population of these objects is difficult. Sawicki and Yee (1998) argue that these objects are dominated by very young stars with ages $\lesssim 0.2$ Gyr.

4.1. MODELS OF GALAXY FORMATION

A very important challenge for astrophysics is to build reliable and robust models of galaxy formation. The goal is to model the evolution of galaxies with, as the main input, the physical conditions as they existed in the very early universe. The model should then predict the gross properties of galaxies including their number density, masses, luminosities and stellar content. Although this is a difficult problem, during the last 10 years important progress has been made. Here I will briefly give a general overview of the basic ingredients of the present day models. The aim is to give a flavour of what is possible with present day techniques. For a thorough account I refer to many excellent articles and reviews that have been written (see for example: Rees and Ostriker 1997; White and Rees 1978; Baron and White 1987; White and Frenk 1991; Cole 1991; Lacey and Silk 1991; Kauffmann et al. 1993; Lacey et al. 1993; Cole et al. 1994; Kauffman, et al. 1994; Baugh et al. 1996; Baugh et al. 1997).

Most of the current simulations assume that the dynamics of the large scale mass distribution in the early universe is driven by the gravity exerted by some form of dark matter. This dark matter is supposed to be collisionless and of pre-galactic origin. Furthermore it is assumed that at early times there are density fluctuations within this medium with a pre-described distribution depending on the physics in the early universe and the nature of the dark matter. The distribution of these fluctuations is normally taken to be a power-law. This, plus a suitable choice of cosmological parameters, determine the initial conditions for modelling of structure formation in the universe. The evolution of the dark matter distribution can be studied with the help of N-body simulations. In Fig. 3 an example of the result of such a simulation is given, showing how halos of dark matter merge and form progressively larger and larger structures. The panels in Fig. 3 are at 4 different time intervals ($z = 3.91, 1.15, 0.37$ and 0.0). To the right the distribution of the dark matter particles is shown, and to the left the location of the galaxies as they have been formed in this particular simulation. These simulations have been carried out by Eelco van Kampen and are described in detail in van Kampen and Katgert (1997).

A second step is to include the baryonic gas and follow its hydrodynamic evolution. Gas dynamics, shocks and radiative heating and cooling all need to be part of the simulation to obtain a realistic multi-phase medium. The outcome of this kind of simulation is that a significant fraction of the gas cools and settles at the centres of dark matter halos. It is from that gas that the stars that will make up future galaxies will form. The major problem is now how to proceed with the inclusion of star formation in the simulations. Since there is no generally accepted theory of how stars form, it is not clear how many stars of what mass will form at what rate in such a modelled proto-galaxy. Often it is simply assumed that the rate at which such a galaxy forms stars is proportional to the total amount of gas present and inversely proportional to the dynamical timescale within the dark matter halo.

Figure 3. A simulation of cosmological structure formation by Eelco van Kampen. (right) the distribution of the dark matter particles at 4 different redshifts z, where, from top to bottom, $z = 3.91, 1.15, 0.37$ and 0.0. (left) the location of the galaxies that form in this simulation (for details, see van Kampen and Katgert 1997).

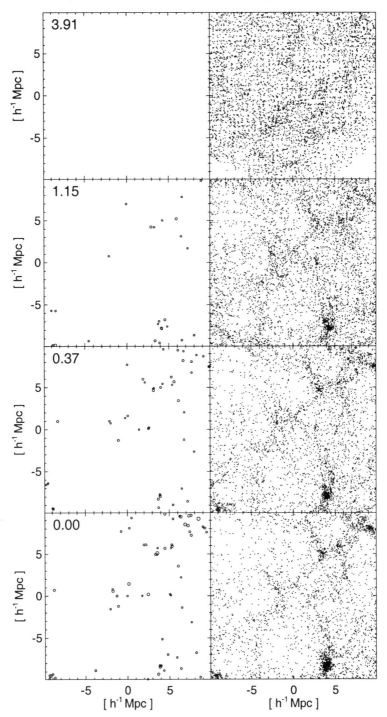

Subsequently, two other aspects need to be incorporated in these models. First, the newly formed set of stars will produce both supernovae and stellar winds. Both of these eject energy into the interstellar medium and, as a consequence, the cold gas present will be heated, and so the rate of star formation will be severely slowed down. Second, we know from observations that the merging of galaxies greatly enhances the combined star formation rate of both galaxies, possibly to a rate whereby a very large fraction of the gas is transformed into stars within a few dynamical timescales. This process is usually taken into account in the models through greatly enhancing the star formation rate during a merging event.

Finally, the result of the star formation has to be taken into account. It is often assumed that when stars are formed that their mass distribution follows a power-law with a lower cutoff around 0.1 solar masses and an upper cutoff around 100 solar masses. There are a number of stellar libraries that give the spectra for a star of given mass as a function of its age. The combination of the star formation rate and the spectral evolution of individual stars will then give the evolution of the integrated spectra of an individual galaxy. As an example of such spectral modelling the results of an instantaneous burst for the models of Bruzual and Charlot (1993) are given in Fig. 4.

With this kind of modelling gross properties of the general galaxy population can be calculated. These properties include the luminosity function of galaxies, the redshift distribution, the relative numbers of ellipticals and spirals, faint galaxy counts, the history of star formation etc. For example, a nice result of this kind of modelling is that the global star formation rate as a function of redshift (see Fig. 2) has nicely been predicted (Baugh et al. 1998).

Although this kind of modelling represents an important step in our understanding of galaxy formation, there are a number of limitations that we need to keep in mind. Several ingredients will have to be added before the global properties can be realistically modelled. These include a more realistic treatment of the star formation, the influence of dust, which will change the colours of the galaxies and the influence of the change of metallicity of the cold gas during the evolution of the galaxy which has an important impact on the characteristics of the stars.

5. Distant radio galaxies

Distant radio galaxies are extremely luminous radio sources that can be observed up to very high redshifts. Since they are associated with very distant massive galaxies, the study of distant radio galaxies can be used to advance our understanding of the formation of massive ellipticals.

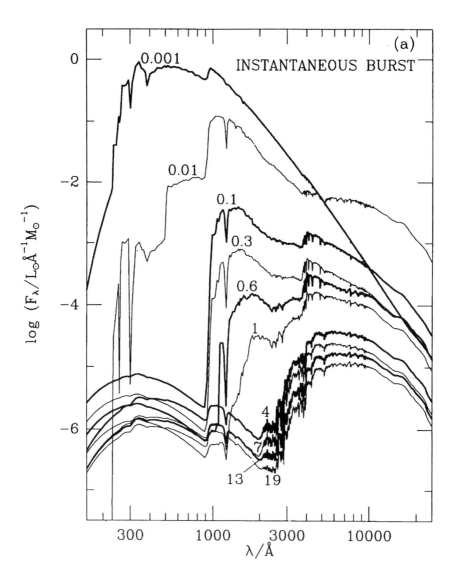

Figure 4. Spectral evolution of the stellar population model of an instantaneous starburst (from Bruzual and Charlot 1993). The ages in Gyr are indicated next to the spectra.

The most luminous local radio galaxy is Cygnus A. Since it has many properties in common with the radio galaxies at large distances, I will briefly review some basic properties of Cygnus A. Subsequently, I will review some basic observed characteristics of distant radio galaxies with the emphasis on their optical morphologies as observed with HST. On the basis of these characteristics scenarios for the formation of the massive hosts of distant radio galaxies will then be discussed.

5.1. CYGNUS A

The detailed radio images that exist of Cygnus A convincingly show what fascinating objects powerful radio galaxies are (for a recent detailed review on Cygnus A, see Carilli and Barthel 1996). In Fig. 5 VLA radio images are shown, illustrating the four basic morphological components of powerful radio sources. At the centre there is a compact component, the *radio core*, with a relatively flat radio spectrum. This core is associated with the very centre of the elliptical galaxy hosting the radio sources. From this compact component two narrow *jets* are emerging connecting this core with the high surface brightness knots, the *hot-spots*, located at the two extremities of the source. Between the two hot-spots the two *lobes* are visible, two regions with filamentary, low surface brightness emission. The overall extent of Cygnus A is 103 kpc, an order of magnitude larger than the hosting elliptical galaxy.

Within the scenario that is normally invoked to explain the structure of these powerful radio sources, the radio core corresponds to the location of the "central engine". This central engine somehow produces energetic outflows of, likely relativistic, particles, the jets. At the locations where the jets terminate, strong shocks form which are observed as the hot-spots. The lobes are then the shocked jet material which expands away from the jet.

The radio emission is due to relativistic particles being accelerated in a magnetic field. With a number of assumptions, the strength of the magnetic field for the various components can be estimated and is given in Table 2. In a number of ways the density of the medium surrounding the radio source can be estimated. Arguably the most reliably way comes from the estimates based on X-ray measurements of the bremsstrahlung that this hot medium is emitting. Typically values of a few times 10^{-3} cm^{-3} are found. The dynamics of the synchrotron emitting plasma can be modelled either through analytic approximations or through detailed hydrodynamic simulations. An important physical ingredient of this modelling is that the thrust of the jet is taken to be balanced by the force due to ram pressure of the medium surrounding the radio source. From this kind of modelling, constraints on a number of important parameters are obtained, including the jet velocity, jet power and age of the whole source (see Table 2).

For a detailed discussion of the assumptions and the uncertainties that go into obtaining the numbers quoted in Table 2, I would like to refer to the review of Carilli and Barthel (1996). There are two further points I would like to make concerning our understanding of the physics of these powerful radio galaxies. First, although the general scenario of how these objects work is probably correct, there are many issues that are not very well understood. What keeps the jet so nicely confined until it reaches its final destination at the hot-spots and why do we find such spectacular

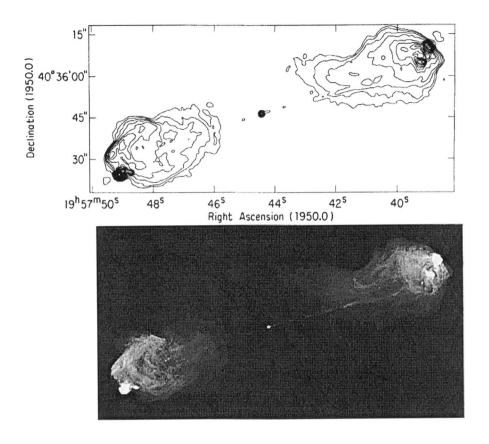

Figure 5. The powerful radio galaxy Cygnus A. (upper) Cygnus A at 5 GHz with 1 arcsec resolution. The contour levels are 0.075, 0.2, 0.5, 1, 2, 3, 4, 5, 7.5, 10,15,20, 30, ..., 80, 90 % of the maximum brightness of 14.5 Jy per beam (from Dreher et al. 1987); (lower) Cygnus A at 5 GHz with 0.4 arcsec resolution (from Perley et al. 1984, courtesy of Chris Carilli).

radio sources only associated with elliptical galaxies and not for example with spirals? Second, I would like to emphasize that the ages derived for the radio sources, of order 10^7 years, are extremely short compared to the age of the universe (e.g. Blundell and Rawlings 1999). These times are so short that it is entirely possible that, while we observe that in our local universe only a tiny fraction of the elliptical galaxies contain a powerful radio source, every elliptical galaxy was once radio active during its life time.

TABLE 2. A number of basic properties of the radio galaxy Cygnus A. (most of the values are taken from Carilli and Barthel 1996 or Begelman and Chioffi 1989)

Monochromatic radio power at 178 MHz	3×10^{35} erg sec^{-1} Hz^{-1}
Total radio power	5×10^{44} erg sec^{-1}
Projected linear size	103 kpc
Magnetic field	
lobes	50 μGauss
jet	130 μGauss
hot-spot	300 μGauss
Minimum total energy in the lobes	10^{60} erg
IGM electron density	6×10^{-3} cm^{-3}
Jet velocity	$\sim c$
Age	10^7 yr
Jet power	10^{45} erg sec^{-1}

5.2. SEARCH FOR HIGH REDSHIFT RADIO GALAXIES

Objects like Cygnus A are extremely rare in our local patch of the universe. This changes drastically towards redshifts one to two. At these epochs the number of such powerful radio galaxies is almost a factor 1000 higher than in the local universe. Normally, the flux of an astronomical objects is a first indication of its distance, ie. fainter (optical) galaxies are in general the more distant. Due to the ubiquity of powerful radio sources at high redshifts, this is not the case for radio galaxies. A randomly selected 1 Jy radio source has an almost equal chance to be located in our local universe or at $z > 1$. This makes radio sources an extremely useful tool for locating very distant galaxies; any sample of fairly bright radio sources will contain a large fraction of $z > 1$ galaxies.

At the moment a number of new radio surveys are available, including the Westerbork survey WENSS at 325 MHz (Rengelink et al. 1997), and the two VLA surveys, FIRST (Becker et al. 1995) and NVSS (Condon et al. 1998), both at 1.4 GHz. These surveys contain many hundred thousand radio sources from which we would like to select the ones with the highest redshifts.

An efficient way of doing that is to select sources with very steep radio spectra, since it has been found that the steeper the spectrum of a radio source, the more likely it is to be at very high redshifts. Realising this, the steps towards obtaining a sample of radio galaxies with $z > 3$

are then straightforward, albeit time consuming. The first step is to select radio sources with ultra-steep radio spectra. These sources should then be imaged to obtain identifications for the radio sources. Typically the R-band magnitude of the identifications are around 24. Subsequent spectroscopy is then needed to obtain a redshift for these faint galaxies. The reason that this is possible with a 4-m class telescope is that the emission lines of these objects are extremely bright, the dominant emission line being Lyα λ1216 which can be as luminous as 10^{44} erg s^{-1}. Other lines that are often present, but with fainter intensities ($< 10\%$ of Lyα) are C IV, He II, C III]. In Fig. 6 a number of examples of spectra of $z > 2$ radio galaxies are given.

After a decade of vigorous search for high redshift radio galaxies, there are now more than 150 radio galaxies with $z > 2$ known (Rawlings et al. 1996, Röttgering et al. 1997, McCarthy et al. 1998). The current record-holder is TN J0924-2201 with a redshift $z = 5.19$ (van Breugel et al. 1999).

5.3. PROPERTIES OF HIGH-Z RADIO GALAXIES

As a probe of the distant universe, high redshift radio galaxies (HzRGs) offer a wide range of distinct components that can be studied. These include stars, AGN, hot (X-ray emitting) gas, ionized gas, neutral (HI) gas, molecular gas, dust and relativistic plasma. Some components have merely been detected (dust, X-ray emitting gas), while others have been extensively studied (ionized gas, neutral (HI) gas, relativistic plasma). In Table 3 we provide an overview of the mass associated with some for the components of HzRGs. Most of these estimates are highly model-dependent and should therefore treated with caution. Radio galaxies have advantages over quasars as cosmological probes in that at least four of these emitting components can be spatially resolved by ground based telescopes. Furthermore, due to the absence of a bright nucleus, the study of components other than the nucleus itself is less challenging.

One of the most remarkable features of distant radio galaxies is that they usually possess giant luminous halos of ionized gas, which can extend to > 150 kpc, with velocity dispersions of typically ~ 1000 km s^{-1}. A spectacular example is the Lyα halo of 1243+036 which has a luminosity $\sim 10^{44.5}$ ergs s^{-1} and extends over $\sim 20''$ (135 kpc, see Fig. 7).

At a spectral resolution of order 1 Å, the Lyα line often shows strong absorption features indicating the presence of regions of neutral gas with sizes of > 10 kpc (Röttgering et al. 1995, van Ojik et al. 1997). Another way of probing relatively cold gas is to observe emission from CO using millimetre or submillimetre telescopes. The many searches that have been carried out have been dramatically unsuccessful. Recently however, two solid detections have been reported for the two radio galaxies 4C60.07 ($z = 3.8$)

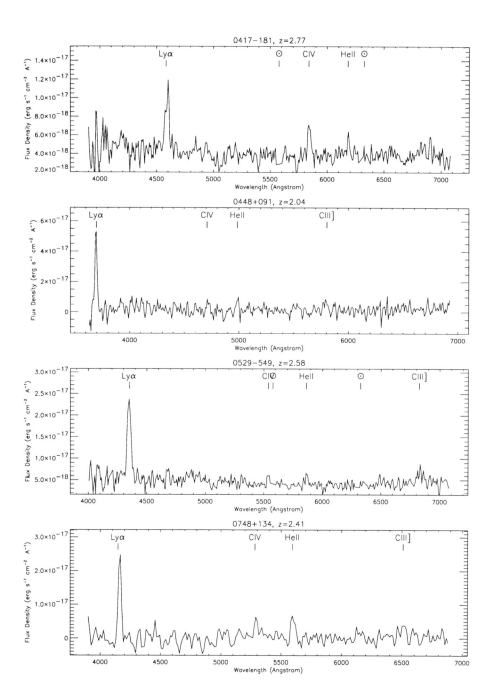

Figure 6. Optical spectra of $z > 2$ radio galaxies (from Röttgering et al. 1997)

TABLE 3. Estimated masses for some of the components associated with distant radio galaxies. A Hubble constant of $H_0 = 50$ km s^{-1} Mpc^{-1} and a density parameter of $\Omega = 1$ are used.

Components	Total Mass $M_\odot$	Reference
Stars	10^{12}	4C41.17, $z = 3.8$, Chambers et al. (1990)
Ionized Gas	10^9	1243+036, $z = 3.6$, van Ojik et al. (1996)
Neutral (HI) gas	2×10^7	0943−242, $z = 2.9$, Röttgering et al.(1995)
Molecular gas	10^{11}	LBDS 53W002, $z = 2.390$ Scoville et al. (1997)
Dust	3×10^8	4C41.17, $z = 3.8$, Dunlop et al. (1994)
Hot (X-ray emitting) gas	4×10^{13}	1138−262, $z = 2.156$, Carilli et al. (1998)

and 1909+722 ($z = 3.6$), indicating that the total amount of molecular gas present in each of these systems has a mass of order 10^{11} M$_\odot$ (Papadopoulos et al. 1999).

The study of the dust content of HzRGs is important for a number of issues including the heavy element production and the role of strong starbursts in the early universe. The best direct way of probing the dust is to measure its (sub)millimetre continuum emission. This is relatively easy at the highest redshifts; there is a favourable K-correction due to the steepness of the dust emissivity at (sub)mm wavelengths, and therefore the expected flux density for a given amount of dust rises between $z \sim 1$ and $z \sim 10$. At the moment 3 HzRGs with $z > 3$ have been detected in the (sub)millimetre continuum, indicating that they have a dust mass of order a few 10^8 M$_\odot$ (Dunlop et al. 1994; Hughes et al. 1997, Ivison et al. Röttgering et al. 1998).

The existence of rich clusters around HzRGs has not been firmly established. There are important observational indications, however, that powerful distant radio galaxies might be in clusters. Recently, the HzRG 1138−262 at $z = 2.156$ (Pentericci et al. 1997, see also Fig. 9), was detected in a 40 ksec ROSAT observation (Carilli et al. 1998). The X-ray luminosity is 7×10^{44} erg sec^{-1} which is similar to that of Cygnus A. This could indicate that the cluster around 1138-262 is as rich as that around Cygnus A. The near-future X-ray missions will answer the important question of whether indeed a large fraction of the HzRGs have similar halos.

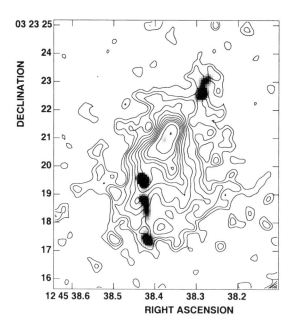

Figure 7. A contour plot of the Lyα halo of the radio galaxy 1243+036 at $z = 3.6$, with a greyscale plot of the 8.3-GHz VLA map superimposed (from van Ojik et al. 1996).

5.4. HST IMAGING AND THE STELLAR CONTENT

One of the most important questions in the field of distant radio galaxies concerns the spatial distribution and the ages of the stars. The study of these issues is a complicated one, mainly for two reasons.

First, the best wavelength region to probe the bulk of the stellar population is around $1-2$ μm rest-frame. For the most distant radio galaxies this region is redshifted towards the thermal infrared, a region in which it is difficult to carry out observations with sufficient sensitivity. Second, in general the optical and radio axes of the galaxies are roughly aligned — the so-called "alignment effect" — showing that the AGN has a dramatic influence upon the optical appearance. One of the prime examples that shows the alignment effect is the radio galaxy 3C368 ($z = 1.132$), which has a highly extended optical morphology with strings of bright knots close to the radio axis (Longair et al. 1995; Best et al. 1998, see also Fig. 8). The three most important models that have been proposed to explain the aligned optical light are that it is due to (i) scattered light from a hidden quasar, (ii) young stars whose formation has been stimulated by the radio jet, and (iii) nebular continuum emission from the ionized gas (see reviews by McCarthy et al 1993; Röttgering & Miley 1996). Although it is clear that, in general, all three mechanisms are contributing to the aligned light, the degree with which they do varies dramatically from object to object (e.g. Simpson et al. 1999).

Hubble Space Telescope WFPC2 images of distant radio galaxies are very valuable for further understanding the nature of the UV-light. The overall impression of the images we have obtained during a project to observe a sample of HzRGs with HST (Pentericci et al. 1999) is that there is a wide variety of structures in the radio galaxy morphology: most objects have a clumpy, irregular appearance, consisting of a bright nucleus and a number of smaller components. The UV-continuum emission is generally elongated and aligned with the axis of the radio sources, however the nature of this "alignment effect" varies from case to case.

To illustrate the variety of structures present in these objects the HST images of three objects, 1345+24 ($z = 2.91$), 1243+036 ($z = 3.6$) and 1138−262 ($z = 2.34$), are shown in Fig. 9 (Röttgering et al. 1998; Pentericci et al. 1998, 1999). The UV-continuum emission of 1345+24 is dominated by a bright compact nucleus. On the eastern side there is a jet-like feature that follows remarkably well the small curvature of the radio jet, suggesting that we might be observing strong interaction of the radiojet with the ISM. The morphology of 1243+036 consists of a nucleus and a narrow and elongated structure emanating from the nucleus which, at the location of bright radio knot in the southern radio jet follows the jet in its bend to the south. There is also a smaller component, about $1''$ beyond the northern radio hot-spot, which is well aligned with the southern UV-emission.

The HST image of the spectacular radio galaxy 1138−262 at $z = 2.2$ shows a very clumpy optical morphology with of order 10 emission clumps (Pentericci et al. 1997, 1998). The clumps have typical sizes in the range of $2 - 10$ kpc and have profiles that can not be well fitted by either an $r^{1/4}$ or an exponential law. Since a significant fraction of the clumps are located away from the radio jet, it seems unlikely that their presence is directly due to the alignment effect and therefore Pentericci et al. suggest that these objects are star forming regions with SF rates in the range of $1 - 5$ $M_\odot$ yr^{-1}. Interestingly, all their characteristics (sizes, profiles, SF rates) are similar to those of the distant normal galaxies at $z \sim 3$ that were discussed at the beginning of this contribution.

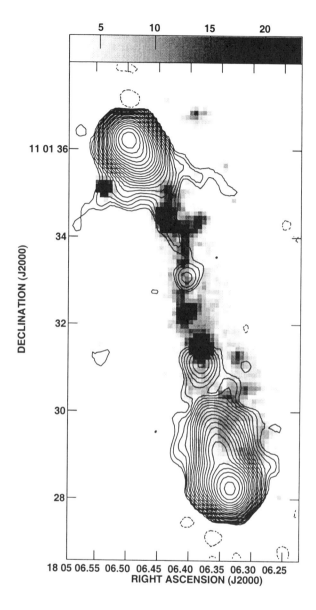

Figure 8. Deep radio observations of 3C368 (z=1.132) with the VLA at 8GHz (contours) with the HST image (greyscale) overlayed (from Best et al. 1998).

5.5. DISTANT RADIO GALAXIES: PROBING THE FORMATION OF CENTRAL CLUSTER GALAXIES?

The observations are putting severe constraints on the way the host of the very distant galaxies are being assembled. From the UV-continuum and dust emission measurements the inferred star-formation rates are often as

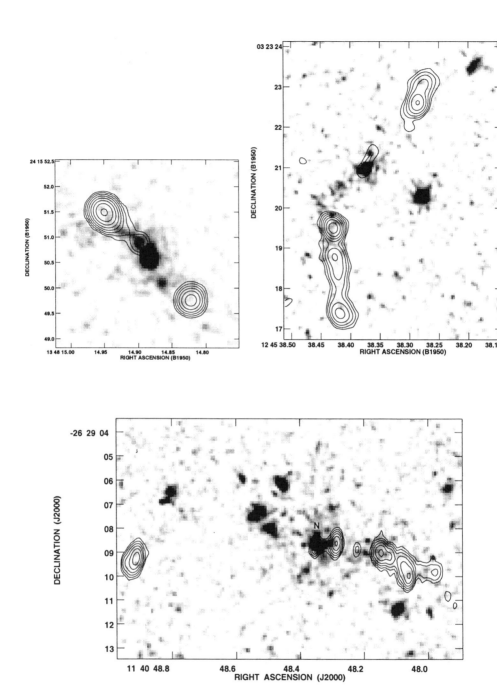

Figure 9. HST WFPC2 images of 3 distant radio galaxies. (upper-left) 1345+24 at $z = 2.91$, (upper-right) 1243+036 at $z = 3.6$ and (lower) 1138−262 at $z = 2.2$ (from Röttgering et al. 1998; Pentericci et al. 1998).

high as 1000 $M_\odot$ yr^{-1} suggesting the presence of a starburst that is so vigorous that a significant fraction of the stars are being formed. The star formation is taking place in clumps that can be spread out over a physical scale of order 50 – 100 kpc. These SF clumps will merge with the central galaxy on dynamical time scales (of order 10^8 years) and at $z \sim 1$ we can observe the fully developed ellipticals. This mode of star formation seems to be different from that as observed in the more classical starburst galaxies such as the ultra luminous IRAS galaxies. In these galaxies the SF is mostly concentrated in a nuclear region with a spatial scale of order 1 kpc (Sanders and Mirabel 1996).

Summarizing, the simple interpretation from the observations indicates that a widespread starburst and the subsequent merging of these SF clumps are both important processes in the formation of the massive ellipticals. Given the – admittedly still some what meager – observational evidence that HzRGs are located at the centers of (forming) clusters of galaxies, a currently favoured and logical conclusion is that the distant radio galaxies are forming brightest cluster galaxies.

6. Issues for high resolution imaging

One of the aims of this school is to discuss the need for high resolution imaging in several areas of astronomy. For the study of distant (radio) galaxies high resolution imaging at a range of wavelength is absolutely critical, and here I will comment on a few issues.

First of all, the relevant angular size scales of the optical/near infrared emission is small. For the UV dropout galaxies, HST has measured typical sizes of 1-10 kpc, corresponding to a few tenths of an arcsec. The overall extent of the optical light observed from distant radio galaxies is more of order 3-4 arcsec. However, many of the clumps that are observed have size scales very similar to those of the UV dropouts.

To be able to study the detailed morphology as a function of redshift, it is important to be able to observe these distant objects at resolutions of order 0.1 arcsec. Issues that can then be addressed include: when do the typical profiles of ellipticals and spirals form?; what fraction of distant galaxies are merging systems?; how many of the distant objects have obvious dust lanes in their centre? It would be very important to also carry out such high resolution imaging at the reddest wavelengths. This would constrain the history of the star formation within the individual components of distant (radio) galaxies and the physical nature of the (radio galaxy) alignment effect.

To study the inter stellar medium (ISM) in these objects, it would be very valuable to obtain high spatial resolution images of the absorption and emission line gas. For the radio galaxies, this would allow important studies of how the radio jets interact with their environment. For the galaxies, the spatial distribution of the ISM can then be compared to that of the stars. This is important for understanding of the origin of the ISM, which on the one hand is the material which the stars are formed from, but on the other hand is the material which is metal enriched through mainly supernova explosions.

Finally, there is the issue that adaptive optic systems can only be used if there is a bright star near a science target. For the study of the general population of distant field galaxies this is not so much of a problem, basically since a field to be observed can be selected to be situated close to a bright star. In the case of the radio galaxies, this requires samples of radio galaxies with spectroscopic redshifts next to bright stars. With the new large radio surveys such samples can indeed be created.

Acknowledgements

I would like to thank the organisers for organising such a pleasant and fruitful meeting. I would like to thank Cedric Lacey for providing me with important input for this contribution. I am grateful to Philip Best, Stephan Charlot, Chris Carilli, Carlos de Breuck, Jim Dunlop, Jaron Kurk, George Miley and Laura Pentericci, for discussions and for allowing me to present material prior to or after publication.

References

1. Baron E., White S. D. M., 1987, ApJ, 322, 585
2. Baugh C. M., Cole S., Frenk C. S., 1996, MNRAS, 283, 1361
3. Baugh C. M., Cole S., Frenk C. S., Lacey C. G., 1998, ApJ, 498, 504
4. Baum S. A., O'Dea C. P., Giovannini G., Biretta J., Cotton W. B., De Koff S., Feretti L., Golombek D., Lara L., Macchetto F. D., Miley G. K., Sparks W. B., Venturi T., Komissarov S. S., 1997, ApJ, 483, 178
5. Becker R., White R., Helfland D. J., 1995, ApJ, 450, 559
6. Begelman M. C., Cioffi D. F., 1989, ApJL, 345, L21
7. Best P., Carilli C., Garrington S. T., Longair M. S., Röttgering H. J. A., 1998, MNRAS, 299, 357
8. Blundell K. M., Rawlings S., 1999, Nat, 399, 330
9. Bruzual A. G., Charlot S., 1993, ApJ, 405, 538
10. Carilli C. L., Barthel P. D., 1996, ARA&A, 7, 1
11. Carilli C. L., Harris D. E., Pentericci L., Röttgering H., Miley G. K., Bremer M. N., 1998, ApJL, 496, L57
12. Chambers K. C., Miley G. K., van Breugel W. J. M., 1990, ApJ, 363, 21
13. Cole S., 1991, ApJ, 367, 45
14. Cole S., Aragon-Salamanca A., Frenk C. S., Navarro J. F., Zepf S. E., 1994, MNRAS, 271, 781
15. Condon J., 1988, in Verschuur G., Kellerman K., eds, Galactic and Extragalactic Astronomy, p. 641, Springer-Verlag New York Inc.

16. Condon J. J., Cotton W. D., Greisen E. W., Yin Q. F., Perley R. A., Taylor G. B., Broderick J. J., 1998, AJ, 115, 1693
17. Dreher J. W., Carilli C. L., Perley R. A., 1987, ApJ, 316, 611
18. Dunlop J., 1998, in M. Bremer, N. Jackson and I. Pérez-Fournon (eds.), Observational Cosmology with the New Radio Surveys, p. 157
19. Dunlop J. S., Hughes D. H., Rawlings S., Eales S. A., Ward M. J., 1994, Nat, 370, 347
20. Efstathiou G., Rees M. J., 1988, MNRAS, 230, 5P
21. Eisenstein D. J., Loeb A., 1995, ApJ, 443, 11
22. Giavalisco M., Steidel C. C., Macchetto F. D., 1997, ApJ, 470, 189
23. Haehnelt M. G., Natarajan P., Rees M. J., 1998, MNRAS, 300, 817
24. Haehnelt M. G., Rees M. J., 1993, MNRAS, 263, 168
25. Hughes D. H., Dunlop J. S., Rawlings S., 1997, MNRAS, 289, 766
26. Ivison R. J., Dunlop J. S., Hughes D. H., Archibald E. N., Stevens J. A., Holland W. S., Robson E. I., Eales S. A., Rawlings S., Dey A., Gear W. K., 1998, ApJ, 494, 211
27. Kauffmann G., Guiderdoni B., White S. D. M., 1994, MNRAS, 267, 981
28. Kauffmann G., White S. D. M., Guiderdoni B., 1993, MNRAS, 264, 201
29. Lacey C., Guiderdoni B., Rocca-Volmerange B., Silk J., 1993, ApJ, 402, 15
30. Lacey C., Silk J., 1991, ApJ, 381, 14
31. Loeb A., 1993, ApJL, 404, L37
32. Longair M. S., Best P. N., Röttgering H. J. A., 1995, MNRAS, 275, L47
33. Madau P., Ferguson H. C., Dickinson M. E., Giavalisco M., Steidel C. C., Fruchter A., 1996, MNRAS, 283, 1388
34. McCarthy P. J., 1993, ARA&A, 31, 639
35. MCCarthy P. J., Kapahi V. K., Van Breugel W., Persson S. E., Athreya R., Subrahmanya C. R., 1996, ApJS, 107, 19
36. Papadopoulos P. P., Röttgering H. J. A., van der Werf P. P., Guilloteau S., Tilanus R. P. J., van Breugel W. J. M., 1999, CO(4-3) and dust emission from two powerful high-z radio galaxies, submitted
37. Peebles P., 1993, Principles of Physical Cosmology, Princeton Univ. Press
38. Pentericci L., Röttgering H., Miley G. K., Carilli C. L., McCarthy P., 1997, A&A, 326, 580
39. Pentericci L., Röttgering H. J. A., Miley G. K., McCarthy P., Spinrad H., van Breugel W. J. M., Macchetto F., 1999, A&A, 341, 329
40. Pentericci L., Röttgering H. J. A., Miley G. K., Spinrad H., McCarthy P. J., Van Breugel W. J. M., Macchetto F., 1998, apj, 504, 139
41. Perley R. A., Dreher J. W., Cowan J. J., 1984, ApJL, 285, L35
42. Rawlings S., Lacy M., Blundell K. M., Eales S. A., Bunker A. J., Garrington S. T., 1996, Nat, 383, 502
43. Rees M. J., Ostriker J. P., 1977, MNRAS, 179, 541
44. Rengelink R. B., Tang Y., de Bruyn A. G., Miley G. K., Bremer M. N., Röttgering H. J. A., Bremer M. A. R., 1997, A&AS, 124, 259
45. Röttgering H., Hunstead R., Miley G. K., van Ojik R., Wieringa M. H., 1995, MNRAS, 277, 389
46. Röttgering H., Miley G. K., 1996, in Bergeron J., ed., The Early Universe with the VLT, pp 285–299, Springer-Verlag
47. Röttgering H., van Ojik R., Miley G., Chambers K., van Breugel W., de Koff S., 1997, A&A, 326, 505
48. Röttgering H. J. A., Best P., Pentericci L., de Breuck C., van Breugel W., 1998, in S. D'Odorico, A. Fontana, E. Giallongo (eds), The Young Universe, ASP Conf. Ser. 146, p. 49
49. Sanders D. B., Mirabel I. F., 1996, Astron. Astrophys. Rev., 34, 749
50. Sawicki M., Yee H., 1998, AJ, 115, 1329

51. Scoville N. Z., Yun M. S., Windhorst R. A., Keel W. C., Armus L., 1997, ApJL, 485, L21
52. Silk J., Rees M. J., 1998
53. Simpson C., Eisenhardt P., Armus L., Chokshi A., Dickinson M., Djorgovski S. G., Elston R., Jannuzi B. T., McCarthy P. J., Pahre M. A., Soifer B. T., 1999, ApJ, in press, astro-ph/9906502
54. Steidel C., Adelberger K. L., Kellogg M., Dickinson Giavalisco M., 1998, to be published in: 'ORIGINS', ed. J.M. Shull, C.E. Woodward, and E H. Thronson, (ASP Conference Series)
55. Steidel C., Hamilton D., 1992, AJ, 104, 941
56. Steidel C., Hamilton D., 1993, AJ, 105, 2017
57. Steidel C., Pettini M., Hamilton D., 1995, AJ, 110, 2519
58. Steidel C. C., Giavalisco M., Pettini M., Dickinson M., Adelberger K. L., 1996, ApJL, 462, L17
59. van Breugel W., de Breuck C., Stanford S. A., Stern D., Röttgering H. J. A., Miley G. K., 1999, ApJL, 518, 61
60. Van Kampen E., Katgert P., 1997, MNRAS, 289, 327
61. van Ojik R., Röttgering H., Carilli C., Miley G., Bremer M., 1996, A&A, 313, 25
62. van Ojik R., Röttgering H. J. A., Miley G. K., Hunstead R., 1997, A&A, 317, 358
63. Weinberg S., 1972, Gravitation and Cosmology, John Wiley & Suns, Inc.
64. White S. D. M., Frenk C. S., 1991, ApJ, 379, 52
65. White S. D. M., Rees M. J., 1978, MNRAS, 183, 341
66. Zombeck M. V., 1990, Handbook of Space Astronomy and Astrophysics -2nd edn, Cambridge University Press

CHAPTER 15
ACTIVE GALAXIES IN THE LOCAL AND DISTANT UNIVERSE

CLÁUDIA S. ROLA
Institute of Astronomy, University of Cambridge
Madingley Road, Cambridge CB3 0HA, United Kingdom

1. WHEN IS A GALAXY ACTIVE ?

1.1. INTRODUCTION

Let's start by defining what we understand by an active galaxy. The term "active" can mean various different things. The first meaning that obviously comes to our minds, is that of a non passive (non-static) object, i.e., one which characteristics may be changing, evolving, transforming in time. Concretely, many phenomena make a galaxy become active. These are for instance, from the smaller scales to the larger scales: star formation, stellar evolution (e.g. supernovas), strong stellar winds, H II regions, formation of bubbles and super-bubbles, jets, etc.

As Bernard Pagel once said at Manchester in 1984: "All galaxies are active, but some are more active than others !". Indeed, all galaxies have stars, and stars evolve. In this context, elliptical galaxies would be the less active ones as they have little or no gas and their spectra presents only continuum and absorption lines. Which galaxies are the most active? This is not a straightforward question as the answer will depend also on the wavelength range one is referring to. Globally speaking, the title of being the most active galaxies, is attributed to "active galactic nuclei" (or AGN). This term AGN, refers to the existence of energetic phenomena in the nuclei, or central regions, of galaxies which cannot be attributed clearly and directly to stars.

The two largest subclasses of AGNs are Seyfert galaxies and quasars, the real boundary between them not being precisely defined, spectroscopically[1]

---

[1] For the case of Seyfert 1s (see later in the text).

nor in terms of the lower luminosity objects[2].

In the context of "activity", situated intermediate between ellipticals and AGNs are the "starburst galaxies", a type of the larger group of HII galaxies (see below). These present also energetic phenomena in the nuclei, but this is clearly associated with the presence of hot, massive main sequence stars. Both HII galaxies and AGNs present what we call an *emission-line spectra* and for this are called *emission-line galaxies* (ELGs).

However, this is just a simple view of the variety of ELGs the Universe has to show us. In reality, AGNs divide in some more subclasses and even HII galaxies whose emission is clearly dominated by massive stars, show different characteristics.

By improving considerably the observing performances of ground based telescopes in the visible and in the near infrared, adaptive optics (hereafter AO) can bring a new insight into the imaging and spectroscopy of astronomical objects. Active galaxies are one out of several interesting targets to gain from its development. AO allows to overcome the limitations induced by atmospherical turbulence, providing stable images, nearly diffraction limited, and improving the spectral resolution and signal-to-noise ratio, making full use of the telescope's capabilities. Details on how does AO works are fully described in this book by other lecturers, and I will therefore, not repeat what was said. Instead, I prefered to stress out the observational requirements which are important to progress the research on active galaxies and that can be achieved by the use and development of AO.

Let us first start by considering in some detail the properties and taxonomy of ELGs.

1.2. TAXONOMY OF EMISSION-LINE GALAXIES

Many galaxies have emission-lines in their spectra and in many of them in particular, the emission-lines are emitted by their nuclei. Many of these nuclei contain large numbers of luminous OB stars, which ionize the interstellar gas surrounding them, producing an HII region-like spectrum. The most luminous of these are often called "starburst galaxies". Other galaxies contain large regions (of the order of hundreds of parsecs in diameter) of gas ionized by OB associations located in their spiral arms, or distributed across the galaxy in the case of irregulars. These are often called "HII region-like galaxies".

The relationship between AGNs and nuclear starbursts is still not clear.

On one hand, there is the idea of a possible evolution between the two types of phenomena. Yet, another view of the problem is that they are

---

[2]Quasars and Seyferts luminosity curves overlap, i.e. the most luminous Seyferts are more luminous than the least luminous quasars).

different manifestations of the same phenomenon. I will comment on this later in the text.

1.2.1. *H II galaxies*

A significant fraction of galaxies show the signature of recent large-scale star formation activity. Such galaxies are known widely as "starburst galaxies". These are generally characterized by relatively blue colours and strong HII-region-type emission-line spectra (due to a large number of O and B-type stars) and relatively strong radio emission (due to supernova remnants). In many cases, the starburst seems to be confined to an unresolved region at the galactic center, which looks very much like an active nucleus. These "nuclear starbursts" are typically around 10 times brighter than the giant HII-region complexes seen in normal spirals and are thus distinct from otherwise inactive late-type spirals. These "nuclear starburst" galaxies (or starburst galaxies, for short) as well as the other emission-line galaxies presenting HII-region-type emission-line spectra make part of the group of H II galaxies.

The most "typical" HII galaxies are dwarf galaxies, undergoing violent star formation, in which thousands of massive stars ($M > 20\ M_\odot$[3]) have recently been formed in a very small volume (of about tens to hundreds of parsecs in diameter) and over a time scale of only a few million years. Their integrated spectra are identical to those of giant extragalactic HII regions associated with normal late Hubble type spirals or irregular galaxies like 30 Doradus in the Large Magellanic Cloud (LMC) and NGC 604 in M 33. Sargent and Searle (1970) called these galaxies "isolated extragalactic HII regions". Other terms like "intergalactic HII regions" or "blue compact galaxies" (BCGs) are also found in the literature. All these make part of the variety of HII galaxies[4]. General statistical spectroscopic properties of HII galaxies can be found, for instance, in the "Spectrophotometric Catalogue of HII galaxies" (Terlevich *et al.* 1991).

HII galaxies are among the most luminous narrow emission-line objects in the sky. The observed Hβ luminosities[5] are frequently in the range of 10^{40} to 10^{41} erg s^{-1} with some galaxies reaching almost 10^{43} erg s^{-1}. The magnitude of the stellar continuum (not including the emission-lines) ranges from $M_B = -14$ to -24 mag.

Despite the fact that they all contain at least one giant active region of star formation whether centered on the nucleus or not, this class of objects

---

[3]$M_\odot = 1.9891 \times 10^{33}$ g.
[4]Note however that not all BCGs have giant HII regions as the dominant components in their integrated spectra. Hence some authors prefer not to call some BCGs, HII galaxies.
[5]$H_0 = 50$ km s^{-1} Mpc^{-1}

presents a wide variety of morphologies of the underlying galaxy. Although HII galaxies tend to be associated with late-type spirals (Sb-Sc's and some Irregulars), recent work (e.g., Telles *et al.* 1997) has revealed that many objects show possible features of interacting or merging systems (wisps, tidal tails or irregular fuzzy extensions). Telles and collaborators also find that one-third of the galaxies in their sample are actually single compact objects with no evidence for extensions or fuzz. Additionally, they point out that the perception of the morphology depends very much on spatial resolution. In fact, while individual star-forming regions within the galaxies may be resolved only for nearby brighter objects, the morphological details of systems at higher redshifts (say, $z > 0.02$) may be smeared out, rendering the galaxies with a smoother compact appearance.

This gives an idea of the variety of questions still needing an answer regarding the morphology of HII galaxies, for which one can add an interrogation on whether it exists any type of evolution between the different types of HII galaxies. In my view, these questions may be answered by AO in the near future, which will also permit to detect any faint interacting companions. AO will be also able to resolve into stars, many star-forming regions in HII galaxies. This will allow to determine the low mass, faint star distribution in these regions. As an example, Fig. 1 illustrates the first result of a programme by Brandl *et al.* (1996) aimed at constraining the distribution of stellar masses in the active region R136 of the 30 Doradus nebula in the LMC.

1.2.2. *Active galactic nuclei*

The first detection of an AGN was done in 1908 by E.A. Fath, which observed the intense emission-lines of NGC 1068, later confirmed in 1917 by V. M. Slipher. However, Carl Seyfert (1943) was the first to realize that there are several similar galaxies which form a distinct class. He observed a group of six galaxies[6], selected on the basis of stellar-appearing cores, for which optical spectra revealed very intense and broad emission-lines. Other types of AGNs, like the radio-galaxies and the quasars, were detected for the first time after the Second World War, thanks to the development of radio-astronomy.

The classification of AGNs tends to be rather confusing as we do not yet understand the physics underlying the AGN phenomenon. Undoubtedly some of the differences we see between various types of AGNs are due more to the way we observe them than to fundamental differences between the various types. Unlike spectra of stars or normal galaxies, AGN spectra cannot be described in terms of blackbody emission at a single temperature, or as a composite over a small range in temperature.

---

[6]These were NGC 1068, NGC 1275, NGC 3516, NGC 4051, NGC 4151 and NGC 7469.

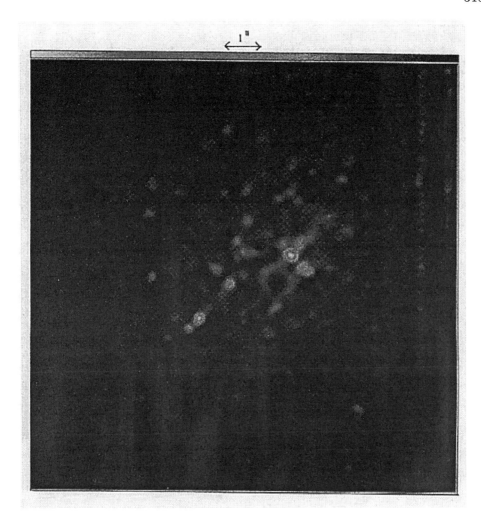

Figure 1. The R136 region in the LMC 30 Doradus nebula, imaged with the 3.6 m ESO telescope at 2.2 μm by adaptive optics/SharpII. The resolution is 0.2" FWHM, north is up and east left. The scale is indicated. Fainter stars in the cluster may be either OB stars or perhaps red supergiants (taken from Beuzit *et al.*, 1994, *The Messenger*, vol. 75, pp. 33).

The fundamental question about AGNs is how the energy that is detected as radiation is generated. Essentially the problem is that an AGN produces as much light as up to several trillion stars in a volume that is significantly smaller than a cubic parsec. One of the models proposed to explain the AGN phenomenon is a "central engine" that consists of a hot accretion disk surrounding a supermassive black-hole. Energy is generated by gravitational infall of material which is heated to high temperatures in a dissipative accretion disk (see e.g. Rees 1984; Begelman 1985). The alternative to the black-hole scenario is the "nuclear starburst" model (see e.g. Terlevich et al. 1992). In this model, the radiated energy is supplied by compact supernova remnants[7] which are responsible for the UV/optical variability, X-ray and radio emission and broad emission-lines.

Globally speaking, an AGN can be defined as a galactic nuclei that emits a large quantity of energy ($\approx 10^{41-48}$ erg s^{-1}), over all the electromagnetic spectrum (from gamma rays to radio waves), from a small dimension source (of the order of some light-years). Furthermore, we should add that the single most common characteristic of AGNs is probably that they are all luminous X-ray sources (Elvis et al. 1978).

Let us now detail about the more common types of AGNs.

Seyfert galaxies A Seyfert galaxy has a quasar-like nucleus, but the host galaxy is clearly detectable. Seyfert galaxies are lower-luminosity AGNs, with $M_B > -21.5 + 5 \times \log h_0$ for the active nucleus[8], which is the criterion used to distinguish them from quasars. This criterion is due originally to Schmidt and Green (1983). The luminosities of Seyfert galaxies range from over 10^{11} up to $\sim 5 \times 10^{12}$ $L_\odot$[9]. The brightest one has less than ten times the luminosity of the brightest spiral galaxies. On the present evidence, there seems to exist some correlation between the degree of Seyfert activity and the absolute magnitude of the underlying galaxy. In fact, the more luminous the galaxy, the more powerful the active nucleus.

The emission-line spectra of Seyferts can be classified into two types, following a scheme first proposed by E. Khachikian and D. W. Weedman (1974), which are distinguished by the presence or absence of broad bases on the permitted emission-lines.

Type 1 Seyfert galaxies have two sets of emission-lines, superposed on one another. One set of lines (forbidden lines) is emitted by a region of ionized gas with electron density, $n_e \approx 10^3 - 10^6$ cm^{-3}. The widths of these lines, like [O III]$\lambda\lambda$4959, 5007, [N II]$\lambda\lambda$6548, 6583 and [S II]$\lambda\lambda$6717, 6731, typically correspond to velocities of the order of $5 \times 10^2 - 1000$ km s^{-1} and are referred to as "narrow lines"[10]. The other set is constituted by permitted "broad lines", originating mainly from hydrogen but also accompanied by lines from species such as HeI (neutral helium), HeII (singly ionized helium) and FeII (singly ionized iron). These lines have Full Width at Half Maximum (FWHM) corresponding to velocities in the range $1-10^4$

---

[7]hereafter cSNRs; cSNRs are ordinary SNRs but evolving in a dense circumstellar medium with $n > 10^6$ cm^{-3}. Such high density is produced by the interstellar medium (ISM) being pressurized by stellar winds and SNRs.

[8]Note that h_0 is the Hubble constant in units of 100 km s^{-1} Mpc^{-1}.

[9]$L_\odot = 3.826 \times 10^{33}$ erg/s.

[10]Note however that they are somewhat broader than the emission-lines in HII galaxies, which have generally widths corresponding to $100 - 200$ km s^{-1}.

km s^{-1}. The absence of broad forbidden-line emission indicates that in the region were the broad lines are emitted the ionized gas has high density ($n_e \geq 10^9$ cm^{-3}).

Besides the information associated with the presence or absence of forbidden lines (which is an indicator of the gas density in the emitting region), these different line widths give clues to the origin of the lines, in terms of the velocity dispersion of the gas that is emitting the line. A simple accepted interpretation is that when lines from the same element or ion have the same width, they are probably formed in the same region. Therefore, in Seyfert 1s the permitted lines and forbidden lines originate from distinctly differing regions.

In this context, the very broad lines delineate what is known as the "broad line region" (BLR) of the emission zone. Within the black hole model, the broad-line clouds are caused by photoionization due to a very hot accretion disk surrounding a supermassive black hole in the nucleus of the galaxy. These clouds are located at about 1 parsec from the central engine. The narrow lines come from what is called the "narrow line region" (NLR). This is believed to lie at a distance from ~10 pc to ~1 kpc from the central engine.

Within the starburst model, cSNRs produce a nuclear spectra similar to a Seyfert 1, after an evolutionary timescale of ~8 Myr, which began by a HII region (or LINER) phase, depending on the density and metallicity). As an example of the "potential" of cSNRs to explain the engine of AGNs, it is worthy to mention the discovery of objects like SN 1987F (Filippenko 1989) and SN 1988Z (Stathakis & Sadler 1991). The spectra of such objects is so similar to that of AGN that led Filippenko to call SN 1987F a "Seyfert 1 impostor".

Type 2 Seyfert galaxies differ from Seyfert 1 galaxies in that only the narrow lines (forbidden and permitted) are present in their spectra and their widths are about equal. These narrow lines are still broader than the ones observed in HII galaxies and their FWHM range up to about 1000 km s^{-1}.

In the case of Seyfert 2s, both permitted and forbidden lines are formed in the same region of the nucleus, in a zone which would be in principle similar to the NLR of Seyfert 1s.

Osterbrock (1981) has introduced the notation *Seyfert 1.5, 1.8 and 1.9*, where these subclasses are based purely on the appearance of the optical spectrum. This subclassification is such that numerically larger subclasses present weaker broad-line components relative to the narrow lines. Therefore, in Seyfert 1.9 galaxies the broad component is detected only in the Hα line and not in the higher-order Balmer lines. In Seyfert 1.8 galaxies, the broad components are very weak, but detectable at Hβ as well as Hα. In Seyfert 1.5 galaxies, the strengths of the broad and narrow components in Hβ are comparable.

Additionally to the strong emission-lines, absorption lines due to late-type giant stars in the host galaxy are also observed in Seyfert spectra. The weakness of most of the observed absorption lines is due to the fact that starlight is diluted by the non-stellar "featureless continuum"[11]. Also, given that the AGN continuum mixes with the stellar continuum, in the case of Seyfert 2s the AGN continuum is so weak that it is very difficult to isolate it from the stellar component unambiguously.

Most of the Seyfert galaxies for which a morphological type could be determined were found to be spirals, and for those that are sufficiently close to be well resolved, they seem to favour barred spirals. Additionally, Seyferts tend to populate the early-type spirals (Sa and Sb) and are rare in Sc and later categories, which have much smaller and less pronounced nuclear zones. Furthermore, about 10% of Seyfert 1s occur in elliptical galaxies.

Another characteristic of Seyfert galaxies is their variability, which was first discovered by Fitch *et al.* (1967) for the continuum emission of NGC 4151. Dibai & Pronik (1967) and Andrillat & Souffrin (1968) observed that the broad component of the Balmer lines in NGC 3516 had disappeared since the observations made by Seyfert in 1943.

Soon after, it was discovered that the variability observed in the continuum and in the emission-lines is related (Lyutyi 1973), as one would expect if the gas which emits the lines is photoionised by a variable continuum. The variability exists for all wavelengths[12] and affects the continuum (in its flux and slope) and the broad emission-lines (in their flux and profile).

LINERs There are many nuclei whose spectra do not show strong enough high-ionization emission-lines to be called Seyfert galaxies, but yet are not HII region-like. Heckman (1981) first clearly recognized these objects and gave them the name of "LINERs" (for Low-Ionization Nuclear Emission-line Regions). His criteria to define this class of objects is given by two line intensity ratios: [O II]λ3727/[O III]λ5007 $\geq$ 1 (as in typical HII region galaxies) *and* [O I]λ6300/[O III]λ5007 $\geq$ 1/3 (much larger than in HII region galaxies). For an ELG to be classified as a "true" LINER, it has *to obey both these criteria* and not only one of them as is sometimes used.

Spectroscopically, LINERs resemble Seyfert 2 galaxies, except that the low-ionization lines, like [O I]λ6300 and [N II]$\lambda\lambda$6548, 6583 for example, are

---

[11] or AGN continuum.

[12] Note, however, that the existence of variability in the far infrared and in radio is still not fully confirmed.

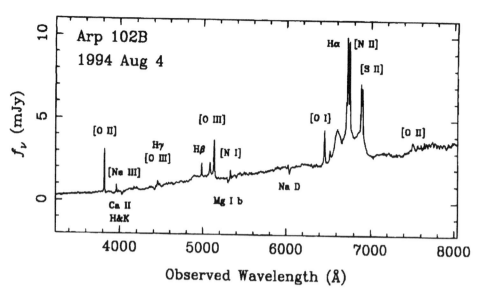

Figure 2. An example of the optical spectrum of Arp 102B, obtained on 4 August 1994 UT (Halpern et al. 1996). The most prominent narrow emission-lines are identified. Residual fringing of the CCD is visible redward of 7500 Å.

relatively strong. Furthermore, they are distinguishable from normal spiral galaxies by frequently having compact flat spectrum radio cores and some show point-like X-ray emission. In the optical however, the nuclear emission is weak compared to the surrounding starlight from the underlying galaxy.

Initially it was thought that the emission-line spectra of LINERs was the result of "shock-wave heating" (Heckman, 1980) by the conversion of kinetic energy of mass motion into heat, causing collisional ionization and excitation. However, later, the comparison of photoionization models, considering a small ionization parameter[13] and a power law type ionizing spectrum (Ferland & Netzer 1983; Halpern & Steiner 1983) with the observed spectra of LINERs has shown a much better agreement. The weak points of shock-heating models are generally related with their non prediction of the emission in He IIλ4686 and their prediction of a strong [O III]λ4363 emission, both contrary to the observations. Additionally, further observations of LINERs have sometimes shown the presence of a weak broad Hα line. A famous case is Arp 102B which spectrum is shown in Fig. 2 (see Halpern et al. 1996). A survey at high angular and spectral resolution of a significant number of LINERs would be important to establish if this is true for all LINERs. Furthermore, it would allow to probe the core of these galaxies to

[13] ratio of density of ionizing photons to the electron density

investigate which type of mechanism is behind the LINER phenomenon. Is it the same type of "engine" that powers Seyfert galaxies? Is a LINER an evolutionary stage in the "life" of an AGN or in galaxy evolution? Here is another thrilling subject, in a still challenging domain, for which AO could help give an important step forward.

Many LINERs were identified in surveys of spiral galaxies by Stauffer (1982) and by Keel (1983). LINERs are very common objects and might be present at detectable levels in nearly half of all spiral galaxies (e.g., Ho, Fillipenko and Sargent 1993). Recently, the analysis of the Canada-France Redshift Survey (CFRS) has suggested the existence of an excess of Seyfert 2 and/or LINERs up to a redshift z=0.3 (e.g., Tresse *et al.* 1996). They obtained that about 17% of all galaxies up to z = 0.3 are narrow-line AGNs (more probably of LINER-type), which contrast with the 2% obtained for the local (z $\approx$ 0) Universe (see e.g. Huchra & Burg 1992). Locally, LINERs are mainly found in the nuclei of Sa and Sb galaxies plus some in Sc and a subset of peculiar galaxies. Indeed, being a common phenomenon in the Universe, it is especially important to discover what is behind their emission.

Quasars These constitute the most luminous subclass of AGNs, with nuclear magnitudes M_B < -21.5 + 5 × log h_0. Quasars present very broad emission-lines with widths of up to 10 000 km s^{-1}. They are distinguished from Seyfert galaxies in that in general they have angular sizes smaller than $\approx$ 7$''$. However, many quasar sources are surrounded by a low surface brightness halo, which appears to be starlight from the host galaxy, and a few sources have other peculiar morphological features, such as optical jets (e.g., 3C 273).

The spectra of quasars are remarkably similar to those of Seyfert galaxies, except essentially in two aspects. Their optical stellar absorption features are very weak, if detectable at all, and the narrow lines are generally weaker relative to the broad lines than is the case in Seyferts.

The overwhelming majority of quasars have weak radio continuum emission compared to the optical and these are often referred to as "radio-quiet quasars"[14]. Radio-loud quasars are less than 10% of the whole class. Both radio-quiet and radio-loud quasars have relatively steep infrared through optical continuum emission, the flux declining to shorter wavelengths, superimposed on very broad emission-lines. This IR-optical continuum follows a roughly power-law form (a straight line, usually of negative slope on a plot of log flux versus log frequency) and like Seyferts, appears to peak

---

[14]In the past, the radio-quiet quasars were called quasi-stellar objects (QSOs). In the text I use the term quasar for the entire class, and refer to a "radio-loud" quasar when it has strong radio emission.

somewhere in the region of 100 μm. By contrast, the radio-loud quasars do not show the prominent peak in the far-infrared/submillimetre region. Some quasars, both radio-loud and radio-quiet, show an excess emission in the blue to UV parts of the spectrum. This has been named the "big blue bump". The quasar flux normally declines through the UV and X-ray regimes, but some have significant X-ray excesses.

Quasars extend in redshift from z < 0.1 to almost z = 5 and have luminosities that go up to 10^{13} $L_\odot$. Therefore, the lower level of luminosity for quasars overlaps that of Seyfert galaxies in that the most luminous Seyferts are more luminous than the least luminous quasars.

In terms of the underlying galaxy, for which the quasar constitutes its nucleus, CCD images taken in the mid-1980s produced evidence of an underlying fuzz, showing the morphology of the host galaxies. Determining whether quasars lie in particular types of galaxies can be done by either investigating the colours or by analysing the profiles of the underlying galaxy. The observations are not fully conclusive yet but many authors suggest that QSO activity might be occuring in the central regions of young massive ellipticals (e.g. Taylor *et al.* 1996; Aretxaga *et al.* 1997). Moreover, spectroscopic studies of the underlying fuzz remain extremely difficult and none have conclusively demonstrated stellar absorption lines typical of starlight from a galaxy. In the near future, observations with AO and the new 8 m and 10 m telescopes will allow a more detailed picture of the quasars' host galaxies.

Other categories I decided to group BL Lacertae objects and optically violent variable quasars (OVVs) together because to a first approximation they show strong similarities of radio-loud flat spectra and variability. This flat radio spectra seems to be produced by a process called synchrotron emission (relativistic electrons in a magnetic field) from a powerful relativistic jet.

BL Lac objects were named after the first member of the class discovered. This was an object previously suspected of being a variable star in our Galaxy and which had been catalogued under the name of BL Lacertae. BL Lac became famous in 1968 because of the identification of a compact and highly variable radio source (VRO 42.22.01) which was coincident with the "star" BL Lac. This previously thought 'star", soon revealed to be quite strange as its optical spectrum was featureless, showing neither absorption nor emission lines. Further work lead astronomers to propose that BL Lac and a handful of other sources were candidates for a new class of extragalactic objects.

Classical BL Lacs are radio-loud objects with a featureless continuum spectrum which shows strong polarization and rapid variability on timescales of days or a few hours. The continuum (synchrotron) emission rises steeply from UV through optical and infrared wavelengths until at a wavelength of around 1mm it turns over and continues to longer wavelengths with a strength which is almost independent of frequency.

With the improvement in detection techniques and more targeted observations, emission-lines have been observed in the optical part of the spectra of a number of BL Lac objects, particularly when they are in a faint phase. These lines, being redshifted, served as a proof that BL Lacs are indeed extragalactic. A number of BL Lacs whose distance has not been determined from the redshift of emission-lines are known to lie at least at a certain minimum distance from us because of the detection of an absorption line due to the light from the BL Lac passing through intervening matter[15] between it and ourselves. In general, the redshifts of these objects tend to be small ($z \leq 0.2$), but some are much more distant.

In OVVs, broad emission-lines are prominent spectral features, except when the continuum is at its brightest, when the emission-line equivalent widths become small as the line flux changes little or not at all while the continuum increases dramatically. Their variability has been found to be very erratic, often consisting of yearly periods of stability followed by rapid flaring of many magnitudes in timescales of weeks or even days. OVVs polarization properties, power-law optical through infrared continuum and flat radio spectra are very similar to the continuum spectra of classical BL Lacs. The redshifts of OVVs tend to be large ($z \geq 0.5$).

Collectively, BL Lacs and OVVs are generally referred to as "blazars"[16]. All known blazars are radio-loud, core-dominated sources and possess a beamed emission from a relativistic jet, produced in the core of the galaxy. An important feature of blazars is that the peak of their spectral energy distribution (SED) occurs at different wavelengths for different sources. Indeed, some emit most of their energy in the IR, while others in the UV/optical.

No completely satisfactory model for the blazar continuum exists. Models that seem to be able to account most successfully for the variability and polarization characteristics involve shocks propagating through an inhomogeneous plasma jet.

---

[15] The intervening matter is probably the halo of a galaxy through which the line-of-sight passes.

[16] The name, suggested by Ed. Spiegel in 1978, derives from a mix of "BL" from BL Lac and "azar" from quasar.

2. UNIFICATION MODELS

In 1985, when Antonucci and Miller showed the Seyfert 2 galaxy NGC 1068 to possess a BLR, albeit only in polarized light (and hence by reflection), astronomers were left with the appealing possibility that the differences we see in the various AGN types might be due to, among other factors, the orientation of the nuclear zone and torus with respect to us[17]. If AGNs were lying more edge-on than face-on then this torus would hide the central zones and at some angle would hide the entire BLR. The much more extended NLR would remain visible, however. This is the foundation of the unified model for AGN.

Within the unification model, the BLR is hidden from direct view and we only see it by reflection (polarized light) off some form of mirror. Presently, it is thought that such mirror is constituted by a thick disk or torus of obscuring material surrounding the central engine, but inside the NLR. This "molecular torus", as it is often called, is believed to be constituted by a combination of dust and molecular gas, making it opaque to the hard X-rays which are observed in Seyfert 1s. The observed polarized spectra can result from scattering or reflection of the AGN continuum, either by dust or by free electrons. Being molecular in composition, the torus cannot lie too close to the "central engine", otherwise the powerful radiation field will immediately dissociate the molecules unless there is extensive dust shielding. Even then, it must lie at a sufficient distance to prevent evaporation of the grains. This gives an inner radius of the order of a parsec in extent, which makes it a very compact phenomenon.

Therefore, in this scheme, Seyfert 1s are distinguished from their counterparts, Seyfert 2s, by the orientation of the obscuring torus. If the torus is seen face-on, our view of the central regions is unobstructed and we detect broad lines. If our view is closer to edge-on, the central regions are not seen directly and no broad lines are detected. BL Lacs and OVVs are the face-on versions of radio sources (i.e., the radio-source axis parallel to the line of sight, or "pole-on"), with the BL Lacs corresponding to the low-luminosity sources and the OVVs corresponding to the more luminous sources. Figure 4 illustrates the basic concept of simple unification models, and shows in a qualitative way how classification might be line-of-sight dependent.

Although the more habitual scenario for the central engine of AGNs within the unification model is the black-hole model, it is worthy to remark that the starburst model is also compatible with the former. In fact, the starburst model contemplates that, at least in some cases, the differ-

---

[17] Since then similar cases have been observed among Seyfert 2s with polarized continua (e.g., Tran *et al.* 1992).

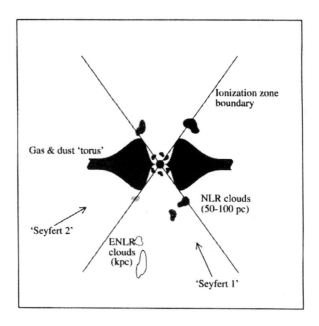

Figure 3. A conceptual scheme for unification of Seyfert 1 and Seyfert 2 galaxies, not to scale. A highly opaque dusty torus surrounds the continuum source (black dot at center) and broad-line region (clouds near center). These cannot be observed directly by an observer close to the torus midplane, although the narrow-line region and extended narrow-line region can be seen directly. This would lead the observer to classify the galaxy as a Seyfert 2. An observer closer to the torus axis would have an unobscured view of the nuclear regions and classify the same galaxy as a Seyfert 1 (Figure by R. Pogge, taken from Peterson, 1997).

ence between Seyfert 2s and Seyfert 1s is not an evolutionary one, but an orientation effect (e.g., Cid-Fernandes, 1997).

Much of the work on unified models is morphological in nature, i.e., based on searches for correlations among observed parameters in the hope that this will lead to physical insight. The assumption is that there is less intrinsic diversity among AGNs than we observe, and that the wide variety of AGN phenomena we see is due to a combination of real differences in a small number of physical parameters (like luminosity) coupled with apparent differences which are due to observer-dependent parameters (like orientation). Undoubtedly, much has yet to come from high-spatial resolution observations of AGNs by the new-generation telescopes.

Just to give some numbers about the spatial resolution required to investigate the core of AGNs, let us consider an example given by M. Ward (1995), of an object at a redshift $z = 0.01$. For this object 1 arcsecond correspond to 300 pc[18]. *For this AGN what will be the angular sizes of its various components?* The size of the accretion disk around the black-hole (if existent) will have about 3 μ arcseconds. The BLR would have the dimension of about 0.1 milli-arcseconds, while the inner region of the NLR would have around 0.03 arcseconds. Then, the angular dimensions of the molecular torus should range between 0.01 and 0.2 arcseconds.

We see immediately that progress in understanding the BLR , for instance, will come essentially from observational spectroscopy, through studies of variability and line profiles, as we will see in the next section.

3. THE PHYSICS BEHIND AN EMISSION-LINE GALAXY SPECTRUM

3.1. INTRODUCTION

The spectra of an ELG is a powerful tool to the understanding of its structure and composition. The spectra of an object so complex as a galaxy has in itself contributions from different sources that one has to be able to isolate in order to understand the whole picture.

A galaxy emission-line spectrum can essentially be divided into a stellar and a nebular component. In the case of HII regions, the nebular component is represented by the emission-lines (permitted and forbidden, depending on the elements and also on the gas density) and by a faint contribution to the continuum, while the stellar component will contribute to the absorption lines and to the greater part of the continuum emission. In the case of AGNs, one has yet to consider the contribution of a "featureless continuum", which can be more or less significant depending on the AGN-type and/or the considered wavelength (see previous sections).

The simplest generalized astrophysical situation for the model of an emission-line region in an ELG, is a photoionized cloud of gas surrounding a cluster of young, hot stars (OB associations). This is what is usually called an HII region. In these gas clouds, hydrogen atoms are ionized by the stellar ultraviolet radiation in the Lyman continuum and then recombine to excited levels; their decay gives rise to observable emission-lines such as the Balmer series. If there is enough hydrogen to absorb all the ultraviolet photons, the HII region is said to be "ionization bounded"[19]. This is usually the case for HII regions, where all Lyman photons (continuum and lines) are eventually degraded into Lyman-α and recombination lines or continua of higher series which eventually escape from the nebula, or may be absorbed by dust. The excess energy of the hydrogen ionizing photons, which

---

[18] for $H_0 = 50$ km s^{-1} Mpc^{-1}
[19] The opposite case is called "matter bounded".

corresponds to the difference between the photon energy and the hydrogen ionization potential (equal to 13.6 ev), supplies heat to the ionized gas. In order to attain a thermal equilibrium, this heating is cooled chiefly by the emission of collisionally excited lines from ions such as O^{++}, O^+, N^+, etc. This leads to a thermal balance at an electron temperature of the order of 10^4 K. The electron temperature depends mainly on the stellar temperatures and on the gas chemical composition, so that the temperature increases with a lowering of abundances because there are, in such case, fewer coolants.

Most of the collisionally excited lines are produced by "forbidden" transitions, i. e. transitions within the ground configuration of atoms or ions having no electric dipole ($\Delta l = 0$), and hence a very low transition probability from magnetic dipole or electric quadrupole interaction, so that they are not seen in the laboratory and have to be identified from spectroscopic term analysis. However for the very low densities in nebulae (typically 10 to 10^4 part. cm^{-3}), collisional de-excitation is so slow that a photon is eventually emitted after a time of the order of seconds.

HII region model calculations for a spherically symmetrical ionization-bounded nebula predict an ionization stratified sphere of gas, bounded by a sharply defined edge. This is known as the "Strömgren sphere". However, the analogy for AGNs breaks down because the BLR is far from being a homogeneous zone, its structure being very filamentary or clumpy. Although astronomers believe the basic physics in AGNs is the same as in HII regions and in planetary nebulae[20] (PNs), AGNs have a much more energetic ionization source and are much more complex entities.

Astronomers have been quite successful to determine the physical characteristics of Galactic and extragalactic HII regions, as their gas densities, temperatures, chemical composition and luminosities have been derived from their observed line spectra. Unfortunately, this is not so straightforward for AGNs due to their complexity and it requires the further development of theoretical photoionization models supplemented by a more precise knowledge of the AGNs core structure. For this reason, still very little is known for example about the metallicities in AGNs. So far there have been some abundance determinations for the NLR of some AGNs (see e.g. the work of Pastoriza et al. 1993 about NGC 3310; Diaz et al. 1985) and also for some Broad Absorption Line QSOs[21] (BAL QSOs) and BL Lacs.

Let us now consider some important aspects one can derive from the analysis of an AGN or a HII galaxy spectrum, in particular from their line intensities.

---

[20] PNs are envelopes of evolved intermediate-mass stars in process of ejection and are ionized by the hot, exposed stellar core.

[21] An atlas of BAL spectra was published by Korista et al. 1993.

3.1.1. *Variability*

The fact that a significant fraction of the luminosity from an AGN can change on timescales of minutes (X-ray emission from Seyfert galaxies) or many hours (emission from quasars), enables firm conclusions to be drawn on the size of the emitting region.

Indeed, if one assumes the existence of a "single" ionizing source, the analysis of continuum variability as well as the profiles of permitted and forbidden emission-lines can give us some informations about the dynamics of these objects, as well as about the sizes of the different emitting regions. How does this works ? Consider that a significant change in the continuum brightness on a short time scale δt requires that a correspondingly significant fraction of the emitting material be contained within a volume limited by light-travel considerations to be $r \leq c\, \delta t$. Translating a time scale for variability into a upper limit on the size of the continuum-emitting region can already give us an idea that, for example, in the case of luminous (L $\approx 10^{44}$ erg s^{-1}) Seyfert galaxies, for which continuum variations occur on time scales of several days, the upper limits on the sizes of emitting regions are $r \leq 10^{16}$ cm.

Variability studies can be done using different approaches that we will not detail here but are well described for example in Peterson (1997). These approches are: monitoring the long-term behaviour of a particular AGN or a sample of AGNs; looking for short-timescale and dramatic activity; and multi-wavelength monitoring searching for links between variability at different wavelegths. Furthermore, in this context, one can investigate the time-delayed response of the emission-lines to variations in the "central engine". This technique is called "reverberation mapping", because it maps the location of the lines that react to the injection of energy by flaring of the central engine. In other words, it measures the response of line emission to a change in the continuum as well as the changes in the profiles of these lines. These studies are quite difficult to undertake as, in addition to the time coordinating and weather problems, the *measurements of the selected lines* and the continuum must be of *high signal-to-noise* to allow the correlation techniques to be meaningful.

Regarding the analysis of the overall optical-ultraviolet continuum emission from AGNs, some complicating factors are associated with line emission and *require high signal-to-noise and high spectral resolution*. Emission from Balmer radiation is a good example of this and occurs in the rest wavelength range of 2700 Å − 3800 Å. This emission arises from complex atomic energy states in the pumped ultraviolet field of an AGN and results in a cascade of electrons moving through their respective energy levels, releasing predominantly Balmer photons. Another source of emission, whose importance has only been recognized over the last 15 years, is the blending of lines of singly ionized iron (FeII). When observed at low wavelength resolution, this blend forms a continuum-like feature predominantly in the 1800 Å − 3500 Å region, but also in the 4400 Å − 4800 Å and 5000 Å − 5500 Å regions. Associated with the Balmer radiation, this forms a pronounced humping in the spectrum which is called the "little blue bump".

Historically, one of the best monitored objects is the Seyfert 1 galaxy NGC 4151. Observation campaigns, undertaken from the 1970s to the 1980s demonstrated that over a period of many years NGC 4151 changed from showing the classical spectral characteristics of a Seyfert 1 galaxy, to those more resembling a Seyfert 1.8 galaxy, and then back again. This apparent metamorphosis was accompanied by a change in the strength of the ultraviolet continuum which was believed to be representative of the ionizing radiation field. In the corresponding paper, Ulrich et al. (1984) have shown a dramatic change in the BLR line shapes, one of the first evidences for a stratification in the BLR.

3.1.2. *Abundance determination*

In contrast to HII regions, PNs and the NLR of AGNs, there are no simple temperature and density diagnostics for the BLR of AGNs. The fundamental reason for this is that the BLR electron densities, n_e, are sufficiently high that virtually all forbidden lines are collisionally suppressed and the emissivity, ϵ, of all these lines is in the high-density limit where $\epsilon \propto n_e$.

These forbidden lines correspond to transitions from metastable levels, i. e., energy levels from which the probability for expontaneous de-excitation is extremely low. An atom or ion reaches such levels by collisional excitation with energetic free electrons (which mean energy is of the order of K T_e, where K is the Boltzmann constant and T_e is the gas electronic temperature), so that if the electronic density, n_e, is sufficiently low (inferior to the so-called "critical density" for collisional de-excitation) the atom or ion finally de-excitates radiatively, emitting a correspondent forbidden line.

How can these forbidden lines give us informations about the physical conditions in the gas of a line emitting region ? I will answer to this question in the text that follows, where I will focus in the three methods generally used to determine chemical abundances from emission-line spectra[22].

The "direct" methods

The most straightforward method to calculate gas metallicities, which is also the more widely used, is called "direct method" because it uses directly

---

[22] For more details on this subject see Aller (1984) or Osterbrock (1989).

the measured line intensities to determine the abundances. The following scheme summarises it:

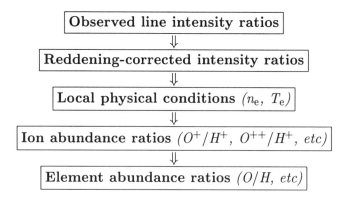

The first step after measuring the relevant emission-line intensities in the nebular spectra, is to correct them from interstellar extinction. This extinction (or reddening) depends on the amount of dust present within the nebula and also on the inclination of the galaxy relatively to the observer. The value of extinction is generally determined from the Balmer decrement[23]. It is a fact that the extinction law can vary in function of the observed line of sight as well as from object to object. This is because, in other objects, the abundance of dust and/or its composition can vary from the ones found in our Galaxy. This can be a real problem, leading to significant uncertainties in line intensity ratios if the wavelengths of the two lines are placed far apart (e. g., the [O III]λ3727/Hβ line ratio). Extinction laws have been derived for other galaxies, as the Magellanic Clouds or for a group of starburst galaxies (Calzetti et al. 1994) which show some deviation from the adopted Galactic laws (e.g., Whitford 1958; Seaton 1979). Even if one uses these as being typical of a certain type of object, one can never be sure, as different galaxies will possibly have slightly different or even quite different extinction laws.

Furthermore, the Balmer decrement itself is subject to uncertainties due to possible underlying stellar absorption. Stellar absorption is increasingly stronger with decreasing wavelength (i.e., it affects more Hβ than Hα) and its contribution depends mainly on the underlying population of A-type stars, which, with the ionised region, were included in the slit aperture. Its consequence is the increase of the measured Hα/Hβ intensity ratio, leading to a "false" increase in the reddening constant C(Hβ).

---

[23]Recombination case B is generally considered. For H II regions the theoretical value of the Hα/Hβ intensity ratio is 2.85 while for the NLR of AGNs it is 3.1 (this increment is due to the collisional excitation of HI (neutral hydrogen) by thermal electrons which produces strong Lα emission and makes a significant contribution to Hα).

The corrected line intensities are then used to determine the physical parameters of the gas. These are the electronic temperature (T_e) and the electronic density ($n_e \approx n_H$, the hydrogen density). T_e is usually determined from the [O III]$\lambda\lambda 4363/4959, 5007$ intensity ratio. This line ratio gives the temperature for the region where the O^{++} ions lie. The line [O III]$\lambda 4363$ ($2p^2\ ^1S_0$) has an energy of 5.3 ev relatively to the ground level, while the line [O III]$\lambda\lambda 4959, 5007$ ($2p^2\ ^1D_2$) has an energy of 2.5 ev. The occupation rate of each level will depend on the speed of the free electrons, therefore on T_e. In this case, the [O III]$\lambda\lambda 4363/4959, 5007$ intensity ratio will be proportional to $\exp[(5.3 - 2.5)/(KT_e)]$, where K is the Boltzmann constant.

If one wishes to determine T_e on a different region in the nebula, like the O^+ region for instance, then another line ratio will have to be used, corresponding to an ion which will co-exist with O^+. This is the case of the [N II]$\lambda\lambda 6548,6583/5755$ line intensity ratio or, in case our spectrum spreads further into the near-IR, the [O II]$\lambda\lambda(7320+7330)/3727$ intensity ratio[24].

The method to determine n_e, is based in comparing the intensities of two lines of the same ion, emitted by different levels but with approximately the same excitation energy. This is usually done from the [S II]$\lambda\lambda 6731/6717$ intensity ratio[25] which depends very little on T_e. In fact this intensity ratio is proportional to $n_e/T_e^{1/2}$ (For a precise expression see e.g. McCall, 1984). Another line ratio that can be used, although it requires a very good spectral resolution to separate the lines, is [O II]$\lambda\lambda 3727,3729$.

Once the electronic temperature and density are determined, the most difficult step in calculating the elemental abundances, is to determine the abundances of ions which have only a few lines observable (this is because usually one has only access to a part of the spectrum and not to the whole UV to IR range).

In HII regions, oxygen is usually an easy case because it presents intense optical lines for the more abundant ions, O^+ and O^{+2}. However, such is not the case for elements like nitrogen and sulfur. In such cases, their total abundances are obtained by adding the abundances of all the observed ions and multiplying it by a ionization correction factor, i_{cf}. This last one, will take into account the non observed ions. The ionization correction factors are generally based in the results of photoionization models but often are also simply based in considering that ions with similar ionization potentials have similar ionic fractions (see e.g. Peimbert & Costero 1969; Stasińska, 1980; Garnett 1995).

What if the T_e determination is impossible ? Empirical methods

[24] For an analytical expression see for example Seaton (1975) or Aller (1984).
[25] Note: [S II]$\lambda\lambda 6731(^4S_{3/2} - {}^2D_{3/2}$ transition); [S II]$\lambda\lambda 6717(^4S_{3/2} - {}^2D_{5/2}$ transition).

In an H II region, the photons from the stellar source impart energy to the gas through photoionization, which liberates energetic free electrons, responsible for the heating. In parallel, the excited ions from metals contribute to the cooling of the nebula through their spontaneous de-excitation from metastable levels, producing forbidden emission-lines. A higher abundance in metallic ions, will make cooling more efficient, diminishing the emissivity of optical forbidden lines. The consequences of this are that higher abundance nebulae will present spectra with very faint optical forbidden lines, as the cooling is transfered to far-infrared lines like [O III] $\lambda 51.8$ μm and [O III] $\lambda 88.4$ μm. Conversely, low metallicity nebulae can easily be spotted by the intense optical emission-lines in their spectra, and in particular for detecting the [O III]$\lambda 4363$ line.

The practical consequences of this behaviour are that a direct T_e determination (which requires observing the weak line [O III]$\lambda 4363$ with reasonably good accuracy) is only possible for nebulae with abundances lower than that of the Orion nebula ($12 + \log(O/H) \approx 8.5$). Therefore, the standard "direct" methods cannot generally be applied to higher metallicity nebulae or to spectra with low signal-to-noise [O III]$\lambda 4363$ line.

In order to estimate abundances in extragalactic H II regions, for such cases, various "empirical" methods, based on ratios of more intense emission-lines, were proposed (Alloin et al. 1979; Pagel et al. 1979). Alloin et al (1979) put forward a calibration of T_e against the readily observable line ratio [O III]/[N II] whereas Pagel et al. (1979) suggested using the R_{23} = ([O II]$\lambda 3727$ + [O III]$\lambda 4959, 5007$)/Hβ ratio, for which they gave calibrations of both oxygen abundance and T_e. This last method proved to be the most useful one.

The R_{23} method was introduced for use in spiral galaxies where abundances range from 20% to 300% of the solar value, and R_{23} increases with decreasing oxygen abundance (Pagel & Edmunds 1981) until, at $\approx$ 20% of the solar abundance value, it starts decreasing with decreasing abundance. Physically, the increase in R_{23} with decreasing abundance (in high abundance H II regions) reflects an increasing gas temperature which is due to a decrease in the presence of the most important coolants. Below $\approx$ 20% of the solar abundance value, the decrease in R_{23} reflects the further diminishing population of oxygen and the increasing importance of Lyα cooling.

Nevertheless, the R_{23} or Pagel method has the inconvenient of leading to a double solution for some R_{23} values. The only way to raise the ambiguity between the higher and lower abundance branches, for which simultaneously correspond certain values of R_{23}, is to use an additional line ratio, like the [O III]/[N II] ratio (see Alloin *et al.* 1979). The calibration of the [O III]/[N II] abundance indicator is uncertain but in most cases can be sufficient to determine which branch to use for a particular object.

The photoionization models

This is the most sophisticated method to determine the physical characteristics of the emitting gas such as, the ionization structure, the chemical abundances, the temperature and density. One can build a model which will reproduce as fair as possible the observed intensities of a nebula, or then build a grid of models, covering a certain domain of nebular parameters. Often, when using "direct" methods, the ionic abundances are generally determined with the help of a grid of models which is used to estimate the effective temperature of the ionizing stars and the i_{cf}s for each ion (e.g., Stasińska 1978; Aller & Czyzak, 1983).

The "ingredients" used to build a photoionization model are generally:
- the effective temperature of the ionising stars, T_{eff}, and the spectral energy distribution of the ionising source
- the chemical abundances for each element (or metallicity, Z), which will control the gas cooling processes
- the ionization parameter[26], U, which represents essentially the ratio between the stellar Lyman continuum and the gas density
- the gas density and its structure in function of the distance to the central source

The following scheme resumes the procedure normally used to model an observed nebula with a photoionization code:

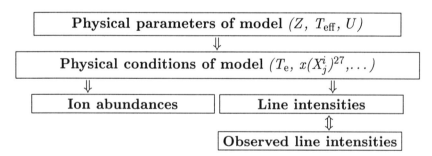

3.1.3. *Stellar populations*
Another important diagnostic that one can obtain from a ELG spectrum is the quantification of the underlying stellar populations. The study of stellar

---

[26]$U = \frac{Q_H}{4\pi n_H R_S^2 c}$, where Q_H is the number of H ionizing photons per second, n_H is the hydrogen density, R_S is the Strömgen radius of the ionized sphere, and c is the speed of light.
[27]Ionic fractions.

populations in AGNs is specially important in order to constrain the models which try to explain the origin of their fundamental continuum source. In the framework of the "starburst model", the study of stellar populations can help to verify if the central cluster is indeed responsible for the activity observed. Alternatively, in the hypothesis of a non-stellar source of activity and as a way to better understand the origin of the continuum emission, it is necessary to account for the stellar contamination, which usually dominates in the optical and IR wavelengths, near the galaxy nucleus. An important application of stellar population studies is, for instance, the correction of the hydrogen recombination lines from the underlying stellar absorption. We have seen earlier that this can affect the determination of the reddening constant, C.

The methods used to analyse the stellar populations in a galaxy are still subject to many uncertainties as they rely partly on our knowledge of stellar evolution, and partly on the existence of a complete stellar data base. There are normally two methods that can be used to analyse the stellar populations behind a spectrum. These are: 1) stellar population synthesis, and 2) stellar evolutionary synthesis.

Evolutionary synthesis determines the stellar populations present in a spectrum based on the following: i) stellar evolution theories; ii) the star formation rates (SFR); iii) the initial mass function (IMF); iv) the metallicity corresponding to each age. The stars are let evolving in an Hertzprung-Russell diagram, accordingly to stellar evolution theories. Using this diagram, one can determine at any moment in the evolution, the corresponding spectrum or the corresponding colour indices. These are then compared with the observations, allowing to determine the history and age of the dominant stellar population (e.g. Bruzual & Charlot 1993).

Many of the stellar spectra used with this method are theoretical ones, calculated from stellar atmosphere models. The advantage of these models is that they permit to take into account phases of stellar evolution which are difficult to observe[28].

The disadvantages of this method are essentially the extrapolations made within the Hertzprung-Russell diagram for the intermediate phases which are not well known and also that the evolutionary tracks are generally calculated only for solar abundances.

The *population synthesis* method consists of trying to reproduce a galaxy observed spectrum (or its colours), from the observed spectra of different types of stars, with different luminosities and metallicities at various evolutionary stages (e.g., Pickles 1985).

---

[28]These are for instance, the pre-main sequence, the asymptotic and post-asymptotic giant branches, the planetary nebula and the white dwarf phases.

The principal, obvious difficulty of this method is the incompleteness of the existent stellar libraries in particular regarding the metallicities of the stars observed.

3.2. NATURE OF AN EMISSION-LINE GALAXY

An important and basic issue, one can extract from the analysis of the spectra of ELGs is the *nature* of the object. Identifying the nature of an ELG, means determining the type of the dominant gas ionizing source, which is therefore mainly contributing to the observed emission-line spectrum. Seyfert 1s are generally easy to identify by their very broad permitted emission-lines. However, active galaxies having narrower lines can be difficult to classify, specially if only a small wavelength range is available. The often used, [O III]λ5007/Hβ, intensity ratio [29], to distinguish Seyfert 2 galaxies from H II galaxies can be misleading because this ratio is also typical of low-metallicity HII regions (or HII galaxies). Moreover, LINERs, Seyfert 2s and HII galaxies cannot be unambiguously distinguished from one another on the basis of any single flux ratio from any pair of lines. However, their nature can be determined based in the relative strengths of pairs of optical emission-line intensity ratios (Baldwin *et al.* 1981; Veilleux & Osterbrock 1987; Rola 1995; Rola *et al.* 1997) or in the equivalent widths of certain optical emission-lines (Rola *et al.* 1997).

4. Active galaxies and Adaptive Optics

Would it be possible to correct in real time the spatial frequencies filtered by Earth's atmosphere of a wave emitted by an astronomical object? The concept of AO, which was first proposed by Babcock (1953), positively answers this question.

AO allows to correct in real time the effects of atmospheric turbulence, providing nearly diffraction limited images. Its principle is simple: firstly, to measure on the pupil, in real time, the distorted wavefront using a wavefront sensor and, secondly, to correct the wavefront for the atmosphere distortion using a deformable mirror. One interesting possibility of this technique is that, when the astronomical source is very faint (like a very distant ELG), and hence does not provide enough signal-to-noise ratio to the wavefront sensor, one can measure the wavefront distortion using a nearby stellar[30] or laser-generated object[31], as far as this is located within the isoplanatic angle[32].

[29] with [O III]λ5007/Hβ > 3 for Seyfert 2s
[30] Natural Guide Star – NGS
[31] Laser Guide Star – LGS
[32] It refers to the distances on the sky over which the wavefront distortions, and hence

AO is much more easy to apply in the near infrared than in the visible, where the deformable mirror needs to have a smaller number of degrees of freedom, N, in order to correct the distorted wavefront, and the timescale of the correction is higher. In fact, for a pupil of diameter D, one has that $N(\lambda) \sim [D/\lambda^{\frac{6}{5}}]^2$ and the timescale of the correction is $\sim \lambda^{\frac{6}{5}}$ (see e.g., Shao & Colavita 1992). Furthermore, the isoplanatic angle increases relatively to the case of visible wavelengths, making it easier to find a NGS sufficiently bright to sense the phase distortions[33]. Thus, using natural sources, AO can be employed over most of the sky at infrared wavelengths (even if the sensing with the reference star is done at visible wavelengths).

Therefore, extending the gains of AO to shorter wavelengths requires a means to increase the availability of reference sources. Synthetic beacons or LGS will significantly increase the sky coverage of AO (see review by Beckers 1993). These provide an artificial reference source near the target using compact laser scattering from the clear atmosphere at an altitude of several kilometers (Rayleigh beacon)[34], or from the mesospheric sodium layer at 92 km (sodium beacon). When using a LGS as reference for wavefront sensing, a NGS will still be required to determine the overall wavefront tilt. Nevertheless, this star can be faint and hence suitable stars can probably be found in the field. Another advantage of using a LGS is that it will eliminate the problem of scattered light from the bright reference star.

However, for the moment it is difficult (even in good seeing conditions) to obtain a good Strehl ratio in the visible with a LGS, although good Strehls are obtained for NGS. This can be a problem for the observations of faint nearby sources in the visible (see Fig. 12 of Le Louarn et al., 1997).

Local AGNs or starburst galaxies have a nucleus in the $8 < m_V < 14$ mag range, which provides a suitable reference for AO guiding. The presence of rings or discs are detectable with AO down to 300 pc at a typical distance of 10 Mpc. As we have seen previously, the existence of a molecular torus surrounding the "central engine" in an AGN could be revelead in images with an angular resolution of the order of 0.1". Data on AGNs resolved with AO may give many answers to several questions like: What is the mechanism which is powering the nucleus emission?; Which type of stellar populations exist within their nucleus?; What is the gas and dust distribution within the nucleus of an AGN?

---

the images, are for all practical purposes the same. The isoplanatic angle increases with wavelength as $\lambda^{\frac{6}{5}}$. For instance, at visible wavelengths is of the order of 2 arcseconds.

[33] However, when the target is an extended bright object, itself can be used as the reference.

[34] This type of laser beacon is not so appropriate for astronomy purposes as, due to its low altitude, it only senses the atmosphere wavefront distortions between it and the telescope and not those of the atmosphere beyond.

Indeed, adaptive optics devices already allow imaging in the near IR of bright AGN with an improved spatial resolution, similar to that achieved with HST at visible and ultraviolet wavelengths. To give an example, the "Come On Plus" device at ESO, has been used to map the nuclear region of NGC 1068 in the J, H, and K bands with an angular resolution of ≈ 0.5 arcseconds[35] in order to test the AGN unified theory. Simultaneous imaging in the K band and in the I band have been performed as well, in order to locate the peaks of emission in these two bands, one with respect to the other (Marco et al. 1997). This data set suggests *the presence of two distinct regions of warm dust and photoionized gas*, extending over 40 to 70 pc. The inner nuclear region seems to be associated with dust heated directly by the central ultraviolet source. The second region, coincident with a peak in the [O III]λ5007 line emission (Evans et al., 1991) and in the 10 μm continuum emission (Cameron et al. 1993), might rather trace *a massive star formation event* (see Fig. 4). The visible and K band continuum peaks are found to be roughly coincident. Marco et al. (1997) conclude that the observed infrared emission peaks outline a dusty torus around a central engine, as expected by the unification models.

Another interesting example of successful application of AO to the study of active galaxies was done by Aretxaga et al. (1998), also using the "Come On Plus" device at ESO, in which they report the first clear detection of the host galaxy of a normal radio-quiet QSO at a redshift $z \approx 2$ in the K-band.

Furthermore, in addition to the increased spatial resolution available via AO, the limiting magnitude for the observed sources will be improved, as the signal-to-noise ratio increases and sky background diminishes. Moreover, with the addition of AO the slit width will decrease, yielding a higher spectral resolution: faint object spectroscopy will be greatly enhanced with the use of AO.

In that case, what are the possibilities of applying AO to the observation of distant ($z > 0.1$) galaxies? High redshift galaxies are very numerous so that the probability of coincidence of a galaxy and a natural reference star in the same isoplanatic field becomes non negligible. For $m_K = 18.5$ (at $\lambda = 2.2$ μm) one has about 10^4 galaxies-mag^{-1}-deg^{-1} which corresponds to 10% probability of finding a galaxy satisfying the above condition around a suitable star. Naturally, the addition of a laser beacon to the system will increase significantly this number.

The current surveys of distant galaxies are unable to establish properly the morphological type of galaxies and miss compact galaxies smaller than the seeing. The improvement in the angular resolution and in signal-to-

---

[35]This corresponds a spatial resolution between 15 to 35 pc.

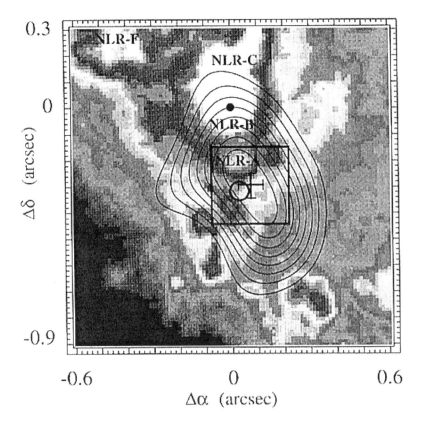

Figure 4. Contour map of NGC 1068 in the K band (linearly scaled from 10% to 100% of the peak intensity), overlaid on an image from the HST in the [O III] line, centered on the HST optical continuum peak (dot symbol). The box indicates position of the 12.4 µm peak, with its error bars. The cross represents the position of the near-infrared peaks location, with its error bars. The circle is the symmetry center of the UV/optical polarization. (Figure taken from Marco et al. 1997).

noise provided by AO will allow to determine the morphology of such yet unresolved objects.

Acknowledgements

Firstly, I would like to thank all lecturers and participants, and most specially to Chris Dainty, Norbert Hubin and Renaud Foy, for an extremely motivating, enjoyable and successful school. I am greatly indebted to E. Terlevich for her comments and suggestions which greatly helped to improve this paper. I am very grateful to I. Lopes for his comments on this paper and help in reproducing the figures here presented. Finally, I would like to thank the "Fundação para a Ciência e Tecnologia" (Portugal), for the fellowship BPD/6064/95.

References

1. Aller, L., 1984, in *Physics of Thermal Gaseous Nebulae*, Reidel
2. Aller, L., Czyzak, S., 1983, *Astrophysical Journal Supp.*, **51**, pp. 211
3. Alloin, D., Collin-Souffrin, S., Joly, M., Vigroux, L., 1979, *Astronomy & Astrophysics*, **78**, pp. 200
4. Andrillat, Y., Souffrin, S., 1968, *Astrophysical Letters*, **1**, pp. 111
5. Aretxaga, I., Le Mignant, D., Melnick, J., Terlevich, R., Boyle, B., 1998, *MNRAS*, **298**, pp. L13
6. Aretxaga, I., Terlevich, R., Boyle, B., 1998, *MNRAS*, **296**, pp. 643
7. Babcock, H., 1953, *PASP*, **65**, pp. 229
8. Baldwin, J., Philips, M., & Terlevich, R., 1981, *Publs astr. Soc. Pacif.*, **93**, 5
9. Beckers, J., 1993, Adaptive Optics for Astronomy: Principles, Performance, and Applications, *Annu. Rev. Astron. Astrophys.*, **31**, pp. 13
10. Begelman, M., 1985, in *Astrophysics of Active Galaxies and Quasi-Stellar Objects*, ed. J. Miller (University Science Books: Mill Valley), pp. 411
11. Bradl, B, Sams, B., Bertoldi, F., Eckart, A., Genzel, R., Drapatz, S., Hofmann, R., Lowe, M., Quirrenbach, 1996, *Astrophysical Journal*, **466**, pp. 254
12. Bruzual, G., Charlot, S., 1993, *Astrophysical Journal*, **405**, pp. 538
13. Calzetti, D., Kinney, A., Storchi-Bergman, T., 1994, *Astrophysical Journal*, **429**, pp. 582
14. Cameron, M., et al., 1993, *Astrophysical Journal*, **419**, pp. 136
15. Cid-Fernandes, R., 1997, *Rev. Mex. Astro. Astrof.*, **6**, pp. 201
16. Diaz, A., Pagel, B., Wilson, I., 1985, in *Active galactic nuclei*, Manchester University Press, pp. 171
17. Dibai, E., Pronik, V., 1967, *Astron. Zh.*, **44**, pp. 952
18. Elvis, M., Maccacaro, T., Wilson, A., Ward, M., Penston, M., Fosbury, R., Perola, G., 1978, *Mon. Not. R. astr. Soc.*, **183**, pp. 129
19. Evans, I. N., et al., 1991, *Astrophysical Journal Letters*, **369**, pp. 27
20. Ferland, G., Netzer, H, 1983, *Astrophysical Journal*, **264**, pp. 105
21. Filippenko, A. V., 1989, *Astronomical Journal*, **97**, pp. 726
22. Fitch, W., Pacholczyk, A., Weymann, R., 1967, *Astrophysical Journal*, **150**, pp. L67
23. Garnett, D., Dufour, , R. Peimbert, M. Torres-Peimbert, S., Shields, G., Skillman, E., Terlevich, E. Terlevich, R., 1995, *Astrophysical Journal*, **449**, pp. 77
24. Halpern, J., Eracleous, M., Filippenko, A. Chen, K., 1996, *Astrophysical Journal*, **464**, pp. 704
25. Halpern, J., Steiner, J., 1983, *Astrophysical Journal*, **269**, pp. L37
26. Heckman, T., 1980, *Astronomy & Astrophysics*, **87**, pp. 152
27. Ho, L., Fillipenko, A., Sargent, W., 1993, *Astrophysical Journal*, **417**, pp. 63
28. Huchra, J. & Burg, R., 1992, *Astrophysical Journal*, **393**, pp. 90
29. Le Louarn, M., Foy, R., Hubin, N., Tallon, M., 1998, *MNRAS*, **295**, pp. 756
30. Lyutyi, V. M., 1973, Soviet Astronomy
31. McCall, M., 1984, *MNRAS*, **208**, pp. 253
32. Marco, O., Alloin, D., Beuzit, J., 1997, Positioning the near-infrared versus optical emission peaks in NGC 1068 with adaptive optics, *Astronomy & Astrophysics*, **320**, pp. 399
33. Osterbrock, D., 1981, *Astrophysical Journal*, **249**, pp. 462
34. Osterbrock, D., 1989, in *Astrophysics of Gaseous Nebulae & Active Galactic Nuclei*, University Science Books
35. Pagel, B., Edmunds, M. Blackwell, D., Chun, Smith, G., 1979, *MNRAS*, **193**, pp. 219
    Pastoriza, M., Dottori, H., Terlevich E., Terlevich R., Diaz, A., 1993, *MNRAS*, **260**, pp. 177
36. Peimbert, M. & Costero, R., 1969, *Bol. Obs. Tonantzintla y Tacubaya*, **5**, pp. 3
37. Penston, M., Pérez, E., 1984, *MNRAS*, **211**, pp. 33p
38. Peterson, B., 1997, in *An introduction to Active Galactic Nuclei*, Cambridge Uni-

versity Press
39. Pickles, A., 1985, *Astrophysical Journal Supp.*, **59**, pp. 33
40. Keel, W. C., 1983, *Astrophysical Journal Supp.*, **52**, pp. 229
41. Khachikian, E., Weedman, D., 1974, *Astrophysical Journal*, **192**, pp. 581
42. Korista, K., Voit, G., Morris, S., Weymann, R., 1993, *Astrophysical Journal Supp.*, **88**, pp. 357
43. Rees, M., 1984, *Annu. Rev. Astron. Astrophys.*, **22**, pp. 471
44. Rola, C. S., 1995, Ph.D. thesis, Université de Paris VII, France
45. Rola, C. S., Terlevich, E., Terlevich, R.J., 1997, New Diagnostic methods for Emission-line Galaxies in Redshift Surveys, *MNRAS*, **289**, pp. 419
46. Schmidt, M., Green, R., 1983, *Astrophysical Journal*, **269**, pp. 352
47. Seaton, M., 1975, *MNRAS*, **170**, pp. 475
48. Seaton, M., 1979, *MNRAS*, **187**, pp. 73P
49. Serote-Roos, M., 1996, Ph.D. thesis, Université de Paris VII, France
50. Shao, M. & Colavita, M., 1992, Long-Baseline Optical and Infrared Stellar Interferometry, *Annu. Rev. Astron. Astrophys.*, **30**, pp. 457
51. Stasińska, G., 1978, *Astronomy & Astrophysics*, **66**, pp. 257
52. Stasińska, G., 1980, *Astronomy & Astrophysics*, **84**, pp. 320
53. Stathakis, R. A. & Sadler, E. M., 1991, *MNRAS*, **250**, pp. 786
54. Stauffer, J. R., 1982, *Astrophysical Journal*, **267**, pp. 66
55. Telles, E., Melnick, J., Terlevich, R., 1997, *MNRAS*, **288**, pp. 78
56. Terlevich, R., Melnick, J., Masegosa, J., Moles, M., Copetti, M., 1991, *Astr. Astrophys. Suppl.*, **91**, pp. 285
57. Terlevich, R., Melnick, J., Moles, M., 1987, *121st IAU Symp.: Observational Evidence of Activity in Galaxies*, eds. Khachikian, E., Fricke, K., Melnick, J., D. Reidel Publs. Comp., 49
58. Terlevich, T., Tenorio-Tagle, G., Franco, J., Melnick, J., 1992, *Mon. Not. R. astr. Soc.*, **255**, pp. 713
59. Tran, H., Miller, J., Kay, L., 1992, *Astrophysical Journal*, **397**, pp. 452
60. Tresse, L., Rola, C. S., Hammer, F., Stasińska, G., Le Fèvre, O., Lilly, S., Crampton, D., 1996, *Mon. Not. R. astr. Soc.*, **281**, pp. 847
61. Ulrich, M., Boksenberg, A., Bromage, G., Clavel, J., Elvius, A., Penston, M., Perola, G., Pettini, M., Snijders, M., Tanzi, E., Tarenghi, M., 1984, *MNRAS*, **206**, pp. 221
62. Veilleux, S. & Osterbrock, D., 1987, *Astrophys. J. Suppl.*, **63**, 295
63. Ward, M., 1995, in *Science with the VLT*, ed. Walsh, J., Danziger, I., Springer-Verlag, pp. 304
64. Whitford, A., 1958, *Astronomical Journal*, **63**, pp. 210

Index

Active galactic nuclei	312
LINERs	316
quasars	288,318
radio galaxies	288
Seyfert	314
Adaptive Optics	
coronography	194
enhancement for astronomy	188,279,332
location	188
photometry	193
polarimetry	**219**
sensitivity	189
speed	189
spectroscopy	196
Adaptive optics results	
asteroids	**243**
binary stars	208
circumstellar shells	209
compact nebulae	209
extragalactic	207
galactic center	207
solar system	206
star clusters	208
young stars	205
ALFA	42,43
AO & LGS astronomical applications	46
Artificial guide star	
(*sodium*) brightness	35
formation	24
Angular anisoplanatism	19
Calibration for sodium LGS	94
auxiliary wavefront sensor	97
dynamic	98
internal sources	96
static	96
Cone effect	107
astrophysical implications	109
parameters	109
Cygnus A	295
Distant (radio) galaxies	**285**
HST imaging	301
properties	298
search	297
stellar content	301
Dual AO	150
ELP-OA	163
Emission line galaxies	
abundance	326
nature	332
physics	**323**
stellar population	330
variability	324
Focal anisoplanatism	30

Fried parameter (r_0)	6
Galaxies	
actives	309
evolution	288
formation	288,290
H II	311
unification models	320
Inner scale	3
Isoplanatic angle	29,192
Kolmogorov turbulence	3,(8?)
Laser	
beacon	67,200
CY dye	74
for sodium beacons	73
mode locking	77
power	159
pulsed dye	75
saturation at sodium layer	72
sum-frequency	76
technology for guide stars	37
Laser guide star	
experiments	41
focus changes	92
operational issues	41,**89**
return flux	60,62,83
Multiple LGS	112
3-D mapping	114
inversion	119
modal approach	119
spot geometry	119
spot field	119
stitching	112
Natural guide star	
focus	198
static aberrations	198
tilt	199
Optical pumping	84
Outer scale	3
PASS	159
Point Spread Function	15,192
Polychromatic laser guide star	5
principle	151
required photon flux	153
Post-processing techniques	**230**
blind deconvolution (IDAC)	238,244
CLEAN	232
Knox-Thompson	233
maximum entropy	231
Richardson-Lucy	232
triple correlation	234
with a known PSF	230
with a poor PSF	236
with wavefront sensor data	237

Pulse format	
LLNL	62
sum frequency	64
Rayleigh scattering	90,154
Redshift	286
Safety for LGS systems	99
aircraft	101
fire	100
laser coordination	103
laser eye	100
spacecraft	102
Sky coverage	**169**
computation	170
object-counts	177
statistical	175
Sodium	
atmospheric layer	32,51,93
atom	32
atomic physics	52
excitation of $4P_{3/2}$	154
optical pumping	57
radiation pressure	57
saturation	55,157
Strehl ratio	7,226
Transfer function	cf *Point Spread Function*
Tilt	25,**126**,**147**
von Karman spectrum	3
Young stars	
disks	277
early years	263
Herbig Ae/Be	268
HH flows	271
T Tauri	268
Zernike polynomials	8